Population and Community Ecology

Principles and Methods

Population and Community Ecology

Principles and Methods

E. C. Pielou
Killam Research Professor of Biology
Dalhousie University

GORDON AND BREACH SCIENCE PUBLISHERS

NEW YORK   PARIS   LONDON

Gordon and Breach, Science Publishers, Inc.
One Park Avenue
New York, New York 10016

Gordon and Breach Science Publishers Ltd.
42 William IV Street
London, WC2N 4DF

Gordon & Breach
7-9 rue Emile Dubois
75014 Paris

First published December 1974
Second Printing July 1976
Third Printing (with corrections) February 1978

Preface

THE MOUNTING CONCERN felt by scientists and public alike about the environmental crisis has made everybody aware of ecology—of what it is and why the study of it is of vital importance. Increasing numbers of students are choosing ecology as their field, both because of its relevance to human affairs and for its great theoretical interest.

This book has been written as a textbook for a two-semester course for undergraduates who want to know more about population and community ecology than a cursory survey of the subject can provide. It covers a far wider field than my earlier book* and is at a much lower level of mathematical difficulty; the reader is assumed to have had no more than a first course in each of calculus and statistics. There is obvious need for a book at this level that gives equal attention to theoretical principles and practical details, to the abstract and the concrete. A working ecologist, in carrying out a research project, engages in a whole sequence of operations. The process starts with the formulation of hypotheses to be tested and questions to be asked. Next one must decide what observations should be made, or experiments performed, to provide direct and unambiguous answers to these questions. The practical work must then be planned and executed. Finally the observations and results have to be analysed and interpreted. Therefore, if a research project is to yield concrete conclusions, every link in this chain of operations must be strong and no ecologist is fully trained who cannot handle all of them.

This is the belief that has led to my choice of the ingredients for this book. I have tried to give due attention to every one of the links just mentioned. The search for general theories, and the success so far attained, are fully discussed. So also are the various ways of testing the theories. Of course, since the organisms ecologists study range from protozoa to whales, precise instructions on what to do in every conceivable set of circumstances can never be given. I have therefore tried to present full and explicit accounts of

* *An Introduction to Mathematical Ecology*, Wiley-Interscience, 1969.

v

a varied array of research methods which the student can adapt to his special needs. A study of particular cases should also develop his capacity to judge exactly how much (and how little) a given body of data can be expected to contribute to the solution of a problem. The commonest mistake of beginning research workers is to overestimate the usefulness of data (however abundant) of a single kind, and to neglect the need for corroborative evidence of different kinds, when hypotheses are tested.

Familiarity with certain elementary mathematical and statistical techniques is a necessity for anyone who wishes to learn more about modern ecology than can be covered in an elementary survey course. To accomodate a wide range of mathematical experience among readers, while at the same time avoiding lengthy digressions in the text, I have discussed a number of mathematical topics in appendices to several of the chapters. These appendices are of different levels of difficulty. Some are elementary and will be useful mainly as memory refreshers; others are more advanced and can be safely ignored by students willing to accept certain statements without proof. The text makes clear which category each appendix belongs to.

I have used the material in this book in teaching a course in population and community ecology at Dalhousie University, and it has benefited greatly from the feedback I have received from my students.

Lastly, I am very grateful to my husband D.P. Pielou for his help and encouragement at all times.

E. C. Pielou

Halifax, Nova Scotia
Canada

Contents

CHAPTER 1

Simple Population Growth

POPULATION DYNAMICS, the study of the fluctuations in numbers of natural populations of animals and plants, is of fundamental importance in ecology. No natural population maintains its size unaltered for very long and any population which an ecologist is investigating will certainly change. Its size may increase or decrease, changes may occur quickly or slowly, and the effects on other populations may be marked or slight. Whichever of these things happens one would like to know the reason for it; if current happenings are explicable, future happenings are more likely to be correctly predictable. And well-argued predictions, with a high probability of being right, are the most important products of an ecologist's labors.

Before discussing the changes in a population, it is obviously necessary to define the word *population*. We shall use it to mean all the animals or plants of one species that inhabit an area sufficiently small to enable them to interbreed freely.

1 EXPONENTIAL POPULATION GROWTH

Any account of population dynamics should begin with a description of the kind of population growth known as geometric, exponential, or Malthusian. This last name is derived from that of Thomas Robert Malthus (1766–1834) who realised that all species had, theoretically, a capacity to increase that ultimately would exceed any conceivable increase in the means of subsistence of those species. The writings of Malthus had a great deal of influence on Charles Darwin and led to Darwin's concept of "survival of the fittest" as a central theme in the theory of natural selection.

Unchecked exponential growth of a population leads to a "population explosion". It is easy to see that if, on average, every member of a population is replaced at its death by more than one successor, the population must inevitably grow ever larger; in other words, a population explosion must occur. If the population doubled at each generation, then a single individual would be replaced by over a million individuals after 21 generations.

Let us suppose that in general, at the end of any unit of time there are always λ times as many individuals as there were at the beginning of the unit of time, and let

$$N_0, N_1, N_2, N_3, \ldots$$

denote the size of a population at times

$$t = 0, 1, 2, 3, \ldots .$$

Then

$$N_1 = \lambda N_0$$

$$N_2 = \lambda N_1 = \lambda^2 N_0$$

$$N_3 = \lambda N_2 = \lambda^2 N_1 = \lambda^3 N_0$$

and so on. Thus the size of the population at the sequence of times

$$0, \quad 1, \quad 2, \quad 3, \quad \ldots, \quad t,$$

is

$$N_0, \lambda N_0, \lambda^2 N_0, \lambda^3 N_0, \ldots, \lambda^t N_0,$$

and we see that these numbers form a geometric series. When we consider the population growth of a mammal that bears one litter per year, or of a plant that produces a single crop of seeds per year, it is clear that the size of the population changes with time in the manner shown in the graph in Figure 1.1. The vertical steps in the graph show the increments to the population that occur at a succession of breeding seasons; and each gently descending slope shows how the population decreases, owing to deaths, between one breeding season and the next. But since the number of births exceeds the number of deaths each year, the population as a whole increases. In the example shown the time unit is one year and we have put $\lambda = 1.2$; thus the population size at the end of each breeding season is 1.2 times its size one year earlier. The constant λ is known as the finite rate of natural increase. Rephrasing the whole argument, we could say that the population is growing like a sum of money earning compound interest with the interest compounded annually. Then N_t, the size of the population after t years is

$$N_t = N_0(1 + r)^t$$

where r is the interest rate expressed as a fraction.

Next let us predict the growth of a population in which breeding is continuous rather than seasonal. Consider again the compound interest law

$$N_t = N_0 (1 + r)^t$$

If interest were compounded n times per year instead of annually, this equation would be replaced by

$$N_t = N_0 \left(1 + \frac{r}{n}\right)^{nt}.$$

If we now allow n to become very large indeed it may be shown (see Appendix 1.1) that in the limit

$$N_t \to N_0 e^{rt} \tag{1.1}$$

where $e = 2.71828\ldots$ is the base of the natural (or Napierian) logarithms.

Figure 1.2 shows the growth curve for a population growing in this manner, with $r = 0.1823$. The constant r is known as the instantaneous rate of

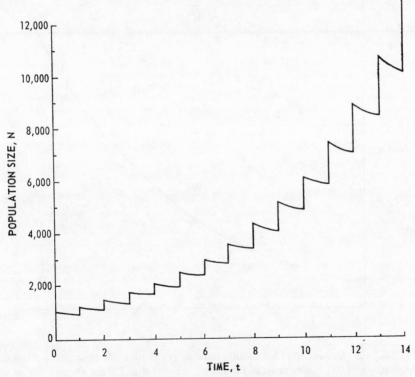

FIGURE 1.1 Growth of a population with one short breeding season each year. The finite rate of increase is $\lambda = 1.2$

1*

natural increase.* It has dimension number/time and if we assert that a particular species, for example man, has an r value of 0.0277 say, we must also state the time unit, in this case years. If the time unit were one month the equivalent r would be 0.0277/12 or 0.00231 approximately; if the time unit were one week r would be 0.0277/52 or about 0.000533; and so on. The rate r is an *instantaneous* rate of increase, whereas λ is a *finite* rate. Nor is this the only difference between them; whereas r, as we have seen, is analogous to a rate of interest, λ is the ratio of the population size at any moment to its size one unit of time earlier. Thus the relation between λ and r is this: since the equations

$$N_t/N_0 = \lambda^t \quad \text{and} \quad N_t/N_0 = e^{rt}$$

are equivalent, it follows that $\lambda = e^r$ or $r = \ln \lambda$ where $\ln \lambda$ denotes $\log_e \lambda$, the natural logarithm of λ. Thus the stepwise growth (with $\lambda = 1.2$) shown in Figure 1.1 is equivalent in the long run to the continuous growth shown in Figure 1.2 (with $r = 0.1823 = \ln 1.2$).

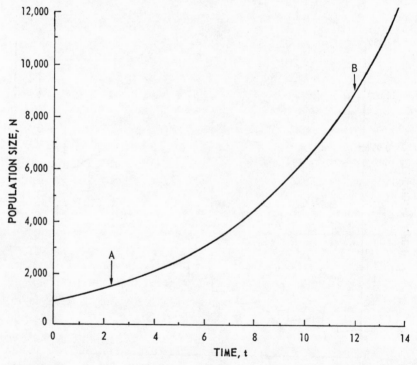

FIGURE 1.2 Growth of a continuously breeding population. The instantaneous rate of increase is $r = 0.1823$. Segments of the growth curve at A and B are shown enlarged in Figure 1.3

---

* It is sometimes called the "*intrinsic* rate of natural increase".

In Figure 1.2 the growth of the population is shown as a smooth curve. In reality, of course, the events causing changes in population size are births and deaths and between any two consecutive events the size of the population remains constant. When an event occurs the population size, N, increases to $N + 1$ if the event is a birth or decreases to $N - 1$ if the event is a death. The instantaneous rate of natural increase r is, in fact, the difference between the birth rate and the death rate. That is $r = b - d$ where b and d are the average numbers of births and deaths respectively per unit of time. It follows that a particular rate of increase can equally well result from a combination of high birth and death rates or of low birth and death rates. Thus a population of mice and a population of elephants could in principle have the same instantaneous rate of natural increase; but the mice would have more frequent births, and correspondingly, more frequent deaths.

2 DETERMINISTIC AND STOCHASTIC EVENTS

Now consider the actual sequence of events represented by the smooth curve of Figure 1.2. In all that follows we shall assume the unit of time is one year. Figure 1.3 shows two extremely small segments of the curve enormously magnified; (to appreciate the degree of magnification, contrast the scales of the axes in the two figures). We have supposed that the birth rate is $b = 0.2434$ births per individual per year; and that the death rate is $d = 0.0611$ deaths per individual per year. Therefore $r = b - d = 0.1823$ as before. The rate at which events (births and deaths) occur is $b + d = 0.3045$ events per individual per year. Therefore, when the size of the population is $N = 1500$, for example, the average number of events per day is $(0.3045 \times 1500)/365 = 1.25$; and when $N = 9000$, the average number of events per day is six times as great, or 7.5. In both cases, however, the birth rate is almost four times the death rate (the exact ratio in this example is $0.2434/0.0611$ or 3.984). Now births and deaths are "stochastic", as opposed to "deterministic", events. That is to say, they are the outcome of chance and succeed one another, not at fixed intervals, but after intervals of varying length. The *average* length of the intervals between events is given by $1/N (b + d)$ time units; when $N = 1500$ (as in Figure 1.3a) the average time is therefore $365/(1500 \times 0.3045)$ or 0.8 days, approximately; and when $N = 9000$ (Figure 1.3b), the average time is 0.13 days, approximately. Moreover one can never predict with certainty whether the next event will be a birth or a death, but can only give the probabilities. In the example we are considering, the probability that an event will be a birth is

$$b/(b + d) = 0.2434/(0.2434 + 0.0611) = 0.799$$

and that it will be a death is

$$d/(b + d) = 0.0611/(0.2434 + 0.0611) = 0.201.$$

Roughly, therefore, there are four births for every death. We now see why the smooth curve of Figure 1.2 rises gently when N is small and becomes progressively steeper as N increases. The relative proportions of births and deaths do not change but, as population size increases, the intervals between

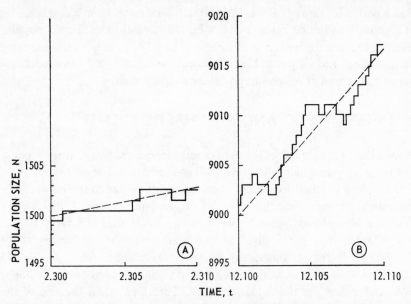

FIGURE 1.3   Enlarged segments of the growth "curve" shown in Figure 1.2. They are centered, respectively, on the points marked A and B. In this figure the scale is so large that the unit increases and decreases in population size, resulting from individual births and deaths, appear as distinct steps

events become progressively shorter so that events follow one another at an ever-increasing rate.

Let us next consider what predictions we can make about the future size of a population if we know its instantaneous rate of increase, r, and its present size N_0. The "expected" size of the population after t units of time is denoted by $\mathscr{E}(N_t)$ and is

$$\mathscr{E}(N_t) = N_0 e^{rt}.$$

We should, however, be extremely surprised if, after the t time units had elapsed, the population size had exactly this predicted value. As already remarked, births and deaths are chance, or stochastic, events and any predictions as to the future size of a population are bound to be imprecise; indeed, their precision will get steadily less the farther into the future we attempt to prophesy.

As a measure of this imprecision we may take the variance of N_t, denoted by var (N_t) (see Appendix 1.2). It is given by

$$\text{var}(N_t) = N_0 (b + d) e^{rt} (e^{rt} - 1)/r. \tag{1.2}$$

Then, using arguments based on probability theory, we can say that the probability is 1 in 20 (or 95%) that after t time units the size of the population will lie between the limits

$$\mathscr{E}(N_t) - 1.96\sqrt{\text{var}(N_t)} \quad \text{and} \quad \mathscr{E}(N_t) + 1.96\sqrt{\text{var}(N_t)}.$$

These limits are known as the 95% confidence limits and define the 95% confidence interval of N_t. Obviously the greater the value of var (N_t) the wider the confidence interval, or, in other words, the less precise the prediction.

Returning to the example described earlier and shown in Figure 1.2, for which $b = 0.2434$, $d = 0.0611$ and $r = b - d = 0.1823$ with one year as the unit of time. Let us take 1970 as the zero'th year, at which time the population's size is $N_0 = 1000$; we wish to predict the size, N_{12}, of the population in 1982, twelve years hence. The expected size of the 1982 population is

$$\mathscr{E}(N_{12}) = 1000\, e^{(0.1823 \times 12)} = 8914.$$

Also, the variance of N_{12} is, from equation (1.2)

$$\text{var}(N_{12}) = 117,831.$$

Therefore the 95% confidence interval for the predicted size of the population in 1982 is given by $\mathscr{E}(N_{12}) \pm 1.96\sqrt{\text{var}(N_{12})}$, or 8242 to 9586 individuals. Not only can we say that the probability is 95% that the 1982 population size will lie between these limits, we can also assert that there is a $2\frac{1}{2}\%$ probability that it will be smaller than 8242; and a $2\frac{1}{2}\%$ probability that it will be larger than 9586. Thus by using arguments based on probability theory we can state not only the most likely, or expected, outcome of population growth but also the probabilities of specified outcomes that, though unlikely, are by no means impossible.

Next consider the growth of a population with the same instantaneous rate of natural increase, r, and the same initial size, but suppose now that the birth rate is $b = 0.1900$ and the death rate is $d = 0.0077$; that is, although the difference of b and d is the same as before, their sum is less. Then the rate of occurrence of "events" (births and deaths together) is less, and as the population grows there will be fewer opportunities during a given period for the population size to diverge from its theoretical expectation. As a result, although the predicted population size for 1982 is still $\mathscr{E}(N_{12}) = 8914$, the 95% confidence interval is narrower: it is 8371 to 9457.

As a final numerical example, consider again the population having $b = 0.2434$, $d = 0.0611$ and $N_0 = 1000$, for which we predicted in 1970 that the 95% confidence interval for the 1982 population was 8242 to 9586. Next imagine that in 1976 the population size is actually found to have the unexpectedly low value of 2500 instead of the predicted 2986. We must now (1976) revise our forecast of the population size in 1982 in the light of the latest available knowledge and of the fact that 1982 is only six instead of twelve years distant. Writing N'_6 for the newly predicted 1982 population we find its expected value to be

$$\mathscr{E}(N'_6) = 2500 \, e^{(0.1823 \times 6)} = 7464$$

with 95% confidence limits 7156 and 7772.

This demonstration of the way in which predictions must be repeatedly altered to accord with the most up-to-date information will come as no surprise to anyone who follows the weather forecasts.

In the foregoing examples we have considered increasing populations, those exhibiting positive growth. "Negative growth" or a decline in the size of a population is also possible of course; then r is negative and λ is less than 1.

3 EXPONENTIAL GROWTH IN NATURE

It is often asserted that exponential population growth cannot go on indefinitely in nature. This statement is too vague to be accepted or denied. It is quite true that for any given positive r we can always find a value of t for which N_t is so large that it would be absurd to suppose it could ever be realized. But conversely, if we prescribe a maximum size, N_{max} say, as an absolute upper limit for the realizable size of a population, we can always find a value of r so small that $N_t = N_0 e^{rt}$ shall be less than N_{max} for any given value of t.

This raises a problem to which no satisfactory, universally acceptable answer is yet available: how far into the future can ecologists reasonably hope to make predictions? Their prophetic powers are thwarted by two quite different factors. In the first place there is a continual accumulation of *chance* divergences from theoretically predicted population sizes. As is clear from the arguments above, the farther into the future we try to make predictions, the wider become the confidence intervals for predicted population sizes. Different people hold different opinions about how vague a prediction can be and still be useful; even so it is obvious that for any population there is a time limit beyond which predictions are so imprecise as to be wholly worthless. The second source of difficulty is this: the instantaneous rate of increase, r, on which predictions are based, depends not only on the species of organism concerned but also on the conditions in which it is living. Therefore predictions about the growth of a population are only possible for as long as these conditions remain unchanged.

Now, as we said earlier, it is often maintained that exponential growth cannot occur in nature; the argument is that if it did population size would soon attain such a high level that the environment would be "saturated", that is it could not support any further increase. However, as we have shown, exponential increase may be so slow that the *feasibly* predictable future is much shorter than the time that would be taken for the environment to become saturated.

It is highly probable that true exponential growth takes place, for a limited period, when species that are very sparse in winter go through a rapid build-up in numbers in spring. This is especially likely to be true of outbreaks of pest species—often the rampant growth of such species is what makes them pests. Examples are the algal blooms of eutrophic lakes, mouse plagues, and outbreaks of insect pests. These populations may increase exponentially until changing weather halts the process; the change in weather may often occur long before the environment has had time to become saturated. Indeed, in any period when the growth of a population is proceeding unchecked by any extrinsic factors, it is likely to be exponential. Of course, if the rate of increase is high, these periods of exponential growth will necessarily be short since saturation of the environment and a consequent reduction in the growth rate must soon ensue.

In descriptions of the growth of human populations, a number often given is the number of years in which the population would double in size given continued increase at the rate obtaining when the forecast is made. For example in 1798 Malthus wrote as follows: "In the United States of America, where the means of subsistence have been more ample, the manners of the people more pure, and consequently the checks to earlier marriages fewer, than in any of the modern states of Europe, the population has been found to double itself in 25 years."

A more recent figure for the doubling period is obtainable from data given by Keyfitz and Murphy (1967) who give the finite rate of increase for 1960* as $\lambda = 1.11$ (using a 5-year interval as the unit of time). Let us find D, the number of years in which the population would be expected to double in size. Writing t for the number of 5-year intervals (so that $D = 5t$), we require the value of t for which $N_t/N_0 = 2$.

But

$$N_t = N_0\lambda^t \quad \text{or} \quad N_t/N_0 = \lambda^t.$$

Therefore we must solve

$$\lambda^t = 2 \quad \text{or} \quad t = \frac{\log 2}{\log \lambda} = 6.64 \quad \text{when} \quad \lambda = 1.11.$$

Then $D = 5t \doteqdot 33$ years is the answer sought. (Notice that in this calculation the base of the logarithms is immaterial.) This result applies to the United States; the doubling period for the population of England and Wales, also based on 1960 data, is 71 years.

---

* The value of λ used here is rounded. The more precise values given by Keyfitz and Murphy are: for females $\lambda = 1.10919$; and for males $\lambda = 1.10934$.

CHAPTER 2

Predicting the Growth of a Population

IN THE PREVIOUS CHAPTER we showed, in equation (1.1), that a population initially of size N_0 undergoing steady exponential growth would, after an interval of length t, be of size

$$N_t = N_0 e^{rt}$$

where r is the instantaneous rate of natural increase.

Another way of looking at exponential population growth, using elementary calculus, is to differentiate both sides of this equation with respect to t giving

$$\frac{dN_t}{dt} = rN_0 e^{rt} = rN_t$$

or

$$\frac{1}{N_t} \frac{dN_t}{dt} = r.$$

This shows that $(1/N_t) \, dN_t/dt$, which is the growth rate *per individual*, is constant and equal to r at all times and whatever the size of the population. Thus r is an extremely important biological property of a species living under given environmental conditions. It is the rate of increase possessed (on average) by every individual of the species when living in those conditions; it defines the innate capacity of the species to increase when population growth is not slowed down by competition. In colloquial terms, it is the best the species can do in the given conditions. Also, the environmental conditions for which r is a maximum are, by definition, the optimum conditions for the species. Therefore it is natural to inquire how r may be determined.

It must first be emphasized that for a population to grow exponentially, one or other of the following two pre-conditions must hold. Either the birth and death rates of the individuals must be independent of their ages; obviously this is most unlikely to be true. Or, and this is the realistic alternative, the population must have a *stable age distribution*. That is to say, the relative proportions of individuals in the different age classes must persist

11

unchanged. This happens when every age class, considered as a single entity, is growing at the same rate. Any given age class gains members as young individuals grow older, and loses members as some die and others grow older still. But if the proportion of the total population in each age class remains constant, then the age distribution is stable.

1 CALCULATION OF *r*, THE INSTANTANEOUS RATE OF INCREASE

Now let us consider how r is determined in practice, a thing that can only be done for a population that has a stable age distribution. It is calculated for the female part of the population only, assuming that males are always sufficiently numerous to ensure that no female goes unmated.

We must begin with careful definitions of the symbols to be used in the mathematical arguments. It cannot be denied that these arguments are often confusing, even in the somewhat simplified version given here. We shall be considering what happens to individuals (females) as time passes and as they grow older, and the two processes necessarily go hand in hand. Thus one cannot, even conceptually, imagine an individual growing older while time stands still or vice versa. This frequent source of difficulty is best overcome by paying very careful attention to the symbols and their subscripts. The subscript x attached to a symbol implies that the number denoted by the symbol (fertility rate, for example) is "age-specific"; that is, it depends on the age, x, of the individuals or group of individuals concerned. Conversely, the subscript t attached to a symbol implies that the number denoted by the symbol (population size, for example) depends on the time, t, at which it is observed. Both age and time are measured in time units of the same duration, which should be short.

The symbols we shall need, and their meanings, are as follows:

l_x is the *proportion* of individuals that survive to age x *at least*. The number l_x is known as the *life table function* or the *age-specific survival rate*.

m_x is the *number* of female offspring born to an individual while her age is in the range x to $x + 1$. This number is the *maternal frequency*, or the *age-specific fertility* (or *fecundity*) *rate*. Thus suppose it were found that the average number of daughters born by American women between their twenty-second and twenty-third birthdays was 0.071. Then, for Americans, $m_{22} = 0.071$.

c_x is the *proportion* of individuals whose age is in the range x to $x + 1$. The list of c_x values thus constitutes the age distribution. Each proportion remains constant (that is, it does not vary with time) since, as we have already stressed, the population is assumed to have a stable age distribution.

N_t is the *number* of individuals in the population at time t.

B_t is the *number* of individuals born to the whole population in the time interval $t - 1$ to t.

$b = B_t/N_t$ is the birth *rate* per head of the population; i.e. it is the number of daughters born to an individual per unit of time, averaged over all individuals of every age. (The inaccuracy introduced here by using N_t in the denominator is negligible; in fact the population size is N_t only at the end of the interval during which the B_t births take place.) Although birth rates (fertility rates) are age-specific, b is constant because the age distribution is stable.

Now consider the state of the population at time t. The size of the whole population is N_t, and the size of that part of it in the age range x to $x + 1$ is $N_t c_x$. However, these $N_t c_x$ individuals are all survivors of individuals born between $t - x - 1$ and $t - x$ time units ago, of which there were B_{t-x}. But not all the individuals born in that past interval have survived, only a proportion l_x of them. Thus we have

$$N_t c_x = l_x B_{t-x}. \tag{2.1}$$

Notice next that

$$\frac{B_{t-x}}{N_{t-x}} = \frac{B_t}{N_t} = b$$

since, as already remarked, b, the birth rate per head, remains constant.
    Therefore

$$B_{t-x} = bN_{t-x}.$$

However, in an exponentially growing population,

$$N_t = N_{t-x}e^{rx} \quad \text{or equivalently} \quad N_{t-x} = N_t e^{-rx};$$

therefore

$$B_{t-x} = bN_t e^{-rx}.$$

Substituting this last result in equation (2.1) shows that

$$c_x = l_x b e^{-rx}. \tag{2.2}$$

Now recall that c_x is the proportion of the population in the age range x to $x + 1$. The sum of all values of c_x must therefore be unity or, in symbols,

$$c_0 + c_1 + c_2 + c_3 + \cdots = \sum_x c_x = 1.$$

We shall need this equation later.

(For an account of the use of the summation sign Σ, see Appendix 2.1.)

Consider now the newborn individuals that enter the population during the time interval $t - 1$ to t. There are B_t of them altogether of which $N_t c_x m_x$ are the daughters of mothers in the age range x to $x + 1$. Summing over all x in order to take account of mothers of all ages, it therefore follows that

$$B_t = N_t \sum_x c_x m_x.$$

Then, substituting for c_x from equation (2.2)

$$B_t = N_t b \sum_x l_x m_x e^{-rx}.$$

Therefore

$$\frac{B_t}{N_t} \frac{1}{b} = \sum_x l_x m_x e^{-rx}.$$

Now $b = B_t/N_t$ by definition, so $\dfrac{B_t}{N_t} \cdot \dfrac{1}{b} = 1$.

Thus finally

$$\sum_x l_x m_x e^{-rx} = 1 \tag{2.3}$$

and this is the equation required for the determination of r.

Clearly, before r can be calculated for a population, we must have a list of survival rates (l_x) and fertility rates (m_x). Obtaining this information for populations of wild animals calls for considerable ingenuity on the part of field ecologists and we shall mention some of the methods they use in Chapter 3. The rates for domestic animals are, of course, much easier to come by and are likely to be more reliable. As an example of the calculation of r we therefore use data (shown in Table 2.1) pertaining to a population of domestic sheep *(Ovis aries)* quoted by Caughley (1967). The time unit is one year; females (ewes) only are considered and an individual is said to be "of age x" when her age is in the range x to $x + 1$ years.

Using these data, equation (2.3) must be solved by successive approximation; that is, using trial values of r, we must evaluate $\sum_x l_x m_x e^{-rx}$ repeatedly until we find what value of r (of the desired precision) gives this sum the value nearest unity. A rough estimate (usually an underestimate) of r is given by the formula (Laughlin, 1965)

$$\frac{\left\{ \sum_x l_x m_x \right\} \left\{ \ln \left(\sum_x l_x m_x \right) \right\}}{\sum_x x l_x m_x}.$$

This provides a useful starting point for the trial and error process of obtain-

TABLE 2.1   Survival rates, fertility rates, and the stable age distribution for a population of domestic sheep (Data from Caughley, 1967)

Age in years	Probability of surviving to age x	Number of female offspring born to a mother of age x	Proportion of females in age range x to $x + 1$
x	l_x	m_x	c_x
0	1.000	0.000	0.266
1	0.845	0.045	0.184
2	0.824	0.391	0.147
3	0.795	0.472	0.116
4	0.755	0.484	0.090
5	0.699	0.546	0.068
6	0.626	0.543	0.050
7	0.532	0.502	0.035
8	0.418	0.468	0.023
9	0.289	0.459	0.013
10	0.162	0.433	0.006
11	0.060	0.421	0.002
			1.000

ing a precise value of r. Thus for the sheep example, calculation shows that

$$\frac{\left\{\sum_x l_x m_x\right\} \left\{\ln \left(\sum_x l_x m_x\right)\right\}}{\sum_x x l_x m_x} = \frac{2.513 \ln 2.513}{12.825} = 0.181.$$

This suggests that $r = 0.20$ would be a good first trial values to assign to r. With $r = 0.20$ it is found that

$$\sum_x l_x m_x e^{-rx} = 0.9996$$

which is extremely close to 1. In fact 0.20 is the best estimate of r to two decimal places, since further trials show that

$$\sum_x l_x m_x e^{-rx} = \begin{cases} 1.0422 & \text{when} \quad r = 0.19 \\ 0.9591 & \text{when} \quad r = 0.21 \end{cases}.$$

Accepting $r = 0.20$ as the instantaneous rate of increase per annum, we also have $\lambda = e^r = 1.221$ as the finite rate of increase per annum. In other words, if the population of sheep were left to itself in constant environmental conditions, the size of the population would be multiplied by the factor 1.221 every year. This growth would ultimately slow down, of course, when population size became so large that the population itself altered the environment and negative feedback started to operate. We are asserting only that exponential growth would go on *until* this stage was reached.

Having derived r we are now in a position to find the stable age distribution. Recall equation (2.2), namely

$$c_x = bl_x e^{-rx}.$$

Treating b simply as a constant of proportionality, this could be written

$$c_x \propto l_x e^{-rx}$$

so that

$$\frac{c_x}{\sum\limits_x c_x} = \frac{l_x e^{-rx}}{\sum\limits_x l_x e^{-rx}}.$$

Then, since

$$\sum_x c_x = 1$$

we have

$$c_x = \frac{l_x e^{-rx}}{\sum\limits_x l_x e^{-rx}}. \tag{2.4}$$

It is found that

$$\sum_x l_x e^{-rx} = 3.757.$$

Equation (2.4) now permits calculation of the proportions c_x of the stable age distribution. For example:
when

$$x = 3, \quad l_x = 0.795 \quad \text{and} \quad e^{-rx} = e^{-0.60} = 0.5488.$$

Therefore

$$c_3 = \frac{0.795 \times 0.5488}{3.757} = 0.116.$$

All the c_x values for the sheep population are given in the last column of Table 2.1.

So far the arguments in this chapter apply to all exponentially growing populations regardless of whether breeding is continuous or is confined to an annual breeding season. Continuous breeding, going on at a rate that is unaffected by the seasons, seems to be uncommon though a few examples are known. Sadleir (1969) quotes reports of continuous breeding in some tropical bat species, and almost continuous breeding in White-tailed deer *(Odocoileus virginianus)* in Florida; it is also reported among lizard species living in equatorial rain forest in Borneo (Inger and Greenberg, 1966). In the great majority of species breeding is probably seasonal, at least to some extent.

In low latitudes breeding may go on all through the year but with a peak rate at one season; or there may be a true breeding season of several months duration. At higher latitudes, or where the climate shows marked annual fluctuations, births occur only in one comparatively short annual breeding season when young are born to parents one year old or older. In the rest of this chapter we shall consider only these seasonal breeders. It is necessary to make a mild simplifying assumption: that the breeding season is infinitesimally short instead of being spread over a month or two as it usually is in nature. The effects of the simplification are negligible.

For seasonal breeders, the instantaneous rate of natural increase r, the finite rate of natural increase λ, and the stable age distribution c_x may be obtained as already described. Observe that for continuous breeders the stable age distribution is the same all the time. But for seasonal breeders the age distribution varies cyclically during the course of the year and to say that it is "stable" implies that the same sequence of distributions is repeated in an endless cycle. The distribution given by the proportions c_x occurs each year just at the completion of the breeding season. This is true of the sheep population discussed above, since sheep are seasonal breeders.

Another measure of population growth that is often used is R_0, the *net reproductive rate*. It is defined as

$$R_0 = \sum_x l_x m_x$$

and is the average number of daughters that will be born to a female in her lifetime. To see this, consider a group of N females all starting life at the same instant; such a group is known as a *cohort*. In the first unit of time following their births, as they age from 0 to 1, they will produce Nm_0 daughters (this will be zero if $m_0 = 0$ as is usually the case); in the next time unit the Nl_1 females that still survive will age from 1 to 2 and will produce Nl_1m_1 daughters; in the next time unit there will be Nl_2 survivors and they will produce Nl_2m_2 daughters; and so on. By the time the whole cohort is dead its total output will be $\sum_x Nl_xm_x$ daughters; that is $\sum_x l_xm_x$ per head of the original cohort. For the sheep population of Table 2.1, $R_0 = 2.513$ ewe lambs per ewe per generation.

Two other ways of wording the definition of R_0 are the following: it is the finite rate of increase per individual *per generation;* (contrast λ which is the finite rate of increase per individual per arbitrarily defined unit of time). Secondly, R_0 is the ratio of the total number of female births in one generation to the number in the previous generation.

R_0 is an interesting property of a population but it cannot readily be converted to an absolute measure of population growth unless we know the mean duration of a generation. The "mean duration of a generation" is not an intuitively clear notion and there are various ways of defining it mathematically which we shall not discuss in this book. The subject is dealt with in, for example, Slobodkin (1962), Laughlin (1965), Leslie (1966) and Caughley (1967).

2 PREDICTING WITH THE AID OF MATRICES

A very instructive way of following, conceptually, the course of events in a growing population requires the use of matrix algebra. To follow the argument in this section the reader need only know how matrices are multiplied; this is explained in Appendix 2.2.

In what follows we shall, as before, consider only females. Imagine, for simplicity, an animal species that never reaches the age of four years. Taking one year as time unit, all the members of a population of this species could be classified into four age classes: those between 0 and 1; those between 1 and 2, those between 2 and 3 and those between 3 and 4 years old. Suppose a census is taken just before each breeding season and at the census taken at time $t = 0$ let the population consist of a total of N_0 individuals of which n_{00}, n_{10}, n_{20} and n_{30} belong to the respective age classes. In these symbols with two subscripts, the first digit in the subscript refers to age and the second to time; in general n_{xt} denotes the number of x-year old individuals present at time t. Also

$$N_0, N_1, ..., N_t, ...$$

denote the total number of individuals (all ages together) in the population at times

$$0, 1, ..., t, ...$$

so that

$$\sum_x n_{x0} = N_0, \quad \sum_x n_{x1} = N_1, \quad ..., \quad \sum_x n_{xt} = N_t, \quad ....$$

Now consider what will have happened after one year to the individuals that were present at time $t = 0$. Of the n_{00} initially in the youngest age class, some will have died and the survivors will have grown a year older so that they now belong in the next age class. And likewise up through the age classes except that all n_{30} three-year olds present initially will have died; (since, for this imaginary species, the fourth birthday is unattainable). Also, during the breeding season that came soon after the start of the year we are considering, numbers of newborn individuals were added to the population. Their mothers belonged to several different age classes, and at $t = 1$ the year's newborn are still all of age zero.

If we represent the composition of the population at any time by a column vector, the course of events can be shown schematically as follows:

At $t = 0$ At $t = 1$

$$\begin{pmatrix} n_{00} \\ n_{10} \\ n_{20} \\ n_{30} \end{pmatrix} \qquad \begin{pmatrix} n_{01} \\ n_{11} \\ n_{21} \\ n_{31} \end{pmatrix}.$$

The solid arrows show how some individuals in each age class survive into the next older age class. The dashed arrows show how members of some or all of the age classes present initially contribute, by bearing young, to the youngest age classes at time $t = 1$.

To represent precisely what is shown pictorially by arrows, more symbols are needed. Let F_x be the number of offspring born by a mother of age x that survive to the next census. And let P_x be the proportion of x-year old individuals that survive to age $x + 1$. Then the number, n_{01}, of individuals in the youngest age class when the population is censused at time $t = 1$ is given by

$$n_{01} = F_0 n_{00} + F_1 n_{10} + F_2 n_{20} + F_3 n_{30}.$$

This is the total of the year's newborn individuals alive at time $t = 1$. The individuals in the older age classes at $t = 1$ are simply those who have survived through the preceding year, growing a year older as they did so. Thus

$$n_{11} = P_0 n_{00}, \quad n_{21} = P_1 n_{10} \quad \text{and} \quad n_{31} = P_2 n_{20}.$$

Therefore, summarizing these results in the form of a single equation showing the equality of two vectors, we have

$$\begin{pmatrix} n_{01} \\ n_{11} \\ n_{21} \\ n_{31} \end{pmatrix} = \begin{pmatrix} F_0 n_{00} + F_1 n_{10} + F_2 n_{20} + F_3 n_{30} \\ P_0 n_{00} \\ P_1 n_{10} \\ P_2 n_{20} \end{pmatrix}.$$

This equation implies that each element in the column vector on the left is equal to the corresponding element in the column vector on the right.

Next, consider the matrix product

$$\begin{pmatrix} F_0 & F_1 & F_2 & F_3 \\ P_0 & 0 & 0 & 0 \\ 0 & P_1 & 0 & 0 \\ 0 & 0 & P_2 & 0 \end{pmatrix} \begin{pmatrix} n_{00} \\ n_{10} \\ n_{20} \\ n_{30} \end{pmatrix} = \begin{pmatrix} F_0 n_{00} + F_1 n_{10} + F_2 n_{20} + F_3 n_{30} \\ P_0 n_{00} \\ P_1 n_{10} \\ P_2 n_{20} \end{pmatrix}. \quad (2.5)$$

The right hand side is identical with the vector given above which represents the composition of the population at time $t = 1$. The left hand side shows the vector representing the composition of the population at $t = 0$ *pre*-multiplied by the square *projection matrix*.

Or, in words:

$$\text{(Square projection matrix)} \times \text{(age composition vector at time } t_0)$$

$$= \text{(age composition vector at time } t_1).$$

We see that the projection matrix "operating on" (i.e. premultiplying) the age distribution vector for the time $t = 0$ yields as product the age distribution vector for time $t = 1$.

We may write the foregoing more compactly by representing a whole matrix by a single letter printed in boldface. Following the usual custom, a matrix with several rows and columns is denoted by a capital letter and a column vector (which is a matrix with only one column) by a lower case letter. Thus, putting

$$\begin{pmatrix} F_0 & F_1 & F_2 & F_3 \\ P_0 & 0 & 0 & 0 \\ 0 & P_1 & 0 & 0 \\ 0 & 0 & P_2 & 0 \end{pmatrix} = \mathbf{M},$$

the square projection matrix,

$$\begin{pmatrix} n_{00} \\ n_{10} \\ n_{20} \\ n_{30} \end{pmatrix} = \mathbf{n}_0 \quad \text{and} \quad \begin{pmatrix} n_{01} \\ n_{11} \\ n_{21} \\ n_{31} \end{pmatrix} = \mathbf{n}_1$$

equation (2.5) can be written as

$$\mathbf{M}\mathbf{n}_0 = \mathbf{n}_1.$$

Exactly the same arguments show that

$$\mathbf{M}\mathbf{n}_1 = \mathbf{n}_2, \quad \mathbf{M}\mathbf{n}_2 = \mathbf{n}_3, \quad \text{and so on indefinitely.}$$

In general

$$\mathbf{M}\mathbf{n}_{t-1} = \mathbf{n}_t.$$

In this way, if we know the elements of the projection matrix $\mathbf{M}$ we can predict the future size and age distribution of a population for many years to come. It is not necessary for the population to have a stable age distribution initially. For example we might take, as a starting population, 1000 individuals all of age 1. Then

$$\mathbf{n}_0 = \begin{pmatrix} 0 \\ 1000 \\ 0 \\ 0 \end{pmatrix}.$$

This is obviously not a stable age distribution for even if there were neither births nor deaths the 1000 individuals would inevitably be one year older one year later.

It is usually found, however, that whatever the form of the initial age distribution, if birth and death rates remain constant for a sufficiently long time, the age distribution will change progressively until ultimately it becomes, and remains, stable*. Also, once stability has been reached, the size of every age class (and consequently the size of the population as a whole) is multiplied by the factor $\lambda = e^r$ each year; λ is the finite rate of increase per annum.

Writing this symbolically, let us suppose that the age distribution is stable at time T and also, necessarily, at times $T + 1$, $T + 2$ and so on.

Then

$$\mathbf{Mn}_T = \lambda \mathbf{n}_T = \mathbf{n}_{T+1}$$

where $\lambda \mathbf{n}_T$ denotes the vector $\mathbf{n}_T$ with each of its elements multiplied by λ. Written in full this is

$$\begin{pmatrix} F_0 & F_1 & F_2 & F_3 \\ P_0 & 0 & 0 & 0 \\ 0 & P_1 & 0 & 0 \\ 0 & 0 & P_2 & 0 \end{pmatrix} \begin{pmatrix} n_{0T} \\ n_{1T} \\ n_{2T} \\ n_{3T} \end{pmatrix} = \begin{pmatrix} \lambda n_{0T} \\ \lambda n_{1T} \\ \lambda n_{2T} \\ \lambda n_{3T} \end{pmatrix} = \begin{pmatrix} n_{0,T+1} \\ n_{1,T+1} \\ n_{2,T+1} \\ n_{3,T+1} \end{pmatrix}.$$

We can now demonstrate the process using the data from the sheep population described earlier. First we require the elements F and P of the projection matrix $\mathbf{M}$; the matrix has 12 rows and columns in this example since there are 12 age classes. The F's (the elements in the top row of $\mathbf{M}$) are given by $F_x = \lambda m_x$ with $x = 0, 1, 2, \ldots, 11$. The element P_x which is in the $(x + 2)$th row and $(x + 1)$th column of $\mathbf{M}$ is $P_x = l_{x+1}/l_x$.

The necessary values of m_x and l_x will be found in Table 2.1. Thus, for example, F_4, the fifth element in the first row of $\mathbf{M}$ is

$$F_4 = \lambda m_4 = 1.221 \times 0.484 = 0.591.$$

---

* In some cases the age distribution, instead of becoming stable, oscillates endlessly (see page 75).

And P_5, the element in the 7th row and 6th column of $\mathbf{M}$, is

$$P_5 = \frac{l_6}{l_5} = \frac{0.626}{0.699} = 0.8956.$$

In this way the elements of $\mathbf{M}$ can be quickly calculated and the whole projection matrix is shown in Table 2.2.

Now suppose we start with a flock of 60 sheep (5 in each of the 12 year classes) at time $t = 0$. The initial population vector is thus

$$\mathbf{n}_0 = \begin{pmatrix} 5 \\ 5 \\ \vdots \\ 5 \end{pmatrix}$$

with 12 elements altogether all equal to 5. Repeated premultiplication by $\mathbf{M}$ gives the age distribution in successive years and the results are plotted in Figure 2.1. This shows the changes that take place in the relative proportions of the 12 age classes as the stable age distribution becomes established and

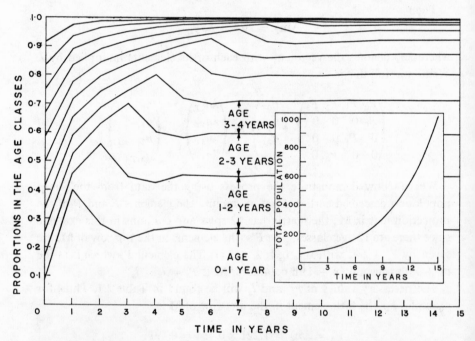

FIGURE 2.1 The changes in age distribution of a growing population of sheep in which, at $t = 0$, the twelve age classes were present in equal proportions. The proportions in every age class in each year are shown by the widths of the strips; (only the strips for the four youngest age classes are labelled). *Inset:* the growth of the total population; $r = 0.20$

TABLE 2.2 The projection matrix for the sheep population

0	0.055	0.478	0.577	0.591	0.667	0.663	0.613	0.572	0.561	0.529	0.514
0.8450	0	0	0	0	0	0	0	0	0	0	0
0	0.9751	0	0	0	0	0	0	0	0	0	0
0	0	0.9648	0	0	0	0	0	0	0	0	0
0	0	0	0.9497	0	0	0	0	0	0	0	0
0	0	0	0	0.9258	0	0	0	0	0	0	0
0	0	0	0	0	0.8956	0	0	0	0	0	0
0	0	0	0	0	0	0.8498	0	0	0	0	0
0	0	0	0	0	0	0	0.7857	0	0	0	0
0	0	0	0	0	0	0	0	0.6914	0	0	0
0	0	0	0	0	0	0	0	0	0.5606	0	0
0	0	0	0	0	0	0	0	0	0	0.3704	0

how the distribution remains constant thereafter. In this example stabilization takes about 11 years. The main figure shows the proportions, not the absolute sizes, of the 12 age classes: the width of each strip represents the proportional size of each class with the youngest (the neonates) at the bottom. The inset figure shows how the size of the whole population increases during the same period. Once the age distribution has become stable the growth of the whole population is exponential with $r = 0.20$; but in the first few years, while the less fertile age classes are overrepresented in the population, growth is slightly slower.

This example shows what a powerful ecological tool the projection matrix is. Its use is not limited to predicting the growth of populations of seasonal breeders. It can also be used to approximate population growth in continuous breeders even though continuously breeding individuals do not belong (as seasonally breeding individuals do) to clearcut, discontinuous age classes.

The method was originally devised in the 1940's, first by Lewis (1942) and independently by Leslie (1945, 1948). Leslie (1945) used the method to explore the growth of a population of rats, *Rattus norvegicus;* later (1966), he applied it to colonies of the Common Murre (guillemot), *Uria aalge*, a sea bird with breeding colonies on coastal cliffs in Scotland. Keyfitz and Murphy (1967) discuss the use of projection matrices in demography, for predicting the growth of human populations. Lefkovitch (1965) has studied population growth in a pest of stored products, the cigarette beetle (*Lasioderma serricorne*).

The method may well prove valuable in predicting the future course of events in a population whose age distribution has been modified by outside interference, of natural or human origin. Thus if some catastrophe decimates a year's crop of newborn animals, for instance, wildlife managers could determine how long a time will be needed for the stable age distribution to be restored, *provided* they have the necessary data to be able to give realistic numerical values to the elements (the F's and P's) in the projection matrix. The practical difficulties of discovering the age-specific birth and death rates of wild animals living in natural conditions are enormous, of course. And this is not the only obstacle to successful prediction: two other difficulties are unavoidable.

In the first place, as we emphasized in Chapter 1 (see page 4), events will not in practice conform exactly to theoretical expectation. The predictions described above are "deterministic"; i.e. they make no allowance for "stochastic", or chance, variation. If, for example, we predict from Table 2.1 that ten years hence the proportion of three-year old sheep in a population having a stable age distribution will be 0.116, we are predicting only the most likely proportion. Ideally we should state a range of values and the probability that the proportion will fall within this range. Pollard (1966) has investigated this approach.

The second difficulty is one that crops up in all branches of ecology. It is one thing to predict the size of a population in nature and quite another to judge whether the prediction is fulfilled. Only in very exceptional cases (the whooping crane, for example) is the size of a wild animal population precisely known. Usually the obstacles in the way of obtaining an accurate census are completely insuperable and one must be content with a very approximate estimate of population size. We shall consider the problems of estimation in Chapter 6.

CHAPTER 3

Life Tables

To DETERMINE the rate of increase of a population it is necessary to know both the age-specific survival rates or life table functions, l_x, and the age-specific fertility rates or maternal frequencies, m_x, of the species. In this chapter, as promised on page 14, we shall describe some of the ways in which numerical values of these rates are obtained.

It will be recalled that the maternal frequency, m_x, is defined as the number of daughters born to a female while her age is in the range x to $x + 1$. To find an m_x value, therefore, two things must be observed: the number of daughters born to a female in a unit of time, and her age. The life table function, l_x, is defined as the proportion of population members that live to age x or older; as will be explained below, l_x values can sometimes be found either by discovering the ages of living individuals, or by judging the ages of naturally occurring carcasses and hence discovering the ages at death of individuals after they have died. In any case, it will be realized that the determination of l_x and m_x values is far from being a routine task; to ascertain them for populations in nature requires detailed knowledge of the habits and natural history of the species concerned. Data are often difficult to obtain and must be eked out be educated guesswork. A few examples will show the variety of difficulties the ecologist must cope with.

1 MATERNAL FREQUENCIES

Maternal frequencies are most easily obtained for small animals that can be reared in the laboratory or in outdoor cages or enclosures. With laboratory populations one may either arrange for experimental conditions to duplicate natural conditions as closely as possible, or one may rear several different populations under different conditions which are held constant for each population. This is done to judge the effects of various factors on the rate of increase. For example, Birch (1948) reared the rice weevil, *Calandra oryzae*, in wheat at temperatures of 23°C, 29°C (the optimum) and 33.5°C. The time unit used was one week. At 29°C it was found that females began to lay eggs when they were four weeks old. On average they laid 40 eggs while they were 4 weeks old (i.e. during the week in which they aged from exactly 4 to exactly 5 weeks old); 46 eggs while they were 5 weeks old; 30 eggs while they were 6 weeks old; and so on. Assuming (in the light of independent evidence) that the ratio of males to females was 1 : 1, one may expect half the eggs laid to yield daughters. The first few m_x values were therefore $m_0 = m_1 = m_2 = m_3 = 0, m_4 = 20, m_5 = 23, m_6 = 15$, etc.

As a contrast to laboratory-reared insects, let us consider large wild mammals. Caughley (1966) describes observations of maternal frequencies made on a population of a horned goat-like ungulate, the Himalayan Thar, *Hemitragus jemlahicus*, living in mountainons country in the South Island of New Zealand. The species was introduced there in 1904 and has become common. In 1963/64 a sample of 623 females was shot and the age of each carcass was determined by counting the annual growth rings on its horns. The hunting was done during the season of births and the average number of births that year per female in a given age class was found by dividing the number that were pregnant or lactating by the total number in the age class. The numbers thus obtained were halved to give the mean number of daughters per female per year for mothers of age 1 year, 2 years, 3 years, and so on. These m_x values were found to be $m_1 = 0.005, m_2 = 0.135, m_3 = 0.440$ etc. female calves per female per annum.

Before accepting these numbers as correct for the whole population, several hidden assumptions must be looked into. In the first place, one would like to be sure that the sample of animals shot was a "representative" sample in terms of representation of the different age classes in the whole population. It might happen that old females were more wary and less likely to be shot than young ones. Or, alternatively, that small, young animals, which present a smaller target to the hunter, were less likely to be shot. In either case a biased, or non-representative, sample of the whole population would be obtained. However, by means of statistical tests, Caughly was able to show that such biases, even if present, were too small to have an appreciable effect on the conclusions. Other assumptions, which were impossible to test, concerned the number of female calves born to the mothers in the sample. Thus it was assumed that the sex ratio of the calves was 1 : 1, that none of them were stillborn, that there was no twinning, and that none of the lactating females were ones that had been barren in the current season but were suckling yearling calves. This account should show how numerous are the possible sources of error that must be thought of and allowed for.

As a final example of the estimation of maternal frequencies, let us take a sea bird, the Thick-billed Murre (*Uria lomvia*) breeding in Canada (see

Tuck, 1960, and Leslie, 1966). The number of young produced by a hen murre in the spring may be taken as the number of fledged chicks, ready to go to sea, reared by a breeding pair of adults. Fifty percent of the young are assumed to be female. The problem of determining the ages of the mothers does not arise if it is assumed that a hen's fertility remains constant throughout her reproductive life which begins when she is 3 years old. This is a reasonable assumption, since one egg is laid per annum per breeding pair of birds; thus the maximum possible value of the maternal frequency is $m_x = 0.5$. The true value must be less than this since not all the eggs laid are successfully incubated, and losses occur between hatching and fledging; on the basis of field observations it was estimated that $m_x = 0.2307$ female chicks per hen per annum when $x = 3, 4, 5$ etc. Since hens younger than 3 do not breed, $m_x = 0$ when $x = 0, 1$ and 2.

2   LIFE TABLE FUNCTIONS

We next consider how values of l_x, the age-specific survival rate or life table function, are obtained.

The problem is comparatively straightforward when the population under investigation is being reared in the laboratory and large numbers can easily be bred. One can then observe the fate of a large group of individuals that all started life simultaneously. Survivors can be counted at fixed time intervals so that a table of l_x values is arrived at automatically. It is exceptional for field ecologists to obtain their data so directly. A case in which it was possible has been described by Deevey (1947) who discussed observations made by Hatton on the barnacle *Balanus balanoides*. At the end of their larval life barnacles attach themselves to rock surfaces and remain there permanently. The attachment season occupies only a short period each spring. One winter Hatton provided clean attachment sites by scraping old barnacles from a rock surface, and the newly adult barnacles that attached themselves to the test surface were therefore known to have begun their adult lives within a very short time period. Survivors were counted every few months and in this way l_x values were obtained that related to the adult lifetimes of the barnacles (the mortality rate among free-living larvae could not be observed).

In a natural population in which all ages are present, there are two ways in which the l_x values may sometimes be deduced: by determining the age distribution among living population members; or by determining the ages at death of naturally occurring carcasses. However, these two methods can be used *only* if the population is "stationary", not otherwise. A *stationary* *population* is one in which the age distribution is stable and at the same time the size of the population as a whole remains constant; i.e. the instantaneous rate of increase, r, is zero, and consequently the finite rates of increase, λ, is unity. The equivalence of the statements $r = 0$ and $\lambda = 1$ is clear from the fact that $\ln \lambda = r$ (see page 4). Since λ is also defined as N_{t+1}/N_t, i.e. as the ratio of population size at any time to its size one time unit earlier, the assertion that $\lambda = 1$ implies that population size remains constant through time*; hence the term "stationary" to describe such a population.

Recall [equation (2.4), page 16] that for any population in which the age distribution is stable the proportion c_x, in the age class x to $x + 1$, is given by

$$c_x = bl_x e^{-rx}.$$

Thus c_x depends on both l_x and x. However, if $r = 0$, then $e^{-rx} = 1$ and consequently $c_x = bl_x$. This shows that if we can discover the c_x values in a stationary population the l_x values can be obtained immediately since they are directly proportional to them.† The c_x values are proportions that sum to unity. The l_x values, however, although they too are proportions, do not sum to unity since they do not measure the relative sizes of *mutually exclusive* portions of the population. Thus the individuals in a population that survive to age $x + 1$, say, are a subset of those that have already survived to age x. Therefore a sequence of l_x values such as l_1, l_2, l_3, etc. is a sequence of steadily dwindling fractions of the l_0 individuals that have survived for at least an infinitesimal period. If we put $l_0 = 1$, then for $x = 1, 2, 3$ etc., l_x is a number less than 1. However, to avoid using small fractions it is often convenient to multiply all l_x values by 1000. Then a statement that $1000l_1 = 700$, for instance, would imply that 700 of 1000 newborn individuals survived for *at least* one unit of time after birth, while the remaining 300 had died off at various times before attaining the age of one time unit. Observe that for a population containing individuals of all ages there is no directly observable number of newborn individuals all present at the same instant such as can be deliberately bred in a laboratory population. Thus the value assigned to l_0 is arbitrary.

To derive l_x values from the ages at death of naturally occurring carcasses we proceed as follows. Suppose, in a large sample of carcasses whose ages at death have been determined, the proportion of individuals that were of

---

* Only a continuously breeding population can be truly stationary, of course. A population of seasonal breeders is bound to show annual fluctuations but such a population is called "stationary" if its size shows no trend over a period of several years.

† Observe that the l_x and c_x values for a population of domestic sheep given in Table 2.1 (page 15) are *not* proportional. This is because that population is not stationary but growing, with $r = 0.20$.

age x when they died is d_x. These individuals must have lived at least to age x, but they did not survive to age $x + 1$. Therefore

$$d_x = l_x - l_{x+1}, \quad \text{for} \quad x = 0, 1, 2, \dots .$$

The d_x values are proportions that sum to unity. That is

$$\sum_{x=0}^{\infty} d_x = 1.$$

This follows since d_x values (like c_x values) are the proportional sizes of mutually exclusive subsets of the population. To derive l_x values from observed d_x values, we use the following equations:

$$l_1 = l_0 - d_0 = 1 - d_0$$

$$l_2 = l_1 - d_1 = 1 - (d_0 + d_1)$$

$$l_3 = l_2 - d_2 = 1 - (d_0 + d_1 + d_2)$$

and, in general,

$$l_x = l_{x-1} - d_{x-1} = 1 - \sum_{j=0}^{x-1} d_j.$$

The foregoing arguments show that, *provided we are dealing with a stationary population*, an observed set of data either of c_x values or of d_x values can be used to obtain l_x values. Values of d_x are not often obtainable in practice since remains of animals that have died naturally are seldom found in sufficient numbers to yield reliable data. Moreover, the bodies of young animals usually decay faster than those of adults, with the result that the value of d_0 is likely to be underestimated.

To obtain a set of c_x values requires that the ages of a representative sample of living population members be discovered; unless the animals can be live-trapped and aged while living, they may have to be killed in the process of capture or of determining age. Care has to be taken to avoid, or allow for, bias (unrepresentativeness) in the sample since animals of different ages may not be equally susceptible to capture. Once the individuals have been caught, their ages have to be determined.

Many animals have organs that show "growth rings" analogous to the annual rings of trees and numerous examples have been given by Taber (1971) for mammals. Thus in the black bear the roots of the canine teeth show layers of dentine which are laid down annually. In bighorn sheep the ridges and furrows on the horns correspond with the alternating periods of fast and slow growth, that is with the seasons. The scales of some fishes show annual rings; fishes are also aged by counting the layers in the otoliths (granules of calcium carbonate in the inner ear). In some animals age can be judged by the condition of structures, such as teeth, that wear away at a steady rate with advancing age. Adult tsetse flies have been aged by noting the degree to which their wings have become tattered with wear and tear. In many species of birds the ages of immature individuals can be judged from plumage characteristics such as color pattern and growth of the wing feathers. Another method of determining age is to infer it from the weight of the lens of the eye which grows continuously throughout life in many animals; the method has been tried on mule deer, foxes, cottontail rabbits, raccoons and rock doves, but there are doubts as to its accuracy (Erickson *et al.* 1970).

Although l_x values for plant populations have rarely been recorded they can be obtained in the same way as for animal populations, provided the population being studied is stationary and its members can be aged. This can often be done for woody plants. Examples have been described by Hett and Loucks (1968). These authors studied three species of tree seedlings in forest stands in Quetico Provincial Park, Ontario, that had been undisturbed for 200 years; the population was thus presumably stationary. We shall use their data to illustrate the construction of a "life table". This is given in Table 3.1 which shows the results for 859 White Pine (*Pinus strobus*) seedlings that were less than 8 years old. The first column gives the age classes. The time unit employed was two years and therefore $x = 0$ denotes the age class 0–2 years; similarly, $x = 1$ denotes the age class 2–4 years; and so on. Ages were judged by counting internodes and terminal bud scars, and were confirmed by counts of annual rings. The second column shows the numbers of tree seedlings in each age class; these numbers are proportional to c_x values and hence, since the population is assumed to be stationary, to l_x values. The 1000 l_x values are shown in the third column; they are obtained by multiplying the corresponding numbers in the second column by 1000/572 which is done merely to standardize the results by making $l_0 = 1$. The fourth column shows successive values of 1000 d_x with $d_x = l_x - l_{x+1}$. Thus 1000 d_x is the number of seedlings, out of an initial population of 1000, that would be expected to die in the 2-year interval between time x and time $x + 1$. The tabulated 1000 d_x values sum to $937 = 1000 - 63$, since on the basis of these observations, 937 of 1000 trees would be expected to die before reaching 8 years of age. The last column shows values of $q_x = d_x/l_x$. Thus q_x is the proportion of the l_x individuals alive at the beginning of the interval from x to $x + 1$ which died during the interval. In other words, q_x is the proportion of the population that died when "of age x" (using the 2-year time units to measure age). We discuss in the next section how tables such as this may be interpreted.

TABLE 3.1   A life table for White Pine seedlings
(Data from Hett and Loucks, 1968)

Age Class	Number of living seedlings	$1000\, l_x$	$1000\, d_x$	$q_x = d_x/l_x$
0–2 years, $x = 0$	572	1000	673	0.673
2–4 years, $x = 1$	187	327	215	0.657
4–6 years, $x = 2$	64	112	49	0.438
6–8 years, $x = 3$	36	63		
	859		937	

4   SURVIVORSHIP CURVES

Hitherto we have assumed that l_x values were to be considered in conjunction with m_x values so that the rate of increase of a population could be calculated by one of the methods described in Chapter 2. Of course l_x values obtained from observations of the ages of living individuals, or of their ages at death, are no use for this purpose since values of l_x so obtained presuppose that $r = 0$ and are reliable only if this is so. However survival rates are interesting in their own right quite apart from their use in calculating rates of increase. A description of the "biology of a population" should certainly contain an account of how long the population members live and whether, for example, the death rate is greatest in infancy, in middle life, or in old age. Observations may be set out in a life table such as Table 3.1 but they are most easily comprehended when they are displayed graphically; there are two ways in which the numbers tabulated in a life table may be graphed: a plot of l_x versus x is called a survivorship curve; and a plot of q_x versus x is called a q_x curve.

Figure 3.1a, b and c give examples of these curves. Figures 3.1a and 3.1b both show survivorship curves and the difference between them is that the l_x axis (the ordinate, or vertical axis) is graduated arithmetically in Figure 3.1a, whereas it is graduated logarithmically in Figure 3.1b. Both figures show two survivorship curves: one is for the domestic sheep population described in Chapter 2 and the data in Table 2.1 have been used to plot it; the other is for a population of lapwings or green plovers (*Vanellus vanellus*) in Britain, and the data are taken from the classic paper by Deevey (1947) in which a large number of ecological life tables have been assembled and compared.

Consider, first, the survivorship curve for the lapwings. When the l_x axis is graduated logarithmically (Figure 3.1b) this appears as a straight line with negative slope. Using natural logarithms, the equation of the line can therefore be written as

$$\ln l_x = \ln l_0 - \delta x$$

or

$$\ln l_x - \ln l_0 = -\delta x$$

where the constant δ is the numerical value of the slope. From this equation we see that

$$\ln \frac{l_x}{l_0} = -\delta x$$

so that

$$\frac{l_x}{l_0} = e^{-\delta x}$$

or

$$l_x = l_0 e^{-\delta x}. \tag{3.1}$$

Thus (3.1) is the equation of the survivorship curve for the lapwings shown in Figure 3.1a in which the l_x axis is arithmetically graduated. The advantage of using $\ln l_x$ rather than l_x on the ordinate is now apparent. The lower curve in Figure 3.1a certainly looks exponential (compare it with the rising exponential curve for population growth shown in Figure 1.2, page 4, which has equation $N_t = N_0 e^{rt}$) but judging the shapes of curves is difficult. The surmise that it is an exponential curve is immediately confirmed, however, when we graduate the ordinate logarithmically and find that the survivorship curve becomes a straight line.

An immediate consequence of equation (3.1) is the following:

$$l_1 = l_0 e^{-\delta} \qquad\qquad \text{whence} \qquad l_1/l_0 = e^{-\delta}$$

$$l_2 = l_0 e^{-2\delta} = (l_0 e^{-\delta})\, e^{-\delta} = l_1 e^{-\delta} \qquad\qquad \text{whence} \qquad l_2/l_1 = e^{-\delta}$$

$$l_3 = l_0 e^{-3\delta} = (l_0 e^{-2\delta})\, e^{-\delta} = l_2 e^{-\delta} \qquad\qquad \text{whence} \qquad l_3/l_2 = e^{-\delta}$$

$$\cdots \cdots \cdots \cdots \cdots \cdots \cdots$$

$$l_{x+1} = l_0 e^{-(x+1)\delta} = (l_0 e^{-x\delta})\, e^{-\delta} = l_x e^{-\delta} \qquad \text{whence} \qquad l_{x+1}/l_x = e^{-\delta}$$

and so on.

Thus

$$\frac{l_1}{l_0} = \frac{l_2}{l_1} = \frac{l_3}{l_2} = \cdots = \frac{l_{x+1}}{l_x} = \cdots = e^{-\delta}.$$

This shows that in every unit of time a constant *proportion*, $e^{-\delta}$, of those birds alive at the start of the time unit survived; and a constant proportion died. The proportion that died is

$$1 - e^{-\delta} = 1 - \frac{l_{x+1}}{l_x} = \frac{l_x - l_{x+1}}{l_x} = \frac{d_x}{l_x} = q_x,$$

using the symbol we have already chosen to denote the proportion of the population that dies at age x.

Next consider the graph for the lapwing population in Figure 3.1c in which q_x is plotted against x (age). Although q_x is not strictly constant it shows no consistent upward or downward trend and it seems fair to conclude that its ups and downs are due only to chance variations in the num-

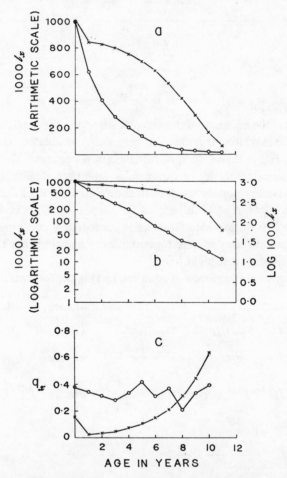

FIGURE 3.1 Survivorship curves, and q_x curves, for lapwings, o——o, (data from Deevey, 1947)); and domestic sheep, x——x, (data from Caughley, 1967). The ordinate is scaled arithmetically in (a) and logarithmically in (b)

bers of deaths at each age. On the basis of this evidence, therefore, it seems safe to assert that if the observed population of lapwings had been very much larger (there were actually 380) the q_x curve would have been much closer to a horizontal straight line, indicating that q_x was constant and did not vary with age.

This assertion can be reworded as follows: for the lapwings, the risk of death was independent of age. The biological implication of this is that none (or negligibly few) of the birds deteriorated physiologically with age: the risk of death was the same for an old bird as a young one. Looked at another way this conclusion amounts to the following: the risk of accidental death was at all times so great that none (or negligibly few) of the birds lived long enough to become physiologically old. These conclusions seem likely to be true of birds in general. In his 1947 paper, Deevey shows curves of $\ln l_x$ versus x for eight other species besides lapwings and all the curves are straight, at least roughly, for most of their lengths.

Now let us contrast the three curves for the lapwings with the corresponding ones for the sheep population. It is clear from Figure 3.1b that the sheep population remained large for a considerable time, after which its size fell off rapidly. The course of events is shown more clearly by the q_x curve in Figure 3.1c. As one would expect, there is a comparatively high risk of death among newborn lambs; this is also apparent, though less obvious, in Figure 3.1a and is scarcely noticeable in Figure 3.1b. Returning to Figure 3.1c we see that the risk of death is smallest for yearling lambs and then, as the animals age, the risk increases at a steadily increasing rate. The curves for the sheep are representative of those for many other mammals (see Caughley, 1966) including man, and also resemble those for such vastly different organisms as rotifers (Deevey, 1947). They are typified by high death rates in infancy and old age and much lower death rates between.

5 INSECT LIFE TABLES AND THEIR PRACTICAL APPLICATIONS

The life tables we have considered up to this point have been those of comparatively long-lived species such as mammals and birds and perennial plants. In populations of this kind the generations overlap; for example it was shown by Leslie (1966) that in the population of Common Murres he studied (see page 24) seven generations could be present simultaneously. It follows that in such a population individuals of many different ages are alive at one time. If the species is a continuous breeder, individuals of any age may be found from neonates (newborn) to the oldest possible for the species. If the species is a seasonal breeder it is clearly impossible for all ages (measured to a fraction of a year) to be present at one particular moment of time but all the year classes are represented. Thus for long-lived species the age distribution that would be revealed by a census taken on a single day will be like that shown in Figure 3.2*a* (for a continuous breeder) or in Figure 3.2*b* (for a seasonal breeder).

Now consider species in which the generations do *not* overlap and there is only one breeding season per year. This is true, for example, of "univoltine" insects and also of most annual plants. Eggs are laid (or seeds are set) at the ends of the parents' lifetimes and the parents are dead before the eggs hatch (or the seeds germinate). When a population of such a species is censused, all the individuals are of roughly the same age; the age difference between the oldest and youngest is no greater than the duration (usually short) of the breeding season. Thus all the individuals belong to a single age class and the age distribution is "degenerate" (see Figure 3.2*c*).

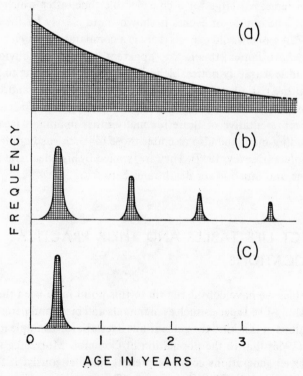

FIGURE 3.2 The three types of age distribution: (*a*) for a continuous breeder; (*b*) for a seasonal breeder; (*c*) for a species that lives for one year and breeds at the end of its life

When a life table is to be constructed for such a species one cannot infer the l_x values (the age-specific survival rates) from the age distribution since there is no age distribution in the ordinary sense. But one can observe the l_x values directly, by determining the size of the population at a succession of times. Indeed, the naturally occurring population is roughly a cohort (i.e. all its members are of roughly the same age) and one can observe the way in which its size dwindles with the passage of time in exactly the same way as one can follow the fate of an even-aged laboratory population. There are often tremendous difficulties in practice of course. Judging the size of a wild population of insects, for example, is always hard to do, but for the moment we shall disregard the problems of ecological sampling which are discussed in Chapter 6.

It is worth emphasizing the following contrast: for species with over-lapping generations one can, at least in principle, infer the l_x values from observations made at one single point in time. Admittedly there may be difficulties in practice; for the method to work the population must be stationary, the sample unbiased, and the individuals in the sample correctly aged. But if all goes well, one need sample the population only once. For species with no overlap of generations, on the other hand, l_x values can be observed only by observing a natural population (which is a cohort) at a succession of times and estimating its size on each occasion.

A modification of this method is used by economic entomologists in compiling life tables for pests. In planning the control of a damaging insect pest, it obviously helps to know all that can be discovered about the natural hazards the pest is exposed to. Repeated sampling of a pest population may enable an investigator to determine not only the size of the surviving population at each inspection, but also the relative importance of the various natural causes of death. Records are kept of the weather so that its effects on survival can be observed; and at each inspection one can attempt to judge the abundance of predators and insect parasites (parasitoids), and the prevalence of different diseases. With some species it is also possible to discover the fate of the pupae by collecting and examining pupal cases; one can judge which of them yielded adults, and which died from parasitism and which from predation (Miller, 1963).

Table 3.2 gives an example of a life table for an insect pest. The pest is a moth, the spruce budworm, *Choristoneura fumiferana;* its larvae (caterpillars) do serious damage to forests in Eastern Canada, especially in New Brunswick. Notwithstanding the insect's name the tree species most seriously affected is Balsam fir (*Abies balsamea*), though spruce are attacked too. The life table was arrived at by taking the means of 80 different sets of observations (see Miller, 1963). The way in which the table is drawn up demonstrates the two chief differences between an insect life table and the sort of table (exemplified by Table 3.1, page 32) that we have discussed hitherto. The differences are:

a)   The left hand column records the "stages", not the ages, of the organism. The x values are merely labels and the time intervals they denote are not of equal length. This is done because judging the ages of insects is extremely difficult, whereas the stages (eggs, the several larval instars, pupae, and adults) are obvious. Nothing is lost by disregarding the unequal durations of the stages, since the values of the life table function are not required for calculating the rate of increase of the population. Their usefulness is in the light they throw on the vulnerability of the pest to its natural enemies. The durations of the stages may be very unequal indeed. In Table 3.2, for example, the "small larva" stage lasts from August to the following May, whereas the adult stage lasts for the last two weeks of July. As a reminder that the numbers of survivors and of deaths, and the proportions dying

TABLE 3.2   A life table for the Spruce Budworm
(modified from Miller, 1963)

Stage	Number of survivors $1000\, l'_x$	Number of deaths $1000\, d'_x$		Proportion of deaths q'_x
Eggs $(x = 0)$	1000	190	Parasites: 90 / Predators: 60 / Other: 40	$\dfrac{190}{1000} = 0.19$
Small larvae $(x = 1)$	810	664	Winter deaths; / Dispersal in / Spring and Fall	$\dfrac{664}{810} = 0.82$
Big larvae $(x = 2)$	146	125	Parasites: 59 / Disease: 33 / Other: 33	$\dfrac{125}{146} = 0.86$
Pupae $(x = 3)$	21	7	Parasites: 2 / Predators: 1 / Other: 4	$\dfrac{7}{21} = 0.33$
Adult moths $(x = 4)$	14	14	(All adults die / in about 2 / weeks)	$\dfrac{14}{14} = 1.0$

in each stage, are not based on values pertaining to equal time intervals, the symbols l'_x, d'_x, and q'_x have been printed with primes. As usual 1000 l'_0 has been set equal to 1000 and the other l'_x values adjusted accordingly.

b)   The 1000 d'_x column shows a breakdown of the deaths, by causes, within each stage. It appears, for example, that parasites cause more deaths than predators. We also see that the greatest numerical loss befalls the small larvae. Some of these are lost by emigration when they are on the move in spring and fall; though they may not necessarily die, they disappear from the population under study. The remainder of the losses in this stage result from the deaths of caterpillars inside the "hibernacula" in which they spend the winter in an inactive state.

The usefulness of a life table as a means of summarizing a mass of observations should now be clear. Other examples will be found in LeRoux *et al.* (1963) and Harcourt (1969).

6   INTERPRETING LIFE TABLES

One would not expect the size of an insect population to be the same year after year. Even if there is no discernible trend, the numbers are likely to vary enormously from one year to the next. We now wish to discover what causes this variation. Of all the many imaginable causes for the observed fluctuations, are there a few (or even only one) whose effects dominate all others? For example, as shown in Table 3.2, the eggs and immature stages of the spruce budworm may be destroyed by a number of different agents: parasites, predators, diseases, winter cold and unidentified "others". What we should like is a method of judging to what extent the size of the budworm population in a particular year is controlled by these several agents. This information is difficult to acquire. What we can do, however, is judge the relative importance of the survival rates during the different stages to the survival of a whole generation. It may then be possible to infer the real causes of low survival from a knowledge of the chief causes of death within each stage.

Let us consider the events that occur in a population of budworm during the course of one year, starting with N_t adults in the summer of year t. A certain proportion, not necessarily one half, of these adults is female; let this proportion be P. Thus there are $N_t P$ adult females and if we denote the mean fertility per female by F, then the females will lay a total of $N_t PF$ eggs in year t. Only a certain proportion, say S_E, of these eggs will survive and

hatch; the remainder will fail for one or other of the causes listed in Table 3.2.
Consequently the number of small larvae will be $N_t PFS_E$. Continuing in the
same way, denote by S_S the proportion of small larvae that survive to become
big, by S_B the proportion of big larvae that survive to pupate, and by S_P the
proportion of pupae that survive to yield adults in the summer of year $t + 1$.
Also, denote by N_{t+1} the number of these adults. Then, clearly,

$$N_{t+1} = N_t PFS_E S_S S_B S_P. \tag{3.2}$$

Each of the factors consisting of S with a subscript represents the proportion
of the population that survives through the stage that has the subscript
letter as initial (E for "eggs" and so on) and all stages together constitute
the whole life cycle.

The ratio $I = N_{t+1}/N_t$ is known as the *trend index*. When there is an
increase in population from one year to the next, I exceeds unity and when
there is a decrease I is less than unity. If we take logarithms of both sides of
equation (3.2) it is seen that

$$\log I = \log P + \log F + \log S_E + \log S_S + \log S_B + \log S_P. \tag{3.3}$$

Thus $\log I$ is the sum of several components, in this case six. Depending on
the species of insect concerned and the ease with which its larval stages can
be recognized, it is sometimes desirable to subdivide the larval stage into
three or four substages instead of only two (small and big); $\log I$ would then
have seven or eight components, but for simplicity in what follows we shall
treat it as having six. We see that $\log I$ is a "dependent" variable; once the
values of the six components (the "independent" variables) are given, the
value of $\log I$ is determined. What we now wish to judge is the relative importance of the independent variables in determining $\log I$. The analysis
described below is due to Mott (1967) (and see, also, Watt, 1961).

In deciding the relative importance of an independent variable, what matters is not its magnitude but the degree to which variation in the dependent
variable is related to, and presumably caused by, that of the independent
variable. Thus if pupal mortality, for example, was always very great but
was known to remain constant from year to year, one would not attribute
a sudden decrease in population size to this high pupal mortality. One would
search for an explanation of the decrease among factors that *had* changed,
not among factors that had remained constant.

To judge the importance of the various components in controlling the
value of $\log I$, we must therefore have data collected over a number of years
or from a number of places; no conclusions can be drawn from a single set
of observations. Assume, then, that we have data for n years (or equi-

valently, for n generations). That is, there are n observed values each of log I, log P, log F, etc. The degree to which log I varies from year to year is measured by its variance, var (log I). Similarly the variances of the independent variables may be calculated.

We must also consider whether any pair of independent variables covary (i.e. vary jointly). The degree to which they do so is measured by their *covariance* (see Appendix 3.1). For example, it might be found that in years with a wet summer survival was poor among both eggs and small larvae, whereas in dry years both stages had high survival. Then the covariance of log S_E and log S_S, namely cov (log S_E, log S_S), would be positive. Such a result is quite likely since weather patterns tend to persist for several weeks. Another example: it could happen that heavy parasitism of the big larvae (and consequently low S_B) would leave the population so small that the birds, especially warblers, which prey on the pupae would stay away from the region *because* of the dearth of pupae; then pupal survival, S_P, would be high. The converse chain of events is: few parasites, high S_B, many birds to eat the pupae, and consequently low S_P. In either case, one stage would experience low survival when the other experienced high survival and cov (log S_B, log S_P) would be negative.

Now it can be shown (see Appendix 3.1) that the variance of the dependent variable is equal to the sum of the variances of the independent variables plus twice the covariance of each pair of independent variables. The equation is:

$$\text{var} (\log I) = \text{var} (\log P) + \text{var} (\log F) + \cdots + \text{var} (\log S_P)$$

$$+ 2 \text{ cov} (\log P, \log F) + 2 \text{ cov} (\log P, \log S_E) + \cdots$$

$$+ 2 \text{ cov} (\log S_B, \log S_P). \tag{3.4}$$

(There are 15 covariances since there are 15 ways in which a pair can be selected from the six independent variables.)

Equation (3.4) is true *exactly*, as are equations (3.2) and (3.3). None has discrepancies due to uncertain prediction or measurement errors; this is because the value of I is calculated directly from equation (3.2), i.e. by multiplying its factors.

The right side of equation (3.4) shows that var (log I) has 21 components (when log I itself has six); when we have an actual example with numerical values in place of the symbols, their relative magnitudes are obvious and it can easily be seen which contributions to log I are the most important. Also, by examining covariances, we can tell when low survival rates in two different stages tend to enhance each other's effects (if their covariance is positive) or tend to cancel each other out (if their covariance is negative).

As a numerical example we shall analyse data from Samarasinghe (1965). The insect concerned is the oystershell scale, *Lepidosaphes ulmi*, a pest of apple trees. The observations were made in Quebec in the 1963/64 season; though data relating to only a single year are used here, the several sets of observations needed for variances and covariances to be calculated are provided for by the fact that data were collected from ten different trees. The numbers of scale insects of each stage on each tree were estimated and in this way ten values of I (the trend index) were obtained, one for each tree. In nine of the trees the 1964 population of scale was much smaller than the 1963 population and I ranged from 0.013 to 0.112; only in the tenth tree did the population of scale increase and in this case I was 1.213.

For this species of insect the trend index, I, is the product of only four factors: S_E, S_S and S_B, the proportions of survivors among eggs, small larvae and big larvae, respectively, and F, the fertility factor. Since in these scale populations every individual is female and reproduction is entirely parthenogenetic (Samarasinghe and LeRoux, 1966), the proportion of females, P, is invariably one. And since the big larvae moult to give adults directly without an intervening pupal stage, there is no pupal survival factor, S_P, as there is for most insects. Thus in the present case the equations analogous to equations (3.2) and (3.3) are

$$I = S_E S_S S_B F$$

and

$$\log I = \log S_E + \log S_S + \log S_B + \log F. \tag{3.5}$$

For compactness, let us now write:

$$\log I = y; \quad \log S_E = x_1; \quad \log S_S = x_2; \quad \log S_B = x_3; \quad \log F = x_4.$$

Thus y is the dependent variable, and the four x's are independent variables. By analogy with equation (3.4) we see that

$$\text{var}(y) = \text{var}(x_1) + \text{var}(x_2) + \text{var}(x_3) + \text{var}(x_4) + 2\,\text{cov}(x_1, x_2)$$

$$+ 2\,\text{cov}(x_1, x_3) + 2\,\text{cov}(x_1, x_4) + 2\,\text{cov}(x_2, x_3)$$

$$+ 2\,\text{cov}(x_2, x_4) + 2\,\text{cov}(x_3, x_4). \tag{3.6}$$

The analysis to be carried out consists in finding the numerical values of each of the ten components on the right side of equation (3.6). Then, by comparing their relative magnitudes, one may judge how important the different factors were in determining the trend index, and to what extent pairs of them reinforced each other (as shown by positive covariances) or acted in opposition to each other (as shown by negative covariances).

The original data (somewhat simplified) and some of the steps in the calculations are given in Appendix 3.2. Here (Table 3.3) we show the sizes of the ten component terms as percentages of var (y) = var $(\log I)$. Before discussing Table 3.3 it must be emphasized that the conclusions drawn from it apply only to the actual data analysed: that is, to the insects on the particular trees studied in the year concerned. Therefore, although a large value for a component probably signifies a true cause-and-effect relationship, small values result merely from chance. For instance, the small negative covariance between $x_1 = \log S_E$ and $x_3 = \log S_B$ cannot be taken to imply any real tendency for low survival of eggs to be partly compensated for by high survival of big larvae or vice versa; we should not be surprised if a comparable set of observations from another group of trees yielded a small positive covariance of these components. Indeed, a value of zero for any of the entries in Table 3.3 is, intrinsically, excessively improbable; thus even when there is no true relationship between two variances (so that their "expected" covariance is zero) a small positive or negative value for their calculated covariance is almost inevitable.

TABLE 3.3   The variances and covariances of the components of $y = \log I$ given as percentages of var (y), the total variance

	Eggs	Small larvae	Big larvae	Fertility
Eggs	var (x_1) 10.16	2 cov (x_1, x_2) 3.67	2 cov (x_1, x_3) −1.89	2 cov (x_1, x_4) 11.15
Small larvae		var (x_2) 13.84	2 cov (x_2, x_3) 10.55	2 cov (x_2, x_4) −7.18
Big larvae			var (x_3) 19.64	2 cov (x_3, x_4) 16.92
Fertility				var (x_4) 23.14

Without doing more elaborate investigations, we cannot decide how large a component would have to be for us to regard it as "significant", i.e. as implying, with high probability, a true relationship that would recur consistently in other investigations of the sáme insect. But the entries in Table 3.3 do tell us, exactly, the relative sizes of the components of var $(\log I)$ in the particular population that was studied, and the conclusions they lead to are as worthy of consideration as the results of any other observations made on the same material. It appears, for instance, that survival rate among big larvae had a greater effect on the trend index than the survival rates in the two earlier stages; that the most important single factor is the fertility of the adults; and that fertility and survival rate of the big larvae are correlated.

Studies like these have so far been carried out only with insect populations; further examples will be found in Mott (1967). Similar investigations on other classes of organisms would probably be very worthwhile.

CHAPTER 4

The Regulation of Populations

WE ARGUED in Chapter 2 that populations of animals and plants can, and probably often do, grow exponentially for limited periods. When a population of size N is growing exponentially, its rate of growth, dN/dt, is proportional to N; and the growth rate *per individual* is therefore constant. That is, we have the relation

$$\frac{1}{N} \frac{dN}{dt} = r \quad \text{(see page 11)};$$

the constant r, which we have called the instantaneous rate of increase, is also known as the "intrinsic" (or "inherent", "innate" or "incipient") rate of increase, or as the "Malthusian parameter".

The rate r depends only on the environmental conditions and for exponential growth to continue at an unchanging rate the conditions must remain constant. Indeed, the value of r under stated conditions is a biological characteristic of the species concerned and its maximum possible value, attained when conditions are optimal, is known as the "biotic potential" of the species.

Now it is obvious that if a population continues to grow for a sufficiently long time, its density will eventually become so great as to cause further growth to be inhibited. In other words, since growth cannot continue forever it must eventually slow down and stop and the "force" opposing continued growth is known as *environmental resistance*. The maximum possible density that a population can maintain for a prolonged period in any given environment is known as the *saturation* density or *equilibrium* density and its magnitude is determined by the *environmental capacity* or the *carrying capacity of the environment*. When a population's density approaches the saturation level, changes in growth rate occur which operate to counteract chance departures from equilibrium. In this way the density of the population is regulated, or could be if the population becomes large enough.

The origins of some of the terms introduced in italics above have been discussed by Cole (1954). The meanings of most are obvious and the concepts underlying them are now so much a part of the thinking of all ecologists that the coining and defining of special terms is no longer necessary. However, the precise consequences of these concepts are not obvious. The purpose of this chapter is to examine some of them; to consider how one may judge whether a population is being constrained to grow at a rate less than r; and to discuss current theories and opinions on population regulation and their evolutionary implications.

Before embarking on technical details, a relevant quotation from Malthus (1798) is apropos. Writing of plants and animals, he commented that "They are all impelled by a powerful instinct to the increase of their species; and this instinct is interrupted by no reasoning, or doubts about providing for their offspring. Wherever therefore there is liberty, the power of increase is exerted; and the superabundant effects are repressed afterwards by want of room and nourishment, which is common to plants and animals; and among animals, by becoming the prey of others."

1 VARIATION IN THE GROWTH RATE PER INDIVIDUAL

Consider a population growing in such a way that its growth rate per individual, instead of being constant, is dependent on population size. A crucial question to ask, if we are to understand the growing population's behavior, is: what is the relationship between growth rate and population size? Before considering how this question might be answered in practice, it is worth exploring the consequences of various hypothetical relationships between growth rate and numbers. It is easy to postulate a variety of plausible relationships; deducing the consequences of the simple ones is a straightforward exercise in ecological model building.

Figures 4.1a, b, c and d show four very simple relationships that are worth looking into. In each of the graphs the population size, N, is measured along the abscissa (the horizontal axis), and the growth rate per individual, $\dfrac{1}{N}\dfrac{dN}{dt}$, along the ordinate (vertical axis). Population size is treated as a continuously varying number though in relality, of course, it must always be a whole number and can change only in discrete, unit steps.

Although in what follows we shall stress the differences among the models, we must not lose sight of five properties they have in common. These are:

i)   In every case it is assumed that the population is confined to some particular limited space so that density is always directly proportional to N; i.e. the effects of crowding cannot be relieved by the individuals' spreading out to broach previously untouched resources.

ii) Population density (or size) is the *only* thing affecting the growth rate per individual.

iii) Changes in N affect growth rate instantly with no time lag.

iv) In all four models the intrinsic* (or instantaneous) rate of increase, r, has the same numerical value, namely 0.5. This is the rate of increase per individual that obtains while the individuals are not interfering with (or competing with) one another. However, most of the time there is mutual interference, with the result that what may be called the *operative rate of increase* is less than r. It is the different relationships between this operative rate, $\dfrac{1}{N}\dfrac{dN}{dt}$, and population size, N, that distinguish the models.

v) In all the models the populations are assumed to be growing in environments with the same carrying capacity so that the equilibrium level is the same in each case. Its numerical value is 500. The conventional symbol for this equilibrium (or saturation) level is K.

Now consider how the four models differ.

In the first (see Figure 4.1a) the operative rate of increase is linearly related to population size. That is, the relationship is given by the equation

$$\frac{1}{N}\frac{dN}{dt} = r - sN \quad \text{for all} \quad N > 0. \tag{4.1}$$

The model entails the assumption that any increment in population size, no matter how sparse the population or how small the increment, is accompanied by a reduction in the operative rate of increase. In other words, the individuals compete or otherwise interfere with one another even at the lowest densities. The smaller the population the greater the operative rate of increase per individual, but the intrinsic rate, $r = 0.5$, is a limiting value never actually attained. Notice that when $\dfrac{1}{N}\dfrac{dN}{dt} = 0$, i.e. when population growth stops

$$N = K \quad \text{(from the definition of } K)$$

and thus

$$r - sK = 0.$$

---

* Inconsistency in the words used to define "r" is unavoidable since the appropriate word depends on the context. When r is contrasted with the *finite* rate of increase, λ, (see page 3) it is natural to describe r as an *instantaneous* rate. When r is thought of as the growth rate the species would exhibit in an unlimited environment (in contrast to the *operative* rate that obtains when there is interference among individuals) it is more appropriate to speak of r as an *intrinsic* rate.

Consequently
$$s = r/K$$

and since we have already put $r = 0.5$ and $K = 500$ we must have $s = 0.001$. The line in Figure 4.1a is therefore

$$\frac{1}{N}\frac{dN}{dt} = 0.5 - 0.001N. \tag{4.2}$$

The second model (Figure 4.1b) is an improvement on the first from a common sense point of view in that it allows for the fact that crowding is not likely to reduce the rate of increase so long as the density is below a certain threshold. Take $N = 100$ as the threshold level below which mutual interference is without effect. Then, for $0 < N \leq 100$, the line is $\dfrac{1}{N}\dfrac{dN}{dt} = r$ $= 0.5$. In other words, over this range of values of N, growth is exponential. Above the threshold level we have, as before, a linear relationship between the operative rate of increase and population size. That is, for $100 \leq N \leq 500$,

$$\frac{1}{N}\frac{dN}{dt} = r_1 - Ns_1.$$

To find the constants r_1 and s_1, observe from Figure 4.1b that the line must go through the two points whose coordinates are $(100, 0.5)$ and $(500, 0)$ respectively.

Therefore
$$0.5 = r_1 - 100s_1$$

and
$$0 = r_1 - 500s_1.$$

Solving these equations yields $r_1 = 0.625$ and $s_1 = 0.00125$. Thus the relationship shown in Figure 4.1b can be summarized as follows:

$$\frac{1}{N}\frac{dN}{dt} = \begin{cases} 0.5 & \text{when} \quad 0 < N \leq 100 \\ 0.625 - 0.00125N & \text{when} \quad 100 \leq N \leq 500. \end{cases} \tag{4.3}$$

The third model (Figure 4.1c) assumes that when population density is low the individuals are so sparse that not all of them find mates; their chances of doing so increase with increasing density and the rate of increase attains its intrinsic value of $r = 0.5$ when $N = 100$. While N is in the range 0 to 100, the rate of increase is directly proportional to N or

$$\frac{1}{N}\frac{dN}{dt} = uN.$$

This line joins the points $(0, 0)$ and $(100, 0.5)$ whence $u = 0.5/100 = 0.005$.

For $100 \leq N \leq 500$ the relationship is the same as in Figure 4.1b. Therefore for the model in Figure 4.1c:

$$\frac{1}{N}\frac{dN}{dt} = \begin{cases} 0.005\,N & \text{when} \quad 0 < N \leq 100 \\ 0.625 - 0.00125N & \text{when} \quad 100 \leq N \leq 500. \end{cases} \tag{4.4}$$

In the fourth model (Figure 4.1d) population growth occurs at the intrinsic rate of $r = 0.5$ when $100 \leq N \leq 400$. That is, when N is in this range growth is uninhibited since the population is neither too sparse nor too dense. When N is in the range 0 to 100, a rise in N improves the chances of successful mating and the growth rate increases linearly. When N is in the range 400 to 500 the rate decreases linearly because of the steadily intensified crowding. The three line segments showing the whole relationship join the four points (0, 0), (100, 0.5), (400, 0.5) and (500, 0). It is easily found that

$$\frac{1}{N}\frac{dN}{dt} = \begin{cases} 0.005\,N & \text{when} \quad 0 < N \leq 100 \\ 0.5 & \text{when} \quad 100 \leq N \leq 400 \\ 2.5 - 0.005\,N & \text{when} \quad 400 \leq N \leq 500. \end{cases} \tag{4.5}$$

When the four models are described in this manner by showing the relationship between the operative rate of increase per individual and population size, the contrasts among them are obvious and striking. But when we look at the integral curves for the models (Figures 4.1a', b', c', d') it is found that they are not nearly so easy to distinguish.

The integral curve for each model is the curve showing the relationship between population size and time for a population whose rate of increase behaves in the way specified by the model. The equations of the integral curves are obtained by integrating the differential equations (4.2), (4.3), (4.4) and (4.5).* Below we give the solutions when N_0, the starting size of the population at time zero, is set equal to 20 in every case. The process of solving the equations is given in full in Appendix 4.1; however, one may check the correctness of the solutions given below by differentiating them with respect to t to confirm that they yield the differential forms once again. This has also been done in Appendix 4.1.

The solutions are as follows. In every case $N_0 = 20$ and $K = 500$.
For equation (4.2) we have

$$N = \frac{K}{1 + Ce^{-rt}} \quad \text{when} \quad 0 < N \leq 500$$

---

* The slopes of these integral curves for any value of N may be found by substituting the desired N in the appropriate expression for dN/dt. The result is the *overall* growth rate of the population (as distinct from the growth rate *per individual*). The student will find it instructive to plot graphs showing the relation between dN/dt and N for the four models shown in Figure 4.1.

with $r = 0.5$ and $C = \dfrac{K - N_0}{N_0} = 24$. This is the equation of the *logistic growth curve*, and it is shown in Figure 4.1a'. It shows the population growing slowly at first, then more rapidly, and then slowing again as its size approaches the saturation level asymptotically. We defer discussion of this result until the equations of all four growth curves have been given.

For equations (4.3) the solution is

$$N = \begin{cases} N_0 e^{rt} & \text{when } \quad 0 < N \le 100 \\[2ex] \dfrac{K}{1 + C_1 e^{-r_1 t}} & \text{when } \quad 100 \le N \le 500 \end{cases}$$

with $r_1 = 0.625$ and $C_1 = 29.298$.

The first of these equations is that of an exponential growth curve and the second that of a logistic growth curve. The reason C_1 has the value given is as follows. While N was small the population was growing exponentially. The time it took to reach a size of $N = 100$, given that at $t = 0$ its size was $N_0 = 20$, is obtained by solving

$$100 = N_0 e^{rt} \quad \text{or} \quad 100 = 20 e^{0.5 t}$$

whence $t = 2 \ln 5 = 3.22$. At this instant growth is assumed to have changed from exponential to logistic but there is to be no discontinuity in the growth curve. Therefore C_1 must be given a value that ensures $N = 100$ when $t = 3.22$ in the logistic equation.

Thus $N = \dfrac{500}{1 + C_1 e^{(-0.625 \times 3.22)}} = 100$ must be solved for C_1. It is found that

$$C_1 = 4 e^{(0.625 \times 3.22)} = 29.928.$$

Figure 4.1b' shows the resultant growth curve.

For equations (4.4) (the third model) the solution is:

$$N = \begin{cases} \dfrac{N_0}{1 - u N_0 t} & \text{when } \quad 0 < N \le 100 \\[2ex] \dfrac{K}{1 + C_2 e^{-r_1 t}} & \text{when } \quad 100 \le N \le 500 \end{cases}$$

with $u = 0.005$ and $C_2 = 593.7$. The growth curve conforms to the first of these two equations until N reaches 100. The time at which this occurs is found by putting $N = 100$ in the equation and solving for t to give $t = 8$. We must then find the appropriate value, C_2, required in the second equation (which is the equation of a logistic curve) to ensure no discontinuity in the whole growth curve. The method is exactly the same as that used in finding the constant C_1 for the second model. For the third model the growth curve is shown in Figure 4.1c'.

Finally, for equations (4.5), the solution is

$$N = \begin{cases} \dfrac{N_0}{1 - uN_0t} & \text{when} \quad 0 < N \leq 100 \\[2ex] C_3 e^{rt} & \text{when} \quad 100 \leq N \leq 400 \\[2ex] \dfrac{K}{1 + C_4 e^{-2.5(t-C_5)}} & \text{when} \quad 400 \leq N \leq 500. \end{cases}$$

Thus the first part (when $N \leq 100$) is as in the third model; the middle section is exponential; and the final section (when $N \geq 400$) is logistic. The whole curve is shown in Figure 4.1d'. The constants $C_3 = 1.8316$, $C_4 = 0.25$ and $C_5 = 10.78$ are such as to ensure smooth continuity of the curve.

We have now constructed four models, simple but possible, to account for the growth of a natural population of organisms. Let us review the outcome. The growth curves (or integral curves) are sigmoid (S-shaped) in form in all

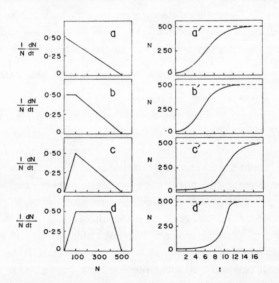

FIGURE 4.1   Four simple models of population growth showing (on the left) the relation between the operative rate of increase per individual, $(1/N)\,dN/dt$, and population size, N. On the right are the corresponding growth curves showing how population size, N, changes with time, t. Observe that the lines in (a) and (b) do not cut the ordinate; this is because $(1/N)\,dN/dt$ is defined as zero unless $N > 0$. For population growth to begin it is necessary that $N \geq 1$ for an asexually reproducing species, or $N \geq 2$ for a sexually reproducing species

four cases. This result is hardly surprising. A sigmoid growth curve is what
we should expect on common sense grounds without any appeal to mathematical argument. A verbal, as opposed to mathematical, description of such
growth, using some of the descriptive phrases introduced on page 45, might
be as follows: while the population is small its overall growth rate (the slope
of the growth curve) increases steadily as the population grows; subsequently,
after its size has reached a certain value which depends on the carrying capacity of the environment, the growth rate begins to decrease as a result of increasing environmental resistance. The rate ultimately decreases to zero and
the population has then reached its saturation density; thereafter no further
growth takes place.

The important point to notice is that this verbal description applies to all
the four growth curves shown in the right half of Figure 4.1 (and, of course,
to a great many other possible growth curves besides). In spite of the marked
contrasts among the four models we have considered, as shown by the entirely
different relationships between growth rate per individual and population
size (see Figures 4.1a, b, c and d), the four outcomes (Figures 4.1a', b', c'
and d') are *not* conspicuously different. If we observed a growing population
of animals in nature and recorded the population's size at a succession of times
we should have the necessary data for plotting a growth curve. But it would
obviously be difficult to argue backwards from an observed growth curve to
the underlying explanatory model, i.e. to the underlying relationship between $\frac{1}{N}\frac{dN}{dt}$ and N. As we have stressed before (on pages 4 and 24)
natural events are stochastic, not deterministic. The size of a natural population is never governed solely by derministic relations such as those given
in equations (4.1) through (4.5). Stochastic, or chance, variations inevitably
cause discrepancies between what actually happens and what had been
expected, on theoretical grounds, to happen. Consequently, since one
sigmoid curve is very like another, it is rarely, if ever, possible to argue
back from a growth curve to a model. An empircal growth curve usually
resembles, more or less, several possible theoretical growth curves and
without additional observations on different kinds of data it is impossible
to choose among them. Guesses as to the model underlying a particular
observed growth curve are no more reliable than guesses as to the contents
of an unopened parcel.

To obtain other kinds of data often entails experiment. A good example
has been described by Smith (1963) who studied the growth of laboratory
populations of the water flea *Daphnia magna*. Instead of attempting to infer
the relationship between $\frac{1}{N}\frac{dN}{dt}$ and N indirectly from an observed growth

curve, he carried out experiments which yielded this relationship directly. The experimental cultures of *Daphnia* were maintained for long periods at each of a number of chosen values of N; population size was held constant at a particular chosen value by removing surplus individuals at frequent intervals. The growth rate per individual, $\dfrac{1}{N}\dfrac{dN}{dt}$, at each of these chosen values of N was determined. Smith thus obtained an empirical example of a "differential" curve (like those in the left half of Figure 4.1) and had no need to infer its form from that of an integral growth curve. It appeared that the relationship was curvilinear; with increasing N the decrease of $\dfrac{1}{N}\dfrac{dN}{dt}$ was very rapid to begin with and much slower when N was large.

2 TESTING FOR DENSITY DEPENDENCE

It should now be clear that arriving at a full and detailed explanation of the manner of growth of a population is difficult. This is especially true of natural populations that are not amenable to experimental manipulation, or if such manipulation itself affects the mode of growth. Indeed, it cannot be too strongly emphasized that to discover the exact relationship between growth rate per individual and population size is an ambitious project. To obtain a complete answer certainly requires a very thorough and careful investigation, if it is possible at all. The data obtained by recording the size of a population at a succession of times, and nothing more, are wholly inadequate as a basis for a "theory" to account for the growth of a real natural population. Failure to uncover a full-fledged "theory" however, need not be discouraging. While it is true that a modest set of observations cannot answer *all* the questions one might ask concerning the growth of a population, it can certainly answer some. And a convincing answer to one clearcut question is a much more satisfactory end-product to an investigation than a large, vague theory.

An attempt to use this strategy that proved unsuccessful is worth describing as it is a very good illustration of the way in which one may be misled by treating population processes as exactly predictable (i.e., deterministic) and ignoring their unpredictable (i.e., stochastic) elements.

Consider the following straightforward question that seems to be answerable when the only datum consists of a record of population sizes at a succession of times. We wish to know whether population growth is density-dependent when density is low. A detailed discussion of density dependence and the mechanisms that bring it about is deferred to sections 3 and 4 of this chapter (see page 58 et seq.). Here it suffices to define* the term: *The growth of a population is said to be density-dependent if the growth rate per individual decreases as population size (and hence density) increases.*

Whenever a growth curve is sigmoid in shape it is natural to conclude that density dependence is operating in the high density part where the curve flattens out; (this conclusion is not logically inevitable, however, unless one can be certain that the observed reduction in growth rate is not due to extrinsic causes). But at low densities it is far from obvious whether or not density dependence is operating. A glance at Figure 4.1 will confirm this. All four models have sigmoid growth curves which are roughly similar; but only in the first model (Figures 4.1a and a') does the growth rate per individual fall off with increasing N for all values of N. In the other three models the growth rate per individual either remains constant (Figure 4.1b) or increases (Figures 4.1c and d) while N increases from 0 to 100. We now enquire how density dependence may be detected in the early stages of population growth.

A method that seems reasonable is the following. Consider the relationship of population size at any time to its size one time unit earlier; in symbols, the relationship between N_{t+1} and N_t. If the growth of the population were independent of its density we should expect the relationship to be

$$N_{t+1}/N_t = \lambda$$

where λ is the finite rate of natural increase. It would follow that

$$\log N_{t+1} = \log \lambda + \log N_t.$$

Therefore, a plot of $\log N_{t+1}$ versus $\log N_t$ would yield a straight line of unit slope.

Now suppose that data from a natural population were found to give a plot of $\log N_{t+1}$ versus $\log N_t$ forming a straight line of slope b, say, with $b < 1$. The equation of the line would then be

$$\log N_{t+1} = \log \lambda + b \log N_t$$

and the relationship implies that

$$N_{t+1} = \lambda N_t^b,$$

---

* This is the customary definition. If the word "increases" were substituted for "decreases", the phenomenon would be called *inverse density dependence*. The terminology is not entirely satisfactory. A separate, more inclusive, word is needed to mean that growth rate *changes* (in either direction) as density increases. The term *density-dependence* is occasionally used in this wider sense, but not in this book.

or equivalently that

$$\frac{N_{t+1}}{N_t} = \lambda N_t^{b-1} \quad \text{or} \quad \frac{\lambda}{N_t^{1-b}}.$$

In this case the ratio N_{t+1}/N_t is not independent of population size. Instead it decreases as population size increases, from which it follows that the population's growth must be density dependent.

The foregoing argument formed the basis of what appeared to be a useful test to judge whether density dependence was operating in nature. It was argued that if an actual plot of values of log N_{t+1} versus log N_t formed (approximately) a straight line with slope less than 1, it could reasonably be inferred that the natural population was regulated by a density-dependent mechanism. An example is shown in Figure 4.2 which is taken from Morris (1963). The data come from populations of spruce budworm larvae in two 25-acre permanent study plots in spruce-fir forest in New Brunswick, Canada; the unit of time was one year and the sizes of the populations were estimated every year for 15 years. Although the points are very scattered, as so often happens with real data, the relationship is roughly linear and the slope of the line that best fits is $b = 0.575$. From the discussion above, it therefore seems reasonable to say that the growth of the spruce-budworm population was density-dependent.

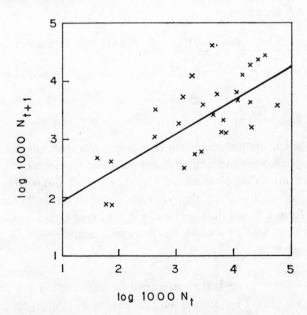

FIGURE 4.2 The relationship between log N_{t+1} and log N_t for populations of spruce budworm larvae in New Brunswick, Canada; (adapted from Morris, 1963)

However, the observation that $b < 1$ could equally well result from the fact that the plotted points are very scattered. This was shown by St. Amant (1970) (and see also Maelzer, 1970). A non-rigorous demonstration is as follows.

For compactness let us put

$$\log N_{t+1} = y_t \quad \text{and} \quad \log N_t = x_t.$$

Assume that population size was estimated at times $t = 0, 1, ..., n + 1$. The scattered points in the graph, to which a straight line is to be fitted, therefore have coordinates $(x_0, y_0), (x_1, y_1), ... (x_n, y_n)$. The slope of the fitted line is calculated from

$$b = \frac{n \sum x_j y_j - (\sum x_j)(\sum y_j)}{n \sum x_j^2 - (\sum x_j)^2}$$

where all the sums are from $j = 0$ to $j = n$.

(A proof is given in Appendix 6.2)

An equivalent formula for b is

$$b = \frac{\sum (x_j - \bar{x})(y_j - \bar{y})}{\sum (x_j - \bar{x})^2} = \frac{\text{cov}(x, y)}{\text{var}(x)}.$$

Multiplying both the numerator and the denominator of this last fraction by $\sqrt{\text{var}(y)}$, it is seen that

$$b = \frac{\text{cov}(x, y)}{\sqrt{\text{var}(x)\,\text{var}(y)}} \cdot \sqrt{\frac{\text{var}(y)}{\text{var}(x)}} = r\sqrt{\frac{\text{var}(y)}{\text{var}(x)}}$$

where

$$r = \frac{\text{cov}(x, y)}{\sqrt{\text{var}(x)\,\text{var}(y)}}.$$

Now r is the correlation coefficient of x and y. We shall not discuss it here (but see page 384) beyond noting that it measures the degree of scatter in a scatter diagram. If the points are but slightly scattered and lie very close to a straight line of positive slope, then r is only slightly less than 1; conversely, if the points form a broad diffuse cloud, r is low, near zero.

Observe, also, that we should usually expect to find

$$\text{var}(x) \simeq \text{var}(y).$$

This follows since var (x) is the variance of the numbers $\log N_0$, $\log N_1$, ..., $\log N_n$ and var (y) the variance of the numbers $\log N_1$, $\log N_2$, ..., $\log N_{n+1}$. Except for the first member of the first list and the last member of the second, the lists are identical.

We thus see that, approximately, $b \simeq r$.

It follows that when a line is fitted to the points in the scatter diagram, its slope will be near 1 if the degree of scattering is slight and considerably less than 1 if the scattering is great. The effect is demonstrated in Figure 4.3. Imagine two populations, A and B, growing exponentially in accordance with the formula $N = 10e^{0.2t}$. The theoretically expected growth curves (the same for both populations) is shown by the continuous curves in Figures 4.3a and b respectively. Now suppose that the inevitable chance deviations from mathematical expectation are much more pronounced in population B than in population A. The actual sequence of population sizes, at times $t = 0$,

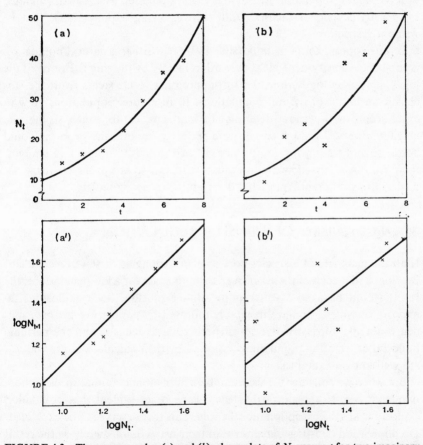

FIGURE 4.3   The upper graphs, (a) and (b), show plots of N_t versus t for two imaginary populations; the smooth curves (identical in the two graphs) are the theoretical growth curves, and the X's are "observed" points. The lower graphs show corresponding plots of $\log N_{t+1}$ versus $\log N_t$, with straight lines fitted to the "observed" points. The line in (b') is less steep than that in (a') merely because the points are more scattered

1, ..., 8, are shown by the X's in Figures 4.3a and b. If we now transform the observed population sizes to logarithms and plot log N_{t+1} versus log N_t the results are as shown in Figures 4.3a' (for population A) and 4.3b' (for population B). The contrast is obvious; the points are close to a straight line in 4.3a' and very diffusely scattered in 4.3b'. The fitted straight lines, which are also shown, are

$$y = 0.148 + 0.954x \quad \text{in Figure 4.3}a'$$

and

$$y = 0.393 + 0.772x \quad \text{in Figure 4.3}b'.$$

As expected, the slope is less (0.772 as compared with 0.954) for population B than for population A. But in neither population was growth regulated by density-dependent factors. Both were growing exponentially, that is, without restriction.

It thus appears that it is impossible to infer whether a natural population is subject to density-dependent regulation merely by judging the slope of the log N_{t+1} versus log N_t line. This disappointing result stems from the stochastic variation of natural populations. If, in nature, populations grew in exact accord with various deterministic mathematical formulas, so that the actually observed growth curves were perfectly smooth, the problem would not arise and testing for density dependence would be simple and straightforward. But stochastic variation causes field data to be not merely fuzzy but misleadingly different from what deterministic theory predicts.

3 MECHANISMS OF DENSITY REGULATION

If all the members of a species bore as many young as they were capable of bearing, and if each of these survived to do likewise for a few generations, the result would be a spectacular population explosion. Such explosions do happen, of course, and are inevitably followed by population crashes: sudden outbreaks of destructive insect pests such as locusts, and the periodic build-up of lemming populations followed by their suicidal dash to the sea, are well-known examples.

But when we consider the vast numbers of plant and animal species whose numbers seem to remain fairly steady, it becomes evident that population explosions are the exception, not the rule. This fact deserves to be explained. Do population densities remain within their usual limits merely as the result of chance, or is density self-regulating in some way? This question has been debated by ecologists for several decades now, and no conclusion is in sight. We shall discuss the controversy in detail in Section 4 (page 68). However, it should be pointed out here that the contentious point is is not whether a population *can* be self-regulating; certainly it can be, since unlimited increase is obviously impossible. Where disagreement arises is over the extent to which self-regulation actually takes place in nature. If chance fluctuations always keep a population's density fairly low, self-regulating mechanisms will seldom be called into play. There is no denying, however, that they do operate at times and in this section we shall examine the various ways in which they function.

Self-regulation is, by definition, a density-dependent process. We have already defined density-dependence (see page 54) and it should be emphasized that the concept entails the notion that the rate of growth of a population is controlled by, and is dependent upon, the population's density. There is obviously an upper limit to the number of individuals of a species that can subsist for long periods within a given area under given environmental conditions, and hence a maximum *sustainable* level of population density. This is known as the saturation density, and it can be exceeded only temporarily. Note that it is misleading to say that growth normally ceases at the saturation density. A population is, in a sense, growing all the time. None of its members is immortal, and there is an unending sequence of births and deaths. If the birth and death rates are equal, growth is still taking place but at zero rate; and the growth rate is positive or negative according as the birth rate exceeds or falls short of the death rate.

Now consider what happens when the density of a population deviates from its saturation value. If the deviation *causes* the growth rate to change in such a way that the saturation density tends to be restored, then the population regulates its own density automatically. There are numerous ways in which this can happen and a list of some of the many possibilities, with a few examples, follows. It should be realized that sometimes two or more processes occurring simultaneously may be required to bring about complete regulation of a population's growth. That in any case, growth may be partly, not fully regulated. And that a self-regulated population does not necessarily maintain a constant, stable size. Indeed, stability would be expected only if effects followed causes with hardly any delay; we shall return to this point later.

Attempts to classify the various mechanisms of density-dependent control have been made (see, for example, Nicholson, 1954) but classification serves no useful purpose. Nor is it helpful to try to distinguish the effects of crowding on the birth rate and the death rate treated separately; thus it is arguable whether the death of a fetus, say, should be thought of as a manifestation of increased mortality or of reduced fertility.

In the list below, a distinction has been made, however, between animal and plant populations. Wherever a density-dependent mechanism of population regulation is described that affects both kinds of populations and for which examples are available, they have been labelled *a* and *b* respectively.

1a Competition for resources among animals

If an animal population increases to a level at which it depletes its food supply (which is itself a growing population of other animals or plants) some of the surplus animals must either starve or fail to reproduce. Only then can the food population recover sufficiently to maintain the survivors. Many carnivores may thus be controlled by the species they prey upon and many herbivores by the plants they graze upon. The fact that a population is limited by food shortage does not *necessarily* mean that its growth is density-dependent, however; the process is density-dependent only if the rate at which food is produced is governed by the size of the population feeding on it. Populations can be food-limited without being density-dependent and an example will be given when we consider density-*in*dependent growth below.

1b Competition for resources among plants

Density-dependent population growth is common among tree species that are intolerant of shade. Consider a one-species forest, for example of Ponderosa Pine, a light-demanding species. As soon as the crowns form a closed canopy excluding sunlight from the forest floor, new individuals can establish themselves only where the death of an existing tree leaves a sunlit gap.

2 Disease

Infectious disease can cause a population to be self-regulating. High population density leads to rapid spread of the disease and consequent high mortality. Then, when population density dwindles, the disease spreads more slowly and mortality is lower. Perhaps human population growth was regulated, at least to some extent, by the Black Death pandemic in fourteenth century Asia and Europe.

3a Predation on animals

If carnivores are regulated by their food supply, then the food supply itself—an animal population of another species—is regulated by the carnivores. An example has been given by Paine (1966) who describes how intertidal populations of mussels, barnacles and goose barnacles were controlled by the predations of starfishes.

3b Predation on Plants

Plants, also, can be the victims of predation. In arid regions, range grasses may be controlled by grazing ungulates. Another example has been mentioned by Harper (1969). He describes how a population of St. John's Wort (*Hypericum perforatum*), a weed of California rangelands, was controlled by the beetle *Chrysolina quadrigemina* which feeds on it. The depredations of the beetles reduced the density of *Hypericum* to such a low level that surviving plants were widely separated. It then became difficult for the beetles to find the food plants they needed, beetle density became reduced, and the surviving *Hypericum* were spared from destruction.

4 Interactions between parasitic animals and their hosts

These often cause density-dependent regulation of both the host and the parasite species. It frequently happens with insects of one species as the host and of another (most often some species of wasp) as the parasite. A parasitic wasp (for example, an ichneumon fly) searches for a larva of its usual insect host and lays an egg in the larva's body. The egg then hatches and the immature wasp lives inside, and feeds upon, the host larva. Finally, if all goes well, the fully-developed adult wasp emerges from the completely consumed body of the host of which only a shell remains. The regulation of pest insect populations by the deliberate introduction of their parasites is a form of "biological control" that permits a pest to be reduced to negligible density levels without the use of insecticides. A famous example (Clausen, 1958) is the successful control of the Coconut Moth (*Levuana iridescens*) in the Fiji Islands by a parasite (in this case a species of fly), *Ptychomyia remota*, which was introduced from Malaya.

5 Accumulation of toxic wastes

If the environment of a population becomes polluted with the by-products of the population's *own* metabolism, and if these metabolites can themselves inhibit further population growth, we have a very clearcut case of density-dependent regulation. It can happen in cultures of yeasts and bacteria. And it now (1972) threatens to be one of the more unpleasant checks on the human population explosion, with the inhibitory by-products in this case being the toxic wastes of civilized life.

6 Thermal pollution

This trouble is not confined to our own species. Solomon (1957) refers to cases in which the metabolic heat generated by dense populations of insects raised their environments to lethal temperatures. It can happen when the environment is a good heat insulator (e.g. stored grain); or when the insects are confined and cannot disperse, as with blowfly maggots in a mammalian carcass.

7a Reduced fertility because of crowding (animals)

This can obviously bring about regulation of a population. In insects (Solomon, 1957) crowding may slow the growth of individuals; sexual maturity is delayed and thus the birth rate reduced. An example of reduced fertility in mammals has been mentioned by Sadleir (1969). A population of swamp rabbits (*Sylvilagus aquaticus*) in Missouri became densely crowded when they were flooded out of most of their territory and restricted to the small areas left above water level. Subsequent dissection of a sample of pregnant does revealed that a high proportion of the embryos, sometimes whole litters, had been resorbed.

7b Reduced fertility because of crowding (plants)

Reduced fertility caused by crowding can be very marked in trees. Intolerant (i.e. light-demanding) species such as lodgepole pine (*Pinus contorta*) can remain alive for many years at excessively high densities; the trees may be far too close for a person to walk among them. But trees crowded in this way usually produce less than a tenth as much seed as isolated trees growing far from competitors (Baker, 1950).

8 Behavioral effects of crowding

There has been speculation for years that a check to the human population explosion may come when densities reach so high a level that crowding affects mating behavior. As for other species, some careful studies were carried out by Calhoun (1962) to observe the effects of overcrowding on laboratory populations of Norway rats (*Rattus norvegicus*). He found that female rats in the dense populations made nests that were sketchy and carelessly constructed compared with those built by uncrowded females. Also, when danger threatened the nest, a mother rat in a sparse population would carry all her pups, one by one, to a safe refuge; a crowded mother, however, would rescue only some of her pups and these she would put in scattered places or simply abandon.

All the foregoing examples of population regulation involve the phenomenon of density-dependence. Next let us consider density-*in*dependent mechanisms that check a population's growth. Anything that kills an animal or plant before it reaches old age, or that impairs its fertility, should be regarded as a mechanism affecting population growth. An exhaustive catalog, even if it could be compiled, would be very long indeed.

In numbered paragraphs below are descriptions and examples of mechanisms which correspond in some way with the density-dependent processes already described. To facilitate cross-referencing, the numbers match those in the previous list, but roman numerals are used in what follows. Some items in the list demonstrate how difficult it can be to judge whether a controlling agent is density-dependent or not; and they also show that some agents may be only partly density-dependent, the degree of density-dependence varying with the circumstances.

Ia *Shortage of resources for animals; e.g. available food or nesting sites*

As remarked in 1a above, a population can be food-limited without its growth being density-dependent. This happens if the food, though present in ample quantity, is hard to obtain. Andrewartha (1961) gives an illustrative example. Tsetse flies (*Glossina* spp.) in East Africa feed by sucking blood, usually from antelopes and other large grazing animals. But a tsetse fly does not remain close to a source of blood between meals and must search for a new supply every time. In bush country where visibility is low and game sparse, a fly can starve or fail to reproduce because of delay in obtaining the next blood meal, although the quantity of blood within reach, if it could be found, enormously exceeds the needs of the whole fly population. The salient feature of this kind of situation is that the needed resource is not appreciably depleted by use.

The same is true of sites for nesting, or for concealment and shelter. An upper limit is set to the size of many bird populations by the availability of nesting sites. A population of Bufflehead ducks (*Bucephala albeola*) described by Erskine (1972) is an example. The buffleheads nested almost exclusively in the old nesting holes of flickers (*Colaptes auratus*) and an upper limit to the number of bufflehead nests was set by the number of flicker holes. Probably individuals of many bird species are barren for a season because no nest site is free.

Ib *Shortage of germination sites for plants*

A lack of suitable germination sites may often affect the growth of plant populations. Harper (1967) has described how very exacting the requirements of a seed may be and mentions experiments by Sagar. Sagar sowed seeds of three species of *Plantago* on smooth surfaces of sifted soil; then he placed objects (such as small glass plates laid flat or on edge) on the seedbed. It was found that seeds of one of the species germinated much more readily under glass; seedlings of another species sprouted in lines around the edges of objects on the seedbed although the seeds had been sown broadcast. Thus, very minor differences in microsite qualities can affect decisively the success of a seed in germinating and becoming established. The special requirements of orchids are even more pronounced. Seeds of most orchids (and probably those of many other plants as well) can germinate only if they chance to touch, on the soil where they fall, fungal hyphae with which they can form mycorhiza. (Infection of an orchid's roots by a symbiotic fungus is obligatory if the orchid is to grow successfully; however, Hadley (1970) has demonstrated by experiment that there is no evidence of species-to-species correspondence between orchid and fungus.) It seems likely that the low densities of many kinds of plants result from their very specialized requirements for germination. Perhaps this explains the surprisingly low density of many plant species notwithstanding the abundant seed they produce and the apparent suitability of large areas of ground for the adult plants.

II *Disease*

Obviously, non-infectious diseases can function as density-independent control agents. With infectious diseases, also, the relation between population density and disease-caused mortality may not be very close. Thus density-dependence with disease as the agent is probably partial at best, and grades imperceptibly into density-independence with diseases that are weakly infectious.

III *Predation*

Cases arise in which a predator population is regulated by the supply of its prey but not vice versa, as an example will show; (it could equally well have been put in paragraph Ia). During an outbreak of spruce budworm in the coniferous forests of New Brunswick, insectivorous birds, especially warblers, became unusually abundant. The strongest and most direct response was shown by the Bay-breasted Warbler (*Dendroica castanea*) whose density increased more than ten-fold in the study area (Mook, 1963). There is no doubt that the warbler population increased because of the abundance of budworm caterpillars for them to feed on, but their predations made no difference to the numbers of budworms since the supply was far too copious for the warblers to deplete it. Thus the increase in predators was insufficient to bring about density dependent regulation in the prey.

IV *Parasitism*

The regulation caused by interaction between insect hosts and their insect parasites is probably always density-dependent to a fairly high degree.

Maturation of a parasite *must* be accompanied by the death of its host. The interaction is very direct: each species controls the other and consequently itself. Other forms of parasitism, for example that of large mammals by intestinal worms, no doubt brings about a certain amount of population control simply by causing ill health in the hosts. But the control is probably only density-dependent to a minor degree since the parasites rarely kill their hosts.

Va   *Pollution and animals*

Whenever population growth is reduced by pollutants *not* produced by the population itself, the process is density-independent. Human actions are by far the most important causes of pollution and the resultant population control may be deliberate or inadvertant. The control of pests with pesticides is an example of the former. An example of the latter was discovered when it was found that peregrine falcons (*Falco peregrinus*) were failing to reproduce because they had ingested DDT and it had caused their eggs to be fragile and almost shell-less. The word "control" is perhaps inappropriate when populations of rare species are involved.

Vb   *Pollution and plants*

Pollution can also have catastrophic effects on vegetation as anyone knows who has seen the many square miles of desert-like conditions that surround big ore-smelters (e.g. in Trail, British Columbia and Sudbury, Ontario). The effect of radioactive pollution on vegetation has been the subject of experiments at Brookhaven National Laboratory in Long Island (Woodwell, 1962, 1963). It is disturbing to speculate whether very dilute pollution, far from its source, has a subtle effect on vegetation that has so far gone undetected.

No parallels suggest themselves for items 6, 7 and 8 of the list of density-dependent controls, and we come now to the more general class of density-independent factors whose effects are all-pervading and basic. They can be classed in a catch-all category as environmental variation due to physical factors, and the list is enormous. These factors, which are chiefly but not exclusively related to weather, range all the way from such things as late frosts or rainy summers to sensational natural phenomena like hurricanes, blizzards, floods, fires, droughts, landslides, avalanches, earthquakes, volcanic eruptions and tsunamis. These catastrophes though conspicuous, are usually local in their influence. The effects of less noteworthy phenomena can be much more far-reaching; for example, a succession of unusually cool, wet summers appears to have been what caused the collapse of a very large outbreak of spruce budworm in New Brunswick in the early 1950's (Greenbank, 1963).

It must not be supposed that the *effects* of changing physical factors are independent of the density of the affected population, even though the population is not itself a *cause* of change in physical factors. Chitty (1960) has emphasized this point. It amounts to saying that a population can exhibit a density-dependent response to a density-independent cause.

Two examples will clarify this:

1. Insect populations are often infected with fungus diseases which function as density-dependent agents. Damp weather, which is obviously independent of the insect population density, favors the spread of these diseases. Here a density-independent agent boosts the effectiveness of a density-dependent one.

2. Neave (1953) provides an exceptionally good example. In British Columbia, mature adults of the pink salmon (*Oncorhynchus gorbuscha*) journey from the ocean up the rivers in September to spawn just upstream of the highest point reached by salt water. The spawning sites are very variable in quality so that when their density is high, many of the salmon are forced to spawn in unfavorable sites where silting is heavy if the rivers flood. Consequently, though flooding and silting are unaffected by salmon density, these physical factors have a more pronounced effect when population density is high, for only then will the salmon have used vulnerable spawning sites.

It should be mentioned here that Neave (1953) also introduced some useful terms that, unfortunately, have not come into general use except among fisheries biologists. Their adoption would be a blessing; it would obviate the need for the hyphenated, similar-sounding expressions *density-dependent* and *density-independent*. As adjectives to qualify "mortality" he proposed:

a) *Compensatory* (when mortality varies directly with density);

b) *Depensatory* (when mortality varies inversely with density); and

c) *Extrapensatory* (when mortality is independent of density).

The effects of the three kinds of mortality are therefore density-dependent, inversely density-dependent, and density-independent respectively.

In the preceding discussion we have assumed that the density of a population could decrease (or the growth rate become negative) only if the death rate exceeded the birth rate for a time. With motile organisms, however, a reduction in density can also be brought about by dispersal. If dispersal is not to result in death for the animals dispersing, there must be some suitable territory for them to disperse into. Thus, dispersal can only serve as a way of relieving population pressure if acceptable environmental conditions for the species are available. These must either be in areas that are contiguous to the areas that have become saturated, or else at a distance not too great to be covered by migrants. If such areas exist, one is immediately faced with the question: why is the population not already at saturation density in these areas also? This question prompts another: are natural populations at their saturation density most of the time or only occasionally? This question will be discussed (though not answered) in Section 4.

Another complication we have not yet considered (it is deferred to Chapter 5) is the problem of delayed responses. If it takes an appreciable time for density-dependent mechanisms to exert their effects, oscillations in population size are usually set up. Thus a growing population may continue to grow even after its saturation density has been attained (i.e. it may "overshoot") because density-dependent effects lag behind their causes; and then, when the growth rate becomes negative and tends to bring the population down to saturation level from above, there may be "undershoot" for the same reason. The complications introduced by time lags do not invalidate any of the earlier discussions in this chapter.

Before concluding this section on the mechanisms of population regulation, we must mention the control exercised on economically important species of plants and animals by human intervention—in other words, resource management (Watt, 1968). The successful practice of resource management requires that populations of pests should be kept as low as possible, and that populations of useful harvestable species should be kept at levels permitting sustained yields. It is the work of applied ecologists and entomologists, wildlife biologists, fisheries biologists, range managers, and foresters.

All practitioners are aware of past blunders. Economic entomologists can give examples of cases when insecticide applications destroyed the natural enemies (parasitic and predatory insects) of a pest and therefore led to subsequent big increases in the population of the pest that was to have been subdued.

The history of wildlife management also has some classic mistakes on record (Dasmann, 1964). There has sometimes been overzealous protection of herbivores from their natural predators and this has resulted in a damaging degree of overgrazing; this acts, of course, to the ultimate detriment of the population (usually of game animals) it was originally intended to foster.

If a population is regulated by density-dependent processes, it can be harvested without being depleted *always provided the harvest taken is not too large*. Exploitation, by hunting or fishing for example, need not act as an addition to natural mortality, but merely as a substitute for some or all of it. This comforting conclusion does not justify overexploitation, of course.

Consider the theory underlying sustained fisheries yields, for example. Fishing usually reduces the size of a fish population to begin with. Then, for

5*

the population to give a constant yield, its growth rate must remain at zero indefinitely. Consequently, one must ensure that when the death rate goes up as a result of fishing, the birth rate shall go up by a corresponding amount. In other words, there must be a faster turnover of individuals. This can be achieved by deliberately modifying the population's age distribution, by selective fishing, in such a way that there is a higher proportion of young fish than in the unfished population. Maximum sustainable yields come from populations that are about half the size of unfished populations, and in which fishing mortality has replaced most natural mortality.

It should be clear that wise resource management hinges on as thorough a knowledge as possible of the natural processes governing a population's density and growth. Until our mastery of population theory greatly exceeds its present level, mistakes are inevitable. The best way to keep them small is to be aware of all the many ways a management program can go wrong, and be constantly alert for the first sign of them.

4 THE "DENSITY-DEPENDENCE" DEBATE

We mentioned on page 58 that there is considerable disagreement among ecologists on the degree to which natural populations are regulated by density-dependent mechanisms. The opinion of one side has been forcefully stated by Nicholson (1954, 1957). For instance in a review article (1957) he wrote: "Logical argument based on certain irrefutable facts shows, not merely that populations may regulate themselves by density-induced resistance to multiplication, but that this mechanism is essential." This view has been supported by many writers especially Solomon (1957) and Lack (1966).

The contrary opinion, that density-dependent processes occur only rarely, has been advanced by, among others, Andrewartha and Birch (1954), Thompson (1956) and Milne (1957). Andrewartha recapitulated his ideas more recently (1961) and gives as his chief criticism of the "theory" [sic] of density-dependent factors that "It is unscientific because its authors have dependend too much on insight and deduction and not enough on experiment and observation." His own field studies, notably on insect populations in parts of Australia where summers are very hot and dry, convinced him that weather changes were the paramount causes of population fluctuations. Populations build up while weather is favorable and dwindle when it becomes unfavorable and only exceptionally does population pressure become so excessive that it *causes* a cessation of growth.

Similarly Thompson (1956) wrote that "climatic and edaphic factors are the basis of the environmental diversity... This is the primary extrinsic factor in natural control .... Populations, therefore, are not truly regulated but merely vary, although indefinite increase is unlikely and *in the long run will become impossible.*" [Italics not in the original.]

It is interesting to note that of the seven authors referred to above, all but one (Lack) are entomologists. It is clear from their writings that each side feels that the obligation to prove its case rests with the opposition.

Nicholson's so-called logical argument is that unless density-dependent processes come into play at some point a population must, in the long run, either become extinct or impossibly big. Solomon (1957) states this argument clearly, as follows: "But to explain why the great irregular increases and decreases from year to year [caused by weather] add up over a period to approximately zero, we seem to have only two alternatives, density-dependence at some point in the population cycle, or pure chance." (The chance is, by implication, negligibly small.) And according to Lack (1966) the question of primary interest is "what determines the levels around which such fluctuations occur, and prevents unlimited increase or decrease."

The various excerpts reproduced above suggest to me that the real bone of contention is not what it is stated to be, and that the debaters' appeal to logic is irrelevant. Those who maintain that density-dependent mechanisms are *not* paramount (Andrewartha, Birch, Thompson, Milne *et al.*) do not deny that they function sometimes, and that when density becomes excessive they must. Thus these workers are not guilty of any logical error; to say that they are is false, and does nothing to support the arguments of Nicholson, Solomon, Lack *et al.* Since both sides concede that density-dependent mechanisms must operate *sometimes*, the argument cannot be resolved by appeal to logic; experiment and observation are required, as Andrewartha (1961) wrote. However, an exchange of anecdotes can never settle the matter, even if it can be settled at all, which I doubt. Every specific example that supports the stand of one group of contenders can probably be met by a counterexample from the other group.

It seems more rewarding to look into the further implications of each group's theory.

If density-dependence operates most of the time in most natural populations, two further conclusions follow. The first is that populations spend most of the time at or close to their current saturation level; (in particular cases this is susceptible to test as we shall discuss below). Note the word *current*; everyone agrees that a population's saturation density varies with varying environmental conditions, and nobody asserts that the density of a natural population necessarily remains constant while it is subject to density-dependent regulation. The second conclusion is that the members of a population are competing with one another (in a broad sense) most of the time.

The word competition as I use it here does not entail any notion of direct confrontation; it means merely that in every generation, some individuals must fail if the rest are to survive and breed, since the population is too large for all its offspring to flourish.

Now let us look at the two chief implications of the opposing theory, which holds that density-dependent mechanisms operate only very occasionally. It follows, firstly, that a population's density is well below saturation level most of the time; it reaches saturation only now and again. And, secondly, that competition among population members happens only occasionally, not all the time.

It now appears that since the whole debate hinges on phrases like "most of the time" and "only some of the time", the deadlock between the two extreme factions is entirely unreasonable. Why should one hold the view that most populations of most species spend most of their time in one mode of growth or the other? To adopt a rigid position is to deprive oneself of the chance to explore all sorts of interesting lines of advance, a few of which will be considered in the remainder of this chapter.

The first of these is the use of experiment to judge whether a particular population is being regulated by a density-dependent mechanism. An experimental test that is very attractive because of its extreme directness was proposed by Nicholson (1957) and has been called the "convergence experiment" (Murdoch, 1970). In principle the method is as follows: a natural population is divided, with as little disturbance as possible, into three subpopulations. The first of these is left in its natural state, the second is reduced in density by removing a proportion of the individuals, and the third is augmented (perhaps, but not necessarily, by adding the individuals removed from the second). The three subpopulations are then left to themselves and if the densities in all three tend to become the same, or converge, it is clear that density-dependence is the regulating mechanism. The advantage of the experiment becomes clear if we consider what could be inferred by following the growth of only one of the subpopulations, the natural, unaltered one. Its density would probably not remain constant throughout a long observation period, but there would be no way of telling whether the changes that occurred were density-independent, or density-dependent adjustments to match changes in saturation density resulting from shifts in environmental conditions. For the convergence experiment to succeed, there is no need for the density of the unaltered subpopulation to remain constant. Evidence for or against density-dependence is obtained according as the subpopulations do or do not converge to a common density.

An example of the experiment has been described by Eisenberg (1966) who subdivided a natural population of the pond snail (*Lymnaea elodes*) by building a row of snailproof enclosures along the margins of a pond. The densities of adult snails were altered to one-fifth of natural density in one group of enclosures, to five times natural density in another group, and left unchanged in the control group. Eisenberg found that the experimentally imposed differences in density of the adult snails were maintained; but when summer came and these adults reproduced, the density differences among the three groups of enclosures entirely disappeared. He wrote: "The regulation of density was far more complete and occurred more rapidly than was thought possible."

For any population, regardless of whether it is controlled continuously or only sporadically by a density-dependent mechanism, it would be interesting to know what factor limits its growth and determines its saturation density. Hairston, Smith and Slobodkin (1960) have discussed whether one can answer this question, in a very general way, for broad categories of organisms. The categories they consider are green plants, herbivores, carnivores, and decomposers (fungi and microorganisms that subsist on, and decompose, dead organic material). Writing of terrestrial communities, they conclude that herbivore populations are usually limited by the attacks of predators, but that among green plants, carnivores and decomposers the limiting factor is usually shortage of resources (food for carnivores and decomposers; and water, light or, occasionally, minerals, for plants).

There has also been speculation on whether populations of certain classes of organisms, again broadly defined, are, in the main, subject to density-dependent controls, while other classes of organisms are not. MacArthur and Connell (1966) consider that large organisms are fairly independent of their physical environment and their populations are probably density-dependent; so also are those of all organisms, large or small, which live in warm, moist climates and are never exposed to adverse weather conditions. In contrast, small organisms in harsh conditions (insects in a desert, for example) are at the mercy of the weather and are probably density-independent. Harper (1967) argues that annual plants of disturbed habitats, notably weeds of arable land, are probably density-independent; whereas perennial plants in stable environments are probably at or near their saturation density most of the time and are subject to density-dependent regulation continuously.

A point we have not yet considered is the following. It is a mistake to regard any large natural population as a homogeneous whole, since the densities of individuals are hardly ever constant from place to place. Most often, density is extremely patchy, being high in some places and low in others. This prompts numerous interconnected questions about how such a population is regulated. Here the problems can only be hinted at. Does the observed patchiness indicate a corresponding patchiness in habitat conditions, or in other words, is the density everywhere as great as it can be? When the densities within the various patches fluctuate, do they do so synchronously over a large area? Or if not, can the seeming stability of an extensive population be actually the resultant of a large number of unsynchronized *local* population explosions and *local* extinctions? These questions suggest several interesting lines of investigation. We shall have more to say about patchiness, and about spatial patterns in general, in Chapter 7.

To end this chapter, consider the evolutionary consequences of density-dependent and density-independent population growth. At times when a population's growth is density-dependent, the struggle for existence must be far more intense than at other times. Thus one would expect advantageous adaptations to become fixed in a population most rapidly during a period when its density was close to saturation and competition was intense; and correspondingly less rapidly in a period when density was low and competition mild. It should be noticed that natural selection must still continue, even in the complete absence of competition. However, the selectively-favored adaptations might then be of a somewhat different kind. For example, if an insect population inhabited a region in which, owing to gradual climatic change, summers were becoming progressively hotter and drier, a life cycle incorporating a long diapause would be selectively favored, regardless of the insects' density and even if they were not competing with one another. If the same population, instead of being exposed to a worsening climate, were limited by the attacks of insectivorous birds, selection would be more likely to favor insects with camouflage patterns. It seems unlikely that speculations like these can lead to testable hypotheses; all the same, nothing is lost by keeping them in mind.

CHAPTER 5

Fluctuating and Oscillating Populations

IT IS PROBABLY rare for any natural population of organisms to remain constant in size for many generations. The rate at which changes take place, however, must obviously depend on the longevity of the species concerned. For example, if the number of trees of one species in a tract of mature forest were to remain approximately the same for three or four human lifetimes, it might still be true that the population was fluctuating "rapidly"; its apparent stability would stem from the long duration, in human terms, of the trees' generations.

No population ever remains constant in a strict mathematical sense, of course; for it to do so would require that every birth be matched by a simultaneous death. The greatest degree of stability attainable in nature is *stochastic* stability; this obtains throughout a period in which the probability of a birth in any short time interval is equal to the probability of a death in the same interval. Even this degree of stability is probably uncommon; in any case it is impossible to judge with certainty whether the minor fluctuations a population undergoes are due to chance or to small changes in intrinsic growth rate. Slight and irregular changes in growth rate, which correspond to changes in environmental factors such as weather, are probably commonplace. Thus a certain amount of irregular fluctuation is the norm for any population; such fluctuations are irregular in the sense that they are aperiodic and not consistent in amplitude.

It is, indeed, difficult to specify what constitutes the norm of population stability and at what stage departures from this norm become remarkable enough to warrant investigation. Judgments as to what is noteworthy are necessarily subjective. However, unusual departures (from an envisaged norm) in *any* direction are interesting. Extreme stability, for example, implies that population growth is both sensitive to, and very prompt in its response to, a density dependent controlling mechanism, and that the saturation level is itself unvarying. Departures from the norm in the direction of instability are more often encountered and can be of different kinds: thus population fluctuations may be surprisingly large, *or* of surprisingly constant amplitude, *or* of surprisingly regular periodicity (or any combination of these).

Very large increases and decreases in population size, that is, irruptions (or outbreaks) and crashes, are not unusual in common species. When a population of an economically important species (a pest or a valuable species) irrupts or crashes, investigation of the underlying cause may be an urgent problem in applied ecology. No doubt outbreaks occur in rare species too. But even if the density of an exceedingly rare species increases ten thousand-fold it may still be rare from the casual observer's point of view; therefore, unless the population happens to be an object of ecological investigation at the time, or is conspicuous and newsworthy, its fluctuations will almost certainly go unnoticed.

Epidemics of infectious diseases are also ecological phenomena; they result from outbreaks of species of pathogenic micro-organisms.

Possibly the most interesting, because most puzzling, form of population fluctuation is regular cyclical change which is not in time with any known inorganic cycle such as the seasons. Natural populations of some mammal, bird and fish species seem to oscillate with great regularity and this has led to much research and speculation on "wildlife cycles" and what causes them. They are discussed in Sections 4 and 5. Borderline cases occur, also, in which it is difficult to judge whether fluctuations should be regarded as truly cyclic or not. In any case, two quite distinct kinds of sequence are defined as cyclic by ecologists. It is obviously appropriate to describe a population's size variations as cyclic if the time intervals between peaks are of approximately constant length; the mean length is the *period* of the cycle. Another form of cyclic variation may be found when we observe populations of two or more species simultaneously. If the peaks of the different species-populations occur in the same order over and over again, the sequence should certainly be regarded as cyclic even though the time intervals between successive peaks may vary considerably. The simplest example is when only two species are involved and the peaks of one alternate with the peaks of the other. Thus the regularity implicit in the concept of cyclic variation may pertain either to the constant intervals between peaks, or to the constant ordering of peaks of different species.

In the following three sections we shall discuss three different mechanisms that can lead to regular fluctuations (i.e. oscillations) in a population's size even though the environment remains constant. That is, the tendency to oscillate is a property of the populations themselves and is independent of any extrinsic factors.

1 OSCILLATIONS CAUSED BY FERTILITY AND SURVIVAL RATES

On page 11 we remarked that for a steadily growing population there is usually a stable age distribution. If an initial population of arbitrary age distribution were left to grow undisturbed, its age distribution would change through succeeding generations until the stable distribution was reached and thereafter this distribution would persist indefinitely. This statement is true of many, perhaps most, populations. It is possible, however, for the fertility and survival rates to have values such that the age distribution and also the total size of the population oscillate indefinitely.

Thus, consider a species which is capable of reproducing only for a short period near the end of its lifetime. Examples are: many insects, all biennial plants, and some fishes, for instance Coho salmon (*Oncorhynchus kisutch*) which spawns in its third year, and Kokanee (landlocked sockeye salmon, *Onchorhynchus nerka kennerlyi*) with populations in which either the three-year-old or the four-year-old individuals spawn just before dying.

The projection matrix (see page 20) for such a species is then of the following form; (we assume, for concreteness, that there are four age classes):

$$\mathbf{M} = \begin{pmatrix} 0 & 0 & 0 & F \\ P_0 & 0 & 0 & 0 \\ 0 & P_1 & 0 & 0 \\ 0 & 0 & P_2 & 0 \end{pmatrix}$$

Since the only fertile individuals in the population are those in the oldest age class, all but the final element in the first row of the matrix are zero.

To see how a population with such a projection matrix would behave, let us assign numerical values to F and the P's and discover by trial what happens to a population growing in this manner. (This is a simple exercise in ecological simulation.)

Thus, put

$$\mathbf{M} = \begin{pmatrix} 0 & 0 & 0 & 8 \\ \frac{3}{4} & 0 & 0 & 0 \\ 0 & \frac{1}{2} & 0 & 0 \\ 0 & 0 & \frac{1}{3} & 0 \end{pmatrix}.$$

Next, calculate the successive powers of $\mathbf{M}$, namely $\mathbf{M}^2$, $\mathbf{M}^3$, $\mathbf{M}^4$, etc. It will be found (see Appendix 2.2) that

$$\mathbf{M}^2 = \begin{pmatrix} 0 & 0 & FP_2 & 0 \\ 0 & 0 & 0 & FP_0 \\ P_0P_1 & 0 & 0 & 0 \\ 0 & P_1P_2 & 0 & 0 \end{pmatrix} = \begin{pmatrix} 0 & 0 & \frac{8}{3} & 0 \\ 0 & 0 & 0 & 6 \\ \frac{3}{8} & 0 & 0 & 0 \\ 0 & \frac{1}{6} & 0 & 0 \end{pmatrix};$$

$$\mathbf{M}^3 = \begin{pmatrix} 0 & FP_1P_2 & 0 & 0 \\ 0 & 0 & FP_0P_2 & 0 \\ 0 & 0 & 0 & FP_0P_1 \\ P_0P_1P_2 & 0 & 0 & 0 \end{pmatrix} = \begin{pmatrix} 0 & \frac{4}{3} & 0 & 0 \\ 0 & 0 & 2 & 0 \\ 0 & 0 & 0 & 3 \\ \frac{1}{8} & 0 & 0 & 0 \end{pmatrix};$$

$$\mathbf{M}^4 = \begin{pmatrix} FP_0P_1P_2 & 0 & 0 & 0 \\ 0 & FP_0P_1P_2 & 0 & 0 \\ 0 & 0 & FP_0P_1P_2 & 0 \\ 0 & 0 & 0 & FP_0P_1P_2 \end{pmatrix} = \begin{pmatrix} 1 & 0 & 0 & 0 \\ 0 & 1 & 0 & 0 \\ 0 & 0 & 1 & 0 \\ 0 & 0 & 0 & 1 \end{pmatrix} = \mathbf{I}, \text{ say.}$$

Because $\mathbf{M}^4 = \mathbf{I}$ has the form shown, it is known as an "identity" matrix and it is easily seen that

$$\mathbf{M}^5 = \mathbf{M}(\mathbf{M}^4) = \mathbf{M}\mathbf{I} = \mathbf{M};$$

$$\mathbf{M}^6 = \mathbf{M}^2(\mathbf{M}^4) = \mathbf{M}^2\mathbf{I} = \mathbf{M}^2;$$

and so on. The numerical values used in the example were, indeed, chosen so that $FP_0P_1P_2 = 1$; this ensures that the population shall not increase or decrease in the long run, although its size will oscillate, as we now show. Suppose at the beginning of the simulation (at time zero) the vector representing the age distribution of the population is

$$\mathbf{n}_0 = \begin{pmatrix} 48 \\ 36 \\ 24 \\ 12 \end{pmatrix}.$$

The total size of the population is the sum of the numbers in the different age classes, i.e. $N_0 = 120$. At a succession of times, $t = 1, 2, 3, ..,$ the population's age distribution becomes

$$\mathbf{n}_1 = \mathbf{M}\mathbf{n}_0 = \begin{pmatrix} 96 \\ 36 \\ 18 \\ 8 \end{pmatrix} \quad \text{so that} \quad N_1 = 158;$$

$$\mathbf{n}_2 = \mathbf{M}\mathbf{n}_1 = \mathbf{M}^2\mathbf{n}_0 = \begin{pmatrix} 64 \\ 72 \\ 18 \\ 6 \end{pmatrix} \quad \text{so that} \quad N_2 = 160;$$

$$\mathbf{n}_3 = \mathbf{M}\mathbf{n}_2 = \mathbf{M}^3\mathbf{n}_0 = \begin{pmatrix} 48 \\ 48 \\ 36 \\ 6 \end{pmatrix} \quad \text{so that} \quad N_3 = 138.$$

Finally,

$$\mathbf{n}_4 = \mathbf{M}\mathbf{n}_3 = \mathbf{M}^4\mathbf{n}_0 = \begin{pmatrix} 48 \\ 36 \\ 24 \\ 12 \end{pmatrix} = \mathbf{n}_0,$$

and the whole cycle repeats itself. Thus the period of the cycle is four time units.* Figure 5.1 shows the result graphically.

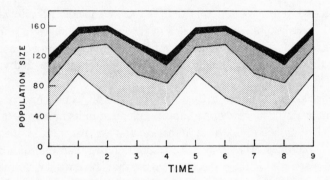

FIGURE 5.1   Oscillations in the total size, and in the age distribution, of a population (artificial) resulting from its fertility and survival rates. The population's projection matrix is given in the text. The numbers of individuals in the four age classes are shown by the widths of the strips. The lowest strip represents the youngest class and the top strip the oldest

The foregoing discussion deals with one of the many ways in which a population can oscillate. Three points deserve comment.

In the first place, the model assumes either that the environment remains constant; or, if it varies seasonally, that there is no long term climatic trend and that the length of the organism's life-cycle is a whole number of years. The implications of these assumptions are as follows: the model could apply to an organism with a life span of several years in which reproduction occurs

---

* Observe that the whole cycle occupies several, say m, time units, and births take place in every time unit. Thus if the time unit is one year and breeding occurs only in spring, the cycle takes m years, in other words m consecutive breeding seasons. Superimposed on this, there may be cyclical change *within* each year as described on page 29n. The distinction between the two situations is this: the age distribution is stable if it repeats itself *every* year; it is oscillatory if it repeats itself every m years with $m > 1$.

at the same season each year (such as the species of *Oncorhynchus*, the Pacific salmon); or to any organism, whatever its life span, living in the unvarying conditions of the humid tropics, of large parts of the ocean, or of caves. But it would not apply to species living in temperate latitudes and having a life span of a few days or weeks, for then successive age classes would live and reproduce in dissimilar weather conditions, some in spring and others in summer. Nor would it apply to univoltine insects or other organisms having no overlap of generations; in such species only a single cohort is present at any one time so that the age distribution is degenerate (see Figure 3.2, page 36).

The second point to notice is that the model makes no allowance for changes in fertility and survival rates in response to changes in population density. Leslie (1959) has shown how a projection matrix can be constructed in which the elements are made to depend on population size, and vary with it.

The third point to notice is this: although the projection matrix we have used as an example has only one non-zero element in its first row, this is not essential. Undamped oscillations in the size and age distribution of a population can sometimes result even when there is more than one non-zero element in the first row of the projection matrix, that is even when the organisms are capable of reproducing in other age classes besides the oldest.*

Although the mechanism just described is theoretically capable of producing cycles in natural populations, it seems unlikely that it would ever be the sole cause of prolonged oscillation. But the process should not be dismissed as unimportant. It may well function as a contributory cause of population oscillation; for instance it might prevent oscillations initiated by some other mechanism from dying out. And, conceivably, the mechanism could impose temporal regularity on relatively irregular fluctuations which were due to a different cause. The age distributions of wildlife populations, and of fish populations, have been observed to vary spectacularly. For example, Mac-Pherson (1969) describes changes in the age distribution of Arctic fox (*Alopex lagopus*) trapped in the Canadian Arctic. In one extreme case it was found that whelps under a year old constituted less than 1 % of the animals trapped in one year, and 100 % in the following year. Although this is an isolated result it demonstrates the importance of taking age distribution into account whenever theories of population growth are devised.

---

* To judge whether a projection matrix operating repeatedly on an age distribution vector will ultimately yield a stable age distribution or not, it is necessary to know the latent roots of the matrix. For details, see Leslie (1945, 1948).

2 OSCILLATIONS DUE TO DELAYED DENSITY DEPENDENT RESPONSE

Oscillations can be set up in a population if its growth rate is governed by a density dependent mechanism and if there is a delay in the response of the growth rate to density changes. When this happens the size of the population alternately overshoots and undershoots its equilibrium level (see page 67).

Let us look at the simplest mathematical model which yields this result. Recall that the simplest possible form of density-dependent population growth is that described by the logistic equation (see page 47). It assumes that at any moment the population's growth rate per individual depends on the size of the population *at that moment*. That is, that

$$\frac{1}{N_t} \cdot \frac{dN_t}{dt} = r - sN_t \quad \text{for all} \quad N_t > 0. \tag{5.1}$$

This is the same as equation (4.1) except that we now write N_t instead of N; the reason for the change will become clear immediately. Now suppose there is a delay, or lag, of length L in the response of the population, so that the growth rate per individual at time t depends on the population's size L time units ago, which was N_{t-L}. The differential equation describing population growth now becomes

$$\frac{1}{N_t} \cdot \frac{dN_t}{dt} = r - sN_{t-L} \quad \text{for all} \quad N_t > 0. \tag{5.2}$$

It can be shown (Pielou, 1969) that an alternative way of writing (5.1), the ordinary lag-free logistic equation, is provided by the *difference equation*

$$N_{t+1} = \frac{\lambda N_t}{1 + \alpha N_t} \tag{5.3}$$

where $\lambda = e^r$ is the finite rate of increase at which the population would grow if its growth were *in*dependent of density; and $\alpha = s(\lambda - 1)/r$. We shall not derive the difference equation here but shall merely draw attention to its implications. Notice that it expresses population size at time $t + 1$, that is N_{t+1}, in terms of its size, N_t, at time t, and two constants (or parameters). These parameters, λ and α, are functions of the parameters r and s which appear in the differential form of the logistic equation. If we choose a value for N_0, the starting size of the population, we can substitute this chosen value in the right hand side of equation (5.3) and calculate N_1; then substitute N_1 in the right hand side and calculate N_2; and so on for as many time units as desired. It is seen that N_{t+1}/N_t, the finite rate of increase of this logistically growing population, is given by $\lambda/(1 + \alpha N_t)$; thus the rate decreases as N_t increases.

Arguing by analogy, it is now evident that population growth with delayed response, described by equation (5.2), can be represented by the difference equation:

$$N_{t+1} = \frac{\lambda N_t}{1 + \alpha N_{t-L}}. \tag{5.4}$$

In this case the finite rate of increase, N_{t+1}/N_t, depends (inversely) on N_{t-L}. In this way a lag in response of L time units duration is allowed for.

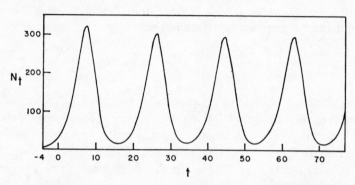

FIGURE 5.2 Oscillations in the size of a population (artificial) caused by delayed response to crowding. The growth of the population is described by equation (5.4') and the way in which successive values of N_t were calculated is shown in Table 5.1

Figure 5.2 shows a theoretical example. A lag of $L = 4$ time units is assumed. The parameters r and s in equation (5.2) were given the values

$$r = 0.5 \quad \text{and} \quad s = 0.005.$$

(As in the lag-free case, the equilibrium population has size $K = r/s$, or 100 in the example). Consequently, the parameters in (5.4) are

$$\lambda = e^{0.5} = 1.6487; \quad \text{and} \quad \alpha = \frac{0.005\,(\lambda - 1)}{0.5} = 0.006487.$$

Equation (5.4) thus becomes

$$N_{t+1} = \frac{1.6487 N_t}{1 + 0.006487 N_{t-4}}. \tag{5.4'}$$

Before the population size at a succession of times can be calculated it is necessary to choose values for $N_0, N_{-1}, N_{-2}, ..., N_{-L}$. Thus $L + 1$ "initial values" must be assigned instead of only one which was all that was required to begin a sequence of calculations based on equation (5.3). The data for

Figure 5.2 were calculated in the following manner. A population originally of size $N_{-4} = 5$ was allowed to grow logistically [in accordance with (5.3)] for four time units and in this way values of N_{-3}, N_{-2}, N_{-1} and N_0 were obtained. With population growth thus started, equation (5.4') could be used in all subsequent calculations. Table 5.1 shows the first few values used in plotting Figure 5.2.

TABLE 5.1   The beginning of the table from which Figure 11 is plotted, to show density-dependent population growth with a delayed response

t	N_t	
−4	5.00	Calculated from
−3	7.98	
−2	12.51	$N_{t+1} = \dfrac{1.6487 N_t}{1 + 0.006487 N_t}$
−1	19.08	
0	27.99	
1	44.70	Calculated from
2	70.07	
3	106.85	$N_{t+1} = \dfrac{1.6487 N_t}{1 + 0.006487 N_{t-4}}$
4	156.76	
5	218.73	

. . . .

Natural populations which oscillate because of a mechanism of this kind are probably common. A very good example of a laboratory population that behaved in this manner was described by Nicholson (1954). In one of many experiments, he reared a population of blow-flies (*Lucilia cuprina*) for about 20 generations. Food and water supplies for both adults and larvae (maggots) were more than adequate at all times but for a single exception: the meat (ground liver) needed by the adults to enable them to lay eggs was rationed to a constant amount per day for the whole population regardless of its size. Consequently, the ground liver was the "governing requisite" of the population. The course of events was as follows. When the density of the adults was low they laid many eggs and the population grew. When the density of the adults became high, competition among them for the available liver was so severe that only a few got enough to allow them to produce any eggs. It takes over two weeks for an egg to develop into an adult and eggs laid before the adults became crowded continued to yield new adults which aggravated the crowding. Population size thus overshot a sustainable level. The adult population inevitably crashed when its only members were those originating from the very few eggs laid during the period of intense crowding. After the crash, adult density was low, the liver supply was ample for the now uncrowded adults so that they produced many eggs, and the whole cycle started again. The resulting oscillations are plotted in Figure 5.3 which shows clearly how peaks in adult numbers present alternated with peaks in the daily output of eggs.

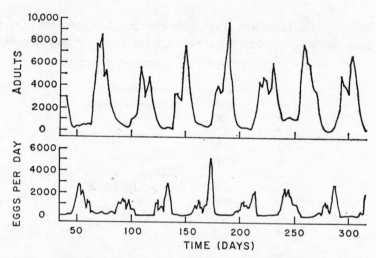

FIGURE 5.3   Oscillations in the numbers of adults, and in the daily rate of egg production, in a laboratory population of the blowfly *Lucilia cuprina*. The peaks in the adult cycle alternate with those in the egg cycle. (After Nicholson, 1954)

This mechanism probably accounts for fluctuations in many field populations, of various species, especially if they cannot relieve overcrowding by dispersal. But outside the constant environment of a laboratory, development times may be variable and it would not be surprising if the periodicity of the fluctuations were much less regular.

3   OSCILLATIONS IN HOST-PARASITE POPULATIONS

A third mechanism that can lead to population oscillations is the interaction between parasites and their hosts or between predators and their prey. The course of events is easy to visualize. Imagine a predator-prey cycle. When the prey is abundant the predators are well fed and breed successfully. The numerous offspring make big demands on their food supply, the prey population, which is consequently reduced. As a result the predators go hungry, their numbers drop because of starvation or lowered fertility, and thereupon the prey population has a chance to recover. Endless cycles can obviously develop.

An entirely analogous course of events is found in host-parasite interactions and is especially clearcut in the case of insect hosts and their insect parasites (see page 61). Since the development of each parasite individual is accompanied by the automatic death of one host individual, the coupling between the numbers of the two species is very close. A great many mathematical models can be thought of which should mimic, at least approximately, the sequence of events in a natural host-parasite set-up. Each involves a pair of differential equations (or of the corresponding difference equations): one equation of the pair expresses the growth rate of the host population and the other that of the parasite population. Here we shall examine only one of the many models described in the literature (see Leslie and Gower, 1960).

Let the numbers of hosts and of parasites be denoted by H and P respectively and suppose that

$$\frac{dH}{dt} = (r_1 - s_1 P)\,H \quad \text{and} \quad \frac{dP}{dt} = \left(r_2 - s_2 \frac{P}{H}\right) P. \tag{5.5}$$

The first equation of the pair implies that the host population would grow at an instantaneous (intrinsic) growth rate of r_1 if there were no parasites; and that the growth rate per individual is decreased by an amount $s_1 P$, which is directly proportional to the number of parasites present. With large enough P, dH/dt becomes negative and the host population dwindles. The second equation shows that the growth rate of the parasites would be unrestricted if there were an unlimited supply of hosts (i.e. if the ratio P/H were negligibly small); but that the rate falls off by an amount proportional to P/H so that as this ratio increases (i.e. as the supply of hosts becomes more and more inadequate for the parasites present) the growth of the parasite population slows and then reverses. There is, indeed, an equilibrium point at which both populations are stationary. This occurs when both $dH/dt = 0$ and $dP/dt = 0$ or when $r_1 - s_1 P = 0$ and $r_2 - s_2 \dfrac{P}{H} = 0$. Solving these equations to obtain the equilibrium sizes of the two populations gives

$$P_{eq} = \frac{r_1}{s_1} \quad \text{and} \quad H_{eq} = \frac{r_1}{s_1} \cdot \frac{s_2}{r_2}.$$

The difference equations corresponding to equations (5.5) are

$$H_{t+1} = \frac{\lambda_1 H_t}{1 + \alpha_1 P_t} \quad \text{and} \quad P_{t+1} = \frac{\lambda_2 P_t}{1 + \alpha_2 P_t/H_t} \tag{5.6}$$

with

$$\lambda_1 = e^{r_1}; \quad \lambda_2 = e^{r_2}; \quad \alpha_1 = (\lambda_1 - 1)\,s_1/r_1; \quad \alpha_2 = (\lambda_2 - 1)\,s_2/r_2.$$

If values for the parameters and for the starting populations, H_0 and P_0, of both species are chosen, it is straight forward to calculate the sequences of population sizes that will result. An example is shown in Figure 5.4a; the numerical values assigned to the parameters are shown in the legend and the starting sizes of the populations were $H_0 = 40$ and $P_0 = 30$. The process was simulated for 25 time units and it is clear that both species oscillate to begin with but fairly rapid damping takes place and the combined population soon approaches equilibrium at which $H_{eq} = 25$ and $P_{eq} = 10$.

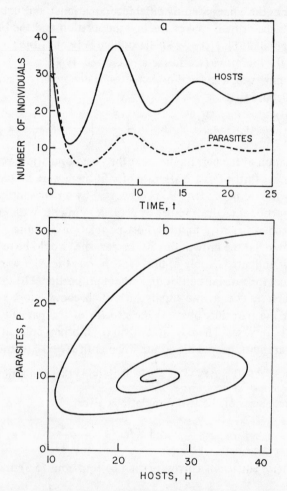

FIGURE 5.4 The behavior of host-parasite populations growing in accordance with the equations: $dH/dt = (1 - 0.1P)\,H$; $dP/dt = (1 - 2.5P/H)\,P$. (a) shows how the sizes of the two populations vary with time; both undergo damped oscillations. (b) shows the trajectory of the combined populations

Another way of looking at the interaction between hosts and parasites is to consider how parasite numbers vary in response to varying host numbers or vice versa. That is, we plot P_t versus H_t for a sequence of values of t. This has been done with the data from which Figure 5.4a was drawn and the result is shown as the continuous spiral in Figure 5.4b. The spiral begins at the chosen starting point $H_0 = 40$, $P_0 = 30$ and then circles inward towards the equilibrium point at (25, 10). The spiral is known as the *trajectory* of the combined population and shows how its species composition progressively changes as equilibrium is approached. This two-dimensional drawing of the trajectory does not show the relationship between species composition and time. To show all three variables (H, P and t) a three-dimensional model would be necessary. Thus the spiral should be thought of as an end view of a tapering coiled spring whose long axis, representing time, comes out of the page perpendicularly.

The spiral in Figure 5.4b, which was arrived at by solving a pair of difference equations, is also an integral curve of the differential equation describing the relationship between H and P. The differential equation required is obtained by dividing the second equation in (5.5) by the first to give

$$\frac{dP}{dH} = \frac{(r_2 - s_2 P/H)\, P}{(r_1 - s_1 P)\, H}. \tag{5.7}$$

Figure 5.4b shows one particular solution of the equation, the one starting at the point (40, 30). In the same way, by choosing other starting values for the host and parasite populations, we could obtain any of the infinitely many particular solutions. There are as many different spirals as there are possible starting points but all converge toward the same equilibrium point.

Figure 5.5 shows an actual example of host-parasite oscillations. The data (from Utida, 1957) were the results of controlled laboratory experiments. The host was the Azuki bean weevil, *Callosobruchus chinensis*, and the parasite a parasitic wasp, *Heterospilus prosopidis*. Only part of Utida's published results are shown here; they are representative, and show how the two populations exhibited very gradually damped oscillations in which host peaks alternated with parasite peaks. Utida found that his experimental populations repeatedly approached equilibrium but that after a few generations of comparative steadiness (for example, from generation 30 to 38, about, in Figure 5.5), "relatively violent oscillations suddenly recurred and were damped in their turn." This is not surprising when we consider that the results come from an actual experiment. Stochastic fluctuations due to chance variations in the numbers of births and deaths of either species are inevitable and prevent an exact realization of the ideal theoretical result. Any chance deviation from equilibrium must start the oscillations going again, and the greater the deviation the larger and more prolonged the subsequent oscillations. Thus even when equilibrium is the theoretical destiny of a combined population, it cannot last for long in practice.

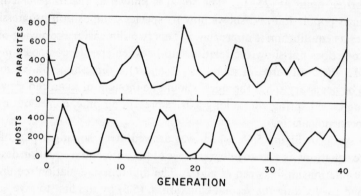

FIGURE 5.5 Host-parasite oscillations over a period of 40 generations in a combined population of *Callosobruchus chinensis* (a bean weavil) and *Heterospilus prosopidis* (a parasitic wasp). (Adapted from Utida, 1957)

4 DIFFERENT TYPES OF FLUCTUATION AND THEIR DIAGNOSIS

We have now considered three theoretical models of population behavior which can bring about *endogenous* or *intrinsic* oscillations. That is to say, the oscillations are caused by the organisms themselves, and are not under environmental control. Fluctuations (whether regularly periodic or not) that are wholly due to environmental effects and owe nothing to biological causes are *exogenous* or *extrinsic*.

We also distinguished (see page 74) between regular fluctuations (oscillations) and irregular fluctuations; regular ones are those which recur with a period of approximately constant length.

When a natural population is found to fluctuate there are therefore two questions to ask concerning it: (i) Are the fluctuations endogenous or exogenous? And (ii) are they regular (periodic or oscillatory) or irregular (aperiodic)? These questions are not logically dependent. Four pairs of answers are possible which may be shown schematically as follows.

| | | Periodicity | |
		Regular	Irregular
Cause of	Endogenous	Type A	Type B
fluctuations	Exogenous	Type C	Type D

Thus, theoretically, four separate types of fluctuation are distinguishable. The examples described in Sections 2 and 3 (blowflies, and bean weevils with parasites, respectively) are clearly of type A. Since the populations were laboratory-reared under constant conditions, their oscillations were certainly endogenous, and their regular periodicity is obvious from Figures 5.3 and 5.5.

Now consider fluctuations of type D. This category includes the behavior of an·enormous number (according to many ecologists, the majority) of natural populations whose sizes fluctuate in response to weather changes. Possibly they respond to density-dependent factors as well. To acknowledge the existence of type D populations is not to take sides in the density-dependence debate (see page 68). Good examples are provided by insect populations that from time to time "explode" and become serious pests.

As to type B fluctuations, it seems likely that species whose populations would oscillate regularly in constant environments, as a result of endogenous mechanisms like those already described, will also oscillate, though perhaps not so regularly, in the field. Whether the fluctuations would become so irregular that they ought to be classified as type B seems not to be known.

Type C fluctuations occur in all fast-breeding species (those with several generations per year) that live in environments affected in any way by seasonal change. The precise annual periodicity of their fluctuations has so obvious a cause that there is no need to investigate it. But the possible existence of type C fluctuations with periods of more than one year has aroused great interest. At one time it was thought that perhaps the many well-documented "wildlife cycles" belonged in this class. These cycles are described in more detail in Section 5 and all that needs to be said about them here is that many warm-blooded animals of high latitudes fluctuate in numbers periodically in a very striking manner. The time intervals between population peaks vary with species and location but are in the range 3 to 12 years. The most puzzling feature of the cycles is that often (though not always) they are in phase (or close to it) over enormous areas. A very famous example is the ten-year cycle in numbers of the lynx (*Lynx canadensis*) in Canada (see Figure 5.6). The following discussion is based on this example and on Moran's (1949, 1953*a, b*) investigations of it. The evidence for the cycles was assembled by

Elton and Nicholson (1942) who consulted the Hudson Bay Company's records on numbers of lynx trapped in Northern Canada. The data used for Figure 5.6 refer to the Mackenzie River District of the Northwest Territories and cover the years 1821 through 1934.

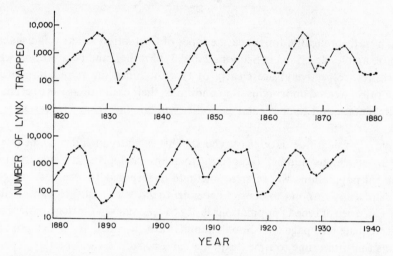

FIGURE 5.6   The ten-year lynx cycle in the Northwest Territories of Canada, over a period of 114 years. (Data from Elton and Nicholson, 1942)

The lynx cycle has been found to be in phase over the whole of Canada. It is obviously out of the question for animals dispersed over such huge distances to be participants in a single endogenously functioning mechanism. This is the reason for supposing that the cycle have an exogenous cause. But then the difficulty arises that there is no known regular cycle in the inanimate world of the required period. It has been suggested that the lynx cycle might be coupled with the 11-year sunspot cycle which (it was argued) must impose changes of the required periodicity on the climate. This theory had to be abandoned when it was shown (Moran, 1949) that the lynx cycle and the sunspot cycle did not stay in phase.

A possible solution to the dilemma was that the wildlife "cycles" were not regular oscillations controlled by an endogenous mechanism but were simply irregular, "random" fluctuations whose regularity, striking though it seems in some cases, was only apparent. If this were so, one could ascribe the fluctuations to an exogenous cause (weather being the obvious candidate) without having to postulate the existence of a *regularly* varying abiotic factor. To test this possibility it is necessary to decide exactly what property distinguishes "true" oscillations from irregular fluctuations and how the property may be recognized in practice. The defining character of a population oscillation is this: whenever the population's size is below the long-term mean it will tend to grow until it exceeds the mean (i.e. it will overshoot); and likewise when it exceeds the mean it will tend to dwindle until it is below the mean (i.e. it will undershoot). By contrast, in a population that is merely fluctuating irregularly, the size "wanders" haphazardly; though peaks and dips must alternate, by definition, there is no consistent tendency for each peak to be above the mean and each dip to be below it.

The method proposed by Moran (1953a) for diagnosing oscillations is as follows. Suppose one has a record of population sizes at a succession of times. Let these sizes be N_1, N_2, ..., N_n respectively. From these data one can calculate a series of s, say, serial correlation coefficients, $r_1, r_2, ..., r_s$ (see Appendix 5.1). Here r_1 is the correlation coefficient between N_t and N_{t+1} for $t = 1, 2, ..., (n - 1)$; r_2 is the correlation coefficient between N_t and N_{t+2} for $t = 1, 2, ..., (n - 2)$; and r_s is the correlation coefficient between N_t and N_{t+s} for $t = 1, 2, ..., (n - s)$.

Thus r_s, for instance, the sth order serial correlation coefficient, provides a measure of the closeness of the relationship between all pairs of observations so chosen that the members of each pair are separated by s steps. It now follows that if the serial correlation coefficients for some values of s are negative, one can conclude that the population is oscillating. The serial correlations will be negative when they measure the relationship between the size of the population when it is increasing (or at a peak) and the size when it is dwindling (or at a dip). Applying this test to the lynx data gave conclusive evidence that the population is truly oscillatory. Figure 5.7 shows the "correlogram" (as it is called) that was obtained when the calculated values* of r_s were plotted against s for $s = 1, 2, ..., 27$. The smoothly oscillating curve of the correlogram shows clearly how population sizes separated by a whole number of wavelengths of the 10-year lynx cycle were positively correlated; and how sizes separated by an odd number of half wavelengths were negatively correlated. The result of the test is scarcely surprising when one looks at the evident regularity of Figure 5.6. But not all suspected wildlife cycles have turned out to be truly oscillatory when subjected to this test. Moran (1952) found that four species of game bird (grouse, ptarmigan, caper and blackgame) in the Scottish Highlands exhibited random fluctuations, not oscillations as had been supposed.

It is now worth enquiring whether a study of the correlogram can cast any light on the question of whether the oscillations are exogenous or endogenous. As Moran (1953a) has shown, it can.

---

* Values of log N_t, not N_t, were used in calculating the correlations.

First, suppose the oscillations are exogenous and that the causative agent has a strictly constant period. Then, even though the oscillating population may have its peaks and dips slightly displaced owing to stochastic variation (or "errors") in population size, these displacements will not be cumulative. The living population does not itself affect the periodicity of the causative environmental factor; and it will always tend to get back into step with the cause in spite of temporary mismatching due to chance.

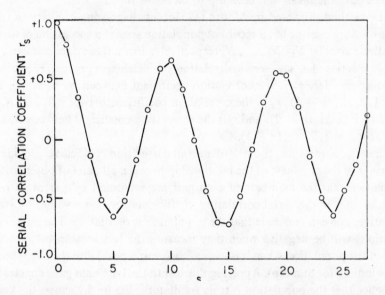

FIGURE 5.7 The correlogram for the lynx data. The curve shows the relationship between r_s, the coefficient of correlation between log N_{t+s} and log N_t, and s. (Data from Moran, 1953a)

Now suppose the oscillations are endogenous. In this case the effect of a chance "error" in population size will persist. The size of the population at some given future time depends on the population's *actual* present size, not on its "expected" present size as specified by some theoretical model. Consequently, chance displacements in population peaks will accumulate, for there is no external pacemaker. The successive population sizes constitute what is called an "autoregressive series".

We now see that if the oscillations are exogenous and their causative agent has a *strictly* constant period of length p, then r_p, r_{2p}, r_{3p} etc. should be of roughly equal magnitude. Because "errors" in population size are not cumulative, there will be no tendency for the correlations to get progressively less. The correlogram will be an undamped wave.

In constrast to this consider data from an endogenously oscillating population (i.e. from an autoregressive series). Again suppose that there is a natural period of length p and that we consider the serial correlation coefficients r_p, r_{2p}, r_{3p} etc. In this case we should expect the correlations to get progressively smaller. The greater the time interval between two observations, the less close will be the dependence of the later on the earlier. As a result, the correlogram will be a damped wave.

The correlogram in Figure 5.7 is evidently damped, leading to the conclusion that the lynx cycle is endogenous. Thus it belongs to type A (see page 87) and we now have to account for the surprising fact that sometimes the oscillations in lynx numbers are synchronized over the whole of Canada. This implies that a great many local populations, each capable of sustaining its own endogenous 10-year rhythm, are all in phase with one another. Moran (1953b) has shown that it is the weather, which is often highly correlated over the whole area, that triggers the several independently oscillating populations. Thus an *irregularly* fluctuating external agent affecting, simultaneously, a number of separate, *regularly* oscillating mechanisms, keeps them all in phase.

A more thorough mathematical analysis of the lynx data has been carried out by Hannan (1960). And Clark (1971), in a recent discussion of the same data, has suggested that the integral of a second order differential equation may perhaps give the best description of the lynx cycle. That is, he assumes that d^2N/dt^2, the *rate of change* of the growth rate, as well as dN/dt, the growth rate itself, depend on N, the population size.

5. WILDLIFE CYCLES

The reality of wildlife cycles and their salient characters if they do, in fact, occur have been debated for many years. As long ago as 1954 the Journal of Wildlife Management (in its Volume 18) published a Symposium on Cycles in Animal Populations, and two contrary opinions on the defining properties of cycles are worth quoting from the Symposium. According to Lack (1954), the "regularity of cycles ... refers primarily to the interval between successive peaks." Rowan (1954), on the other hand, remarked that uniform time intervals are "not essential to cycles" but that there must be fluctuations of large amplitude as opposed to the minor ups and downs of "merely random fluctuations." In terms of the four types of fluctuations defined on page 87, it thus appears that Lack would regard only types A and C as cycles; in other words, Lack treats "cycles" and "oscillations" as synonyms. Rowan would regard any of the four types as cycles if their amplitudes were large enough, but would presumably not treat even perfectly periodic fluctuations as cycles if they were of small amplitude. It seems to me that *all* fluctuations that are believed *not* to be random are best called "cycles", at least until they have been analysed and their type determined. Even "outbreaks" occurring at quite long intervals can be so described. To insist, as some authors do, on restricting the word "cycle" to endogenously caused fluctuations (whether periodic or not) or to periodic fluctuations (whether endogenous or not) creates unnecessary difficulties when one wants to speak of cycles whose cause is unknown. The non-commital word "fluctuations" is all-inclusive and embraces random as well as non-random fluctuations.

We now consider some of the best documented examples of wildlife cycles. Most of the account that follows is based on Lack (1954) and Keith (1963).

The classic examples are all from northern Canada and the animal species exhibiting cyclical behavior can be divided into three distinct groups on the basis of habitat and the duration of the cycle. The three groups are:

1 *Tundra species with a four-year cycle.* The carnivores in the group are Snowy Owl *(Nyctaea scandiaca)*, Arctic fox *(Alopex lagopus)* and Rough-legged Hawk *(Buteo lagopus)*. Their chief prey, which also undergo the four-year cycle, are lemmings *(Lemmus* spp and *Dicrostonyx* spp), voles *(Microtus* spp) and Willow ptarmigan *(Lagopus lagopus)*.

2 *Birch forest species with a four-year cycle.* The habitat in this case is the transition zone, occupied by open, grassy birch woodland between the open tundra to the north and the boreal (coniferous) forest to the south. The carnivores are Red fox *(Vulpes fulva)*, marten *(Martes americana)*, Rough-legged Hawk and Northern Shrike *(Lanius excubitor)*. The herbivores are lemmings, voles and willow ptarmigan as in the tundra cycle, and also Snowshoe hare (also known as Varying Hare, *Lepus americanus*).

3 *Boreal forest species with a ten-year cycle.* The carnivores are lynx *(Lynx canadensis)*, Horned owl *(Bubo virginianus)*, American goshawk *(Accipiter gentilis)*, mink *(Mustela vison)*, fisher *(Martes pennanti)*, marten and red fox. Their herbivorous prey are Snowshoe hare, Ruffed grouse *(Bonasa umbellus)* and Spruce grouse *(Canachites canadensis)*.

It will be noticed that some species, notably the red fox, marten and snow-shoe hare have four-year cycles in the open birch forest and ten-year cycles farther south in the coniferous forest.

The consensus of ecological opinion on the causes of the cycles seems to be roughly as follows (we give it here in the simplest possible form shorn of all qualifications and caveats): the primary cause is a "predator-prey" type cycle (see page 82) in which the herbivorous mammals function as predators and the vegetation as prey. The damage that small herbivorous mammals can do to the plants they feed on is sometimes spectacular; lemmings and voles can strip the ground bare of vegetation and thereby destroy not only their food supply but also the cover which might otherwise conceal them from hunting hawks and owls. Snowshoe hare populations, too, can build up into devouring hordes. Rowan (1954) describes a 23-acre tract of young Jack pines (*Pinus banksiana*) containing more than a million trees, all but 40 of which had been barked, decapitated and stripped of branches by hares whose density he estimated at 47 per acre.

The herbivore cycles cause the carnivore cycles. But as far as the herbivores are concerned, it is shortage of plant food rather than abundance of carnivores that causes their population declines. Thus both herbivores and carnivores are limited by starvation. Evidence for this is provided by the migrations of starving populations in their attempts to find new food supplies. The mass migrations of lemmings, which often terminate in a suicidal march into the sea, are believed to begin as sorties in search of food. And the abundance of snowy owls in temperate eastern North America every fourth year, when lemmings are sparse in the arctic, is well known to thousands of bird watchers; the owls are forced to emigrate from their preferred haunts to avoid starvation.

The cycles in the gallinaceous bird species (grouse and ptarmigan) are believed to result from the predator cycles. The birds are less numerous than voles and hares, and are prey to the same predators. Their cycles do not, therefore, cause carnivore cycles but are caused by them.

Summarizing, the chain of cause and effect seems to be this: The basic cycle at any place results from plant-herbivore interaction. The herbivore cycles cause the carnivore cycles. And the latter cause the grouse and ptarmigan cycles.

The durations of the cycles are presumably linked to the longevities of the most important species. The cycles in which lemmings and voles are the determining species last for four years; the snowshoe hare consequently has a four-year cycle when it occurs with, and is dominated by, the rodents. But where rodents are few and the hare becomes the controlling species, its longer life span causes the cycles to be ten years long.

It is noticeable that the "best" cycles, those with the most constant period and with the most pronounced peaks, occur in high latitudes. Almost certainly this is because in harsh environments species are few and food chains simple and short. The field conditions are such that comparatively uncomplicated interactions among species can occur, and the resulting cycles are almost as regular as those that can be induced in laboratory populations. Pitelka (1967) has given evidence for this belief. He found that lemming populations fluctuated more markedly when only one species was present than when two or more occurred together. Carnivore populations, also, tend to fluctuate more the fewer the number of species available as prey.

In case the foregoing paragraphs have made the problems of cycles seem simple and uncontroversial, at least a few of the puzzles and difficulties should be mentioned. Indeed Keith (1963), confining his remarks to the ten-year cycle, writes: "The belief that some mammals and birds manifest ten-year cycles of abundance is by no means universal among biologists... Many reputable workers [put forward] opposing views." The doubts of some biologists as to the reality of the cycles stem largely from a distrust of the data used as evidence for cycles. Precise records exist for short periods only. Records covering long periods are anything but precise and are often based on hunting statistics, fur returns, and studies of game animals for management purposes.

The estimation of a population's size from the number of skins of that species traded by trappers is obviously open to objection. The number of skins of a species brought to a trading station in a year depends on many things, quite apart from the abundance of the species; the demand for furs of that species, the number of trappers and the chance that they may spend some time in other occupations, and the weather, are all likely to affect the catch. Nevertheless, for such species of the northern forests as lynx, fox, mink, marten, fisher and muskrat, much of the evidence comes from fur

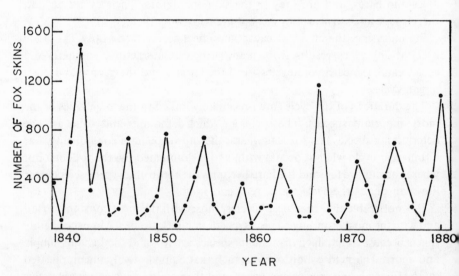

FIGURE 5.8 The four-year cycle for "colored fox" in Labrador, over a period of 41 years. (Data from Elton, 1942)

trading records. As an example, consider the data given by Elton (1942) to illustrate a four-year cycle in "colored fox" (probably *Vulpes fulva*) in Labrador. He obtained these from the published reports of the Moravian Missions of Labrador in which are recorded the numbers of colored fox skins traded to the missionaries by the eskimos during the years 1834 to 1925. (Data covering the period 1840 to 1880 are shown here in Figure 5.8.)

In spite of the admitted defects (from an ecologist's point of view) of records of this sort, they do provide a wealth of data that it would be wasteful to ignore; and they often cover very long periods.

Next we must mention a few of the anomalous observations that do not fit in with the summary given above. The following are simply miscellaneous examples and are intended only to illustrate the kind of fact that is apt to crop up and make generalisation difficult. The ten-year cycle of the boreal forest animals of northern North America is not exhibited by all species, only by some. Also, the cycles occur only in North America, not in Eurasia. Some species are cyclic at some times, but not always; for example Ruffed grouse used to have a ten-year cycle in eastern U.S.A. and New Brunswick, but not in recent years. Some species are cyclic in only part of their range; thus Sharp-tailed grouse (*Pedioecetes phasianellus*) are cyclic in the Canadian prairies but not in North Dakota. The Snowshoe hare whose cycles in the northern forest have been described above does not cycle in the extreme eastern and western parts of North America. (Most of these examples are from Keith, 1963).

Observations on the synchrony of the cycles in different parts of a species' range have given a variety of results. Sometimes the phases of the cycle in neighboring populations of a species appear to be unrelated. And sometimes, as already remarked, synchrony may obtain over an enormous area. Another commonly observed phenomenon is for peak densities to spread out in an expanding wave from an "epicenter" implying interdependence among local populations across thousands of miles. These spreading waves often occur during insect pest outbreaks, as well as with mammal cycles (see Watt, 1968). They are probably caused, at least partly, by successive emigrations from regions of very high population density outward into sparsely occupied territory. Among insects such mass movements can be brought about when swarms of individuals are passively wafted by aerial convection currents (Greenbank, 1957).

Much recent work on cycles has been designed to test the hypothesis that the animals in an increasing population differ *genetically* from those in a declining one. The possibility was first suggested by Chitty; accounts of recent work and many references are given in Chitty and Phipps (1966) and Krebs (1970). Most of the work has been done with voles (*Microtus* spp and

Clethrionomys spp) and has involved laboratory and field experiments as well as the observation of undisturbed natural populations. An individual vole in a sparse population is exposed to quite different environmental conditions from one in a dense population, and the forces of natural selection acting upon it are different. As a consequence, the genetic constitution of a population may undergo cyclic variation in parallel with the variation in population size. An individual animal in a dense population has many more contacts with others of its kind and a strong fighter will be selectively favored. It has been found that voles in expanding populations are larger and more aggressive than those in dwindling populations; there are also differences in litter size, in the hostility of the adults towards the young, in birth and death rates and, possibly, in the susceptibility of the animals to adverse weather. Investigations are going on to determine whether these undoubted qualitative differences are phenotypic or genotypic. In any case, it seems very likely that populations of small mammals can contribute to their own density regulation by agressive behavior and intraspecific strife.

CHAPTER 6

Estimating the Sizes of Plant and Animal Populations

THROUGHOUT THE PRECEDING chapters we have discussed the sizes of populations, that is to say the numbers of individuals comprising them. Populations are the basic entities an ecologist studies and all theorizing in population and community ecology deals, in one way or another, with the sizes of populations, with how they increase or dwindle, become stabilized or fluctuate, and with how they respond to environmental changes and themselves cause changes in the environments of other populations.

It seems fair to say, however, that it is easier to devise hypotheses to account for population phenomena than it is to test them. To test a hypothesis entails comparing its predictions with what is actually happening in nature; and to discover what *is* happening is often extraordinarily difficult, perhaps more so in ecology than in any other natural science. Few ecologists can work for long without being faced with the need to estimate the size of some natural wild population. In all too many cases the precision attainable is disappointingly low; often it is too low to permit discrimination between rival theories.

Estimating the size of a wild population takes planning, effort and ingenuity. The job is often regarded as a distracting chore causing delays and uncertainty in the course of an investigation. Alternatively, it may be treated as an integral (and interesting) part of an investigation. The greater the number of reasearch workers who take the latter approach the better, since ecologists with a flair for population estimation are becoming increasingly necessary to society in the capacity of environmental watchdogs. Monitoring the effects on the environment of new technological developments will certainly require sophisticated methods for measuring population changes in a diverse assortment of living organisms.

In this chapter, therefore, we shall describe some of the ways in which plant and animal populations are determined or estimated. In a few exceptional cases the size of a population can be determined by taking a complete census. This implies that every member of the population under study is seen and counted. Two examples are described in Section 1 below.

Far more often, the size of a population has to be estimated. That is, one must infer it on the basis of the incomplete evidence provided by a sample. It is important to emphasize the distinction between *determining* and *estimating* the size of a population. The size is said to be determined when it is believed that every individual has been seen and tallied and hence that a complete *census* has been taken. Errors may be made of course, but perfect accuracy is attainable in principle if not in practice.

By contrast, when the size of a population is estimated from a sample, from a part rather than from the whole, error is inevitable and the ecologist must accept the fact. No amount of care in counting the individuals that are seen can make up for the fact that part of the population is not seen. To *estimate* is to *infer* and the process is uncertain by definition. When a population's size is estimated, therefore, the result sought is not a single number but a pair of numbers between which the unknown true value is believed to lie. We shall go into details below.

Before the size of a population can be determined or estimated, however, it is necessary to define exactly what constitutes the population under study. The definition must be unambiguous and operational; that is, it must be so clear and explicit that one can say of any individual organism either that it does, or does not, belong to the defined population. Thus one cannot, in the present context, define a population as an assemblage of interbreeding individuals since the breeding behavior of the assemblage one wishes to study may be unknown. Again, an assemblage with clear natural boundaries, whether interbreeding or not (e.g. all the crayfish in a big lake) may be too large an entity to be conveniently studied. One must then so define the population as to make it of manageable size. Another point to notice is that a population, in the sense in which we are using the word here, need not consist of organisms of only one species. For example, one might wish to estimate the number of snails in a marsh, or of fish exceeding a certain size in a river. In such cases, the populations would probably have several species.

The point that must be emphasized is that the definition of a population is a matter of choice. The word "population" connotes the assemblage of organisms that the ecologist has *chosen* to study; he may choose as he likes, but must specify his choice precisely enough for there to be no doubt what is meant. Some examples should make this clear. A forest ecologist might define as the population under study all the sugar maples on a specified tract of land. Is this to include even the smallest seedlings and if not what is the lower limit of height, or diameter, or age, of trees to be treated as population members? Until this question has been answered, the population has not been defined. Another example: an entomologist might wish to estimate the number of grasshoppers present in a delimited area at a particular time. If the estimate were to be made in the breeding season, one would have to decide whether or not eggs were to be counted and until this decision had been made the population would be undefined.

We shall now consider some of the ways in which ecological populations are censused and sampled, and describe actual examples.

1 CENSUSING A WHOLE POPULATION

In a few cases it is possible to count every member of the population chosen for study. An obvious example is the trees in a tract of forest. Provided the area is not too large, once its boundaries have been fixed it is a simple matter to count all the trees that conform to the chosen specifications for population members. Probably some trees will be difficult to classify as to whether they should properly be included in the population; thus severely diseased trees, blowdowns that are still alive, and trees with more than half their branches dead might constitute borderline cases. Whether to treat them as population members or not, however, is a matter of decision, not of uncertain inference.

More interesting problems arise when a population of large active animals is to be censused.

Stephen (1967) has described how the sizes of migrating flocks of sandhill cranes (*Grus americana*) are determined. Large flocks migrate southward from northern Canada in late summer and early fall and pause to feed in prairie grain fields where they damage the ripened crops by feeding and trampling. The numbers may be large. For instance, in one year the peak population was found to be 18,000.

For census purposes a population consisted of all the members of a flock (a natural entity) present on a particular day. The task of censusing them was made possible by the behavior of the cranes, which roost in small areas over night and leave their roosts at daybreak when they fly out to their daytime feeding locations. Two census methods were devised. In one, concealed observers on the ground were stationed near each roost, beginning half an hour before sunrise. Each individual bird was counted as it flew out and the observers maintained their vigil until half an hour after the flight of the last bird seen to leave.

The second method entailed counting the cranes at their roosts from a low-flying aircraft as soon as the light was good enough for the pilot to see the flocks but before the birds had flown out. Determining the numbers of birds in dense flocks on the ground from a plane flying over them required quick judgment; obviously, individual birds could not be tallied one by one. The pilot therefore judged numbers by comparing the appearance of flocks on the ground with reference photographs showing model "flocks" of known size; (the models were constructed by strewing a known number of wheat kernels on a brown background). Perfect accuracy was unattainable of course. Even so, the process should be regarded as an imperfect determination of population size and not as an estimation, since the errors arose from imperfections of technique and not from the inherent uncertainties of drawing conclusions about a population from examination of a sample.

Another example of census-taking has been described by Thomas (1969). The populations censused were herds of barrenground caribou (*Rangifer arcticus*) on their spring migration northward from forest to open tundra. If a herd is concentrated into a small area, it can be photographed from the air. It was found possible to count every individual animal by examining the prints under a dissecting microscope and marking off each animal with ink as it was counted. They showed up especially clearly when there was a background of fresh snow. The sizes of the herds ranged up to 127,000 but even a population as large as this can be censused when its members are all present simultaneously in terrain where concealment is impossible.

2   POPULATION ESTIMATION: SOME GENERAL COMMENTS

In the great majority of cases the size of a plant or animal population must be estimated from a sample. There are a number of ways in which this can be done and in subsequent sections we shall describe some of them. To begin with, a few general comments applicable to all forms of sampling should be made.

Once the precise definition of the population whose size is to be estimated has been decided, the area or volume containing it can often be subdivided, at any rate conceptually, into a large number of parts, or *sampling units*. A few examples should make this clear.

1   Suppose we wish to estimate the number of seedlings of a plant species in an area of 10,000 square meters. A convenient way of doing this would be to count the seedlings in each of several small "sample plots" or "quadrats", of area one square meter, say. The quadrats are the sampling units and the whole population is contained in an area equal to 10,000 of them. If the seedlings are counted in 200 quadrats, for example, then these quadrats constitute the *sample* and the size of the sample is 200 units.*

---

* It is a common error to describe a sampling unit as a "sample". A sample always consists of a number of sampling units.

It is a useful convention to denote by N the number of sampling units containing the whole population, and by n the number in the sample. Thus in the example just given, $N = 10,000$ and $n = 200$. The ratio n/N (1 in 50 in this case) is known as the sampling fraction.

2   A soil ecologist might wish to estimate the number of mites in the top 20 centimeters (a common plough depth) of a hectare (10,000 square meters) of agricultural soil. The population concerned thus lives in a volume of 2,000 cubic meters of soil. To estimate its size one might examine a sample of 50, say, sampling units each consisting of a cylindrical plug of 300 cubic centimeters of soil. Thus the sampling ratio in this example is about 1 in 133,000.

3   In the two examples already given the organisms of the population to be estimated might be found anywhere throughout a continuous area or volume. Subdivision of the area or volume into sampling units was arbitrary; the units did not have natural boundaries. However, cases often arise when natural sampling units do occur. For instance, suppose the population to be estimated consisted of all the caterpillars of a species of cone moth in a tract of spruce forest. The caterpillars are found only inside the spruce cones and consequently a cone is a natural sampling unit. The difficulty arises, of course, that one does not know the number of spruce cones in the whole tract of forest. Though n is known, N is not and it has to be estimated. Thus estimation of the caterpillar population requires *two-phase* or *double* sampling and a simple example is described in Section 5 (see page 113).

In the foregoing examples the organisms were all sedentary or small, so that a number of individuals could be "captured" and counted each time a sampling unit was examined. Obviously this cannot always be done. It clearly will not work when one wishes to estimate the number of squirrels in a wood, or of trout in a large lake. In estimating the sizes of populations such as these, it is not feasible to use sampling units which each contain several individuals of the population concerned. Instead, samples must be taken of the individuals themselves. Details are given in Sections 7 and 8.

We now introduce some of the symbols to be used in Sections 3, 4, 5 and 6 which deal with methods of estimation using sampling units.

Capital letters denote quantities pertaining to the whole population. Thus:

Y is the total number of individuals in the population.

N is the number of sampling units containing the whole population.

$\bar{Y} = Y/N$ is the mean number of individuals per sampling unit, averaged over *all* sampling units.

Lower case letters denote quantities pertaining to the sample. Thus:

y is the number of individuals in a single sampling unit.

n is the number of sampling units in the sample.

$\bar{y} = \Sigma y/n$ is the mean number of individuals per sampling unit *in the sample.*

Symbols surmounted by a "hat" (e.g. $\hat{Y}$) are estimates. Thus $\hat{Y}$ and $\hat{\bar{Y}}$ are the estimates of Y and $\bar{Y}$ obtained by observing the sample. Common sense shows that, in general

$$\hat{\bar{Y}} = \bar{y} \quad \text{and} \quad \hat{Y} = N\bar{y}.$$

Therefore, if all that is required is a single number that is believed to be close to the unknown Y one wishes to discover, the estimation process is mathematically straightforward (even though the field work may not be). Such a single calculated value is called a *point estimate.*

A point estimate by itself is of very little value, however. We have already stressed that estimation is a form of uncertain inference; the probability is negligibly small that an estimate of a population's size is exactly correct and we must therefore give, besides the estimate itself, some measure of its *precision.* The best way of doing this is to calculate an *interval estimate* (as opposed to a point estimate) of the value concerned. That is, one specifies a *confidence interval* for the number that is to be estimated.

The way in which this is done is as follows. The estimate $\hat{Y}$ of the unknown Y is based on counts of the number of individuals in a sample of n sampling units, taken from a total of N sampling units. Now clearly a very great number* of samples of this size could have been drawn; they would have yielded many different values of $\hat{Y}$. Thus $\hat{Y}$ itself is a random variable with its own probability distribution (i.e. the expected relative frequencies of all its possible values). From the observations we can also calculate var $(\hat{Y})$, the *variance* of $\hat{Y}$, which measures the "spread" of the probability distribution; and $s_{\hat{Y}} = \sqrt{\text{var}(\hat{Y})}$, which is known as the *standard error*† of $\hat{Y}$. The necessary formulas for calculating variances and standard errors will be given below.

Now consider the quantity

$$\frac{\hat{Y} - Y}{s_{\hat{Y}}}.$$

---

* A sample of size n from a total of N sampling units can be chosen in $\binom{N}{n}$ $= \dfrac{N!}{n!\,(N-n)!}$ ways. This is the number of possible combinations of N things taken n at a time and is sometimes written $^{N}C_n$ or $_NC_n$.

† The distinction between a *standard error* and a *standard deviation* is explained in the final paragraphs of Appendix 6.1.

On consulting statistical tables (for example Table Q in Rohlf and Sokal, 1969), one can obtain a number, say $t_{0.025}$ such that

$$\Pr\left\{-t_{0.025} \leqq \frac{\hat{Y} - Y}{s_{\hat{Y}}} \leqq +t_{0.025}\right\} = 0.95. \qquad (6.1)$$

(Observe that the subscript in the term $t_{0.025}$ is $(1 - 0.95)/2$.) Here Pr $\{...\}$ denotes the probability that the statement inside the braces is true. Manipulating the inequalities inside the braces it is seen that

$$\Pr\left\{(-t_{0.025}s_{\hat{Y}} - \hat{Y}) \leqq -Y \leqq (-\hat{Y} + t_{0.025}s_{\hat{Y}})\right\} = 0.95$$

or

$$\Pr\left\{(\hat{Y} - t_{0.025}s_{\hat{Y}}) \leqq Y \leqq (\hat{Y} + t_{0.025}s_{\hat{Y}})\right\} = 0.95. \qquad (6.2)$$

In words, the probability is 0.95 (or 95%) that the unknown Y lies somewhere between the two values $\hat{Y} - t_{0.025}s_{\hat{Y}}$ and $\hat{Y} + t_{0.025}s_{\hat{Y}}$. These two numbers are known as the lower and upper 95% *confidence limits* for Y and between them lies the 95% *confidence interval*. If $s_{\hat{Y}}$, the standard error of $\hat{Y}$, is small the interval is narrow and the true value of $\hat{Y}$ has been estimated with high precision. Conversely, if $s_{\hat{Y}}$ is large, the calculated confidence interval is wide and the precision low.

It is now apparent that if $\hat{Y}$ and $s_{\hat{Y}}$ are calculated from the field observations (the appropriate formulas, which depend on the sampling methods used, are given below with the descriptions of the different methods) and if $t_{0.025}$ is taken from statistical tables, the 95% confidence interval for Y can be obtained. Appendix 6.1 gives a somewhat fuller account of the rationale underlying the calculation of confidence intervals and also a table of values of $t_{0.025}$.

The reason that there is a list of values of $t_{0.025}$, instead of a single value, is as follows. Besides depending on $s_{\hat{Y}}$ (the standard error of $\hat{Y}$), the width of a confidence interval depends also on the number of "degrees of freedom" of $s_{\hat{Y}}$ which in turn depends on n, the size of the sample. The number of degrees of freedom* is often denoted by ν and it is always true that $\nu \leqq n - 1$. The exact value of ν depends on the sampling method used and the appropriate value will be given with each of the sampling methods described.

One final point to notice: for convenience, in the examples below 95% confidence intervals have been calculated in every case. That is, the interval

---

* The number of degrees of freedom is the number of observations that are "free to vary". For example, if a list of n numbers is to have a total *given beforehand*, then $n - 1$ of the numbers may have any values at all (positive and negative numbers being allowed), but the nth *must* have whatever value is required to bring the sum to the given total. There-. fore, once the total has been fixed, the number of degrees of freedom of the list is $n - 1$.

given is always such that the probability that the true value it is desired to estimate falls within it is 0.95. This is a common convention but there is no necessity always to use a 95% confidence interval. For example, if one wants to be more confident that a stated interval brackets the true value, a 99% confidence interval could be used. It would, of course, be wider than a 95% interval but this is the price that must be paid for reducing the risk of error. When 99% confidence intervals are required, appropriate values of t are needed and they are somewhat larger than the $t_{0.025}$ values given in Appendix 6.1. They can be found in standard statistical tables.

We come now to some actual examples of different methods of population estimation. In most cases the necessary formulas for $\hat{Y}$ are intuitively obvious or easily derived. However, the formulas for var $(\hat{Y})$, and hence for $s_{\hat{Y}}$, which are required for the computation of confidence intervals, are often difficult to derive and are reached only after fairly intricate mathematical arguments. They are stated here without proof.

Very full elementary discussions of sampling and estimation will be found in Hansen, Hurwitz and Madow, Volume I (1953), Yates (1960), and Sampford (1962). The underlying mathematical theory will be found in Hansen, Hurwitz and Madow, Volume II (1953), Cochran (1963), and Raj (1969). None of these books deals with ecological applications, however.

3 SIMPLE RANDOM SAMPLING

We begin with a very simple example. The population whose size was to be estimated consisted of all the plants of *Balsamorhiza sagittata* (the "sunflower" of British Columbia sagebrush country) growing on a patch of ground 25 meters square (i.e. of area 625 sq. m). The area was sampled with 100 randomly placed meter-square quadrats and the number of plants in each was counted. All quadrats were placed with their edges parallel with the borders of the square area being sampled; their locations were selected by picking two numbers, in the range 0–24, from a random numbers table and using them as coordinates for the south-east corner of each quadrat. If any pair of coordinates chanced to occur more than once it was disregarded and another pair selected, so that the 100 quadrats examined were all different. This is called "sampling without replacement". In effect, therefore, the area was treated as consisting of a grid of 625 square quadrats from which 100, all different, were chosen at random. The number of different samples that could have been obtained was

$$\binom{625}{100} = \frac{625!}{100!\,525!} = 9.5 \times 10^{117} \quad \text{approximately,}$$

and the method of selecting the sample ("simple random sampling") was such that each of these possibilities was equiprobable.

The results obtained are shown in Table 6.1.

TABLE 6.1   Data for estimating the size of a population of *Balsamorhiza sagittata*

Plants per quadrat y	Frequency, f_y, of quadrats with y plants	yf_y	y^2f_y
0	32	0	0
1	40	40	40
2	16	32	64
3	6	18	54
4	4	16	64
5	2	10	50
$\geqq 6$	0	0	0
	100	116	272

From the table it is seen that

$$\bar{y} = \frac{1}{n}\Sigma y = 1.16 \quad \text{plants per sq. m.}$$

Taking $\bar{y}$ as an estimate of $\bar{Y}$, the mean number of plants per square meter throughout the area being sampled, it now follows that

$$\hat{\bar{Y}} = \bar{y} = 1.16 \quad \text{and} \quad \hat{Y} = N\hat{\bar{Y}} = 1.16 \times 625 = 725.$$

Thus $\hat{Y} = 725$ is a point estimate of the number of plants in the defined population.

Let us now obtain a 95% confidence interval for this number. It is first necessary to calculate the variance of $\hat{Y}$ which is

$$\text{var}\,(\hat{Y}) = \frac{N^2}{n(n-1)}\left(1 - \frac{n}{N}\right)\left\{\Sigma y^2 - \frac{(\Sigma y)^2}{n}\right\} \qquad (6.3)$$

$$= 4555.30.$$

Therefore

$$s_{\hat{Y}} = \sqrt{\text{var}\,(\hat{Y})} = 67.5.$$

The number of degrees of freedom is $v = n - 1 = 99$ and (see Appendix 6.1) we therefore take $t_{0.025} = 2$. Thus the 95% confidence interval for Y is

$$725 \pm (2 \times 67.5) \quad \text{or} \quad 590 \text{ to } 860.$$

It is worthwhile to consider the factors influencing var $(\hat{Y})$. The variance of the raw data, the number of plants per quadrat, is

$$\text{var}(y) = \frac{1}{n-1}\left\{\Sigma y^2 - \frac{(\Sigma y)^2}{n}\right\}.$$

Therefore equation (6.3) may be written

$$\text{var}(\hat{Y}) = \left(1 - \frac{n}{N}\right) \cdot \text{var}(y) \cdot \frac{N^2}{n}. \tag{6.4}$$

Clearly the first factor*, $\left(1 - \dfrac{n}{N}\right)$, which is known as the "finite population correction", decreases as the sampling fraction, n/N, increases; it is zero, as we should expect, when the sample contains every sampling unit, for then the population of plants is fully censused and Y is known without error. In the present example the sampling fraction is $100/625 = 0.16$, quite a large proportion. Often in ecological work the sampling fraction is so small that for practical purposes we may take $1 - (n/N) = 1$.

The second factor in equation (6.4) is var (y). As one would expect, the greater the variability among the quadrats in the number of plants they contain, the less precisely can Y be estimated.

The third factor in (6.4), N^2/n, shows that var $(\hat{Y})$ can be reduced by enlarging n. Once again, this accords with common sense.

Now suppose we wished to estimate $\bar{Y}$, the mean number of plants per square meter rather than Y, the size of the total population. A point estimate of $\bar{Y}$ is obviously given by $\hat{\bar{Y}} = \bar{y}$, the sample mean. The 95% confidence interval is

$$\hat{\bar{Y}} \pm t_{0.025}s_{\hat{\bar{Y}}} \quad \text{where} \quad s_{\hat{\bar{Y}}}^2 = \text{var}\left(\hat{\bar{Y}}\right).$$

We therefore require var $(\hat{\bar{Y}})$. Now

$$\hat{\bar{Y}} = \frac{\hat{Y}}{N}$$

and consequently†

$$\text{var}\left(\hat{\bar{Y}}\right) = \text{var}\left(\frac{\hat{Y}}{N}\right) = \frac{1}{N^2}\text{var}(\hat{Y}). \tag{6.5}$$

---

* This factor is to be replaced by 1 if sampling is done "with replacement", i.e. if the fact that a sampling unit has been examined already does not prevent its being picked again. The number of sampling units is then effectively infinite, since choosing a unit does not deplete the supply.

† A proof will be found in any elementary statistics textbook.

In the present example

$$\hat{Y} = \bar{y} = 1.16$$

and, from equation (6.5)

$$\text{var}(\hat{Y}) = \frac{4555.30}{625^2} = 0.01166$$

so that

$$s_{\hat{Y}} = 0.108.$$

As before, there are 99 degrees of freedom and consequently the 95% confidence interval is

$$\hat{Y} \pm 2s_{\hat{Y}} = 1.16 \pm 0.22 \quad \text{or} \quad 0.94 \text{ to } 1.38.$$

In most ecological work an estimate of total population size is of greater interest than an estimate of population mean density (i.e. the mean number of individuals per sampling unit). For some kinds of populations, however, the numbers concerned are so vast that only densities convey any meaning. This is true, for instance, of plankton populations, and an example is given in Section 4.

A final point should be noticed. The precision an estimate is to have can, in theory, be specified in advance; that is, one can stipulate an acceptable width for the confidence interval. A certain amount of preliminary sampling must first be done so that the variance of the observations can be judged roughly. Then the size of the sample required to give a confidence interval of the desired width can be calculated. A sample of the calculated size will *probably* (not, of course, certainly) yield an estimate with the desired precision. Mathematical and computational details will be found in Cochran (1963) and Sampford (1962), for example. However, in ecological work it may be impossible to act with such prescience. Natural populations, especially of small, short-lived organisms, often change so rapidly that there is no time to inspect a preliminary sample and plan the next step in the light of its evidence.

4 STRATIFIED SAMPLING

When sampling units are located at random throughout an area, it seldom happens that the units are evenly spaced. Often, merely as a result of chance, some parts of the area under study contain a disproportionately large share of the sampling units, while other parts contain few or none. If the area is truly homogeneous, in other words if the value of the variate is wholly independent of the location of the sampling unit in which it is observed, this does not matter. Indeed, even if all the units happened to be concentrated in a single compact block, as could conceivably occur, they would be as representative of the area, if it were really homogeneous, as widely scattered units. Therefore, when there is reason to believe that an area is not homogeneous or that individuals, for whatever reason, are more crowded in some parts than in others, it is natural to attempt so-called *stratified sampling*. That is, the area concerned is subdivided into a number of internally homogeneous subareas, or *strata*, and a separate random sample is drawn from each stratum. In this way no part of the area is inadvertently underrepresented.

As an illustrative example we consider estimation of the density of the Chlorophyceae (green algae) forming part of the phytoplankton in the surface waters of Lake Ontario. The data* are from Ogawa (1969) where the methods of obtaining water samples and of extracting and identifying the phytoplankton organisms are fully described. It suffices to note here that each sample of "surface" water (i.e. from a depth of less than five meters) was collected in a Nansen bottle and the phytoplankton was allowed to settle and become concentrated; one milliliter was drawn from the concentrate and part of its contained phytoplankton counted with the aid of a dissecting microscope. These details are given to emphasize that in work such as this it is unreasonable to try to estimate the sizes of whole populations. Average densities are wanted, that can be compared with other average densities obtained at different times and places.

Since inshore waters are generally richer in phytoplankton than more open waters, the total area (i.e. the whole lake) was divided into two strata. The dividing line was defined as a line parallel with the shore and eight kilometers distant from it. Stratum 1, the inshore waters, consisted of all surface waters on the landward side of this line; the rest of the lake surface constituted Stratum 2, the offshore waters. In all that follows the subscripts 1 and 2 will be attached to symbols pertaining to the respective strata.

Water samples (these are the "sampling units") were taken at a total of $n = 23$ observation stations of which $n_1 = 15$ were in Stratum 1, and $n_2 = 8$ in Stratum 2. For the purpose of the present problem each station yielded a single item of data, a count of cells of green algae. The results are shown in Table 6.2.

Let y_i denote the cell count made on one of the sampling units from the ith stratum. Since there are two strata, i can take the values 1 and 2. It is

---

* It should be remarked that the observations were not made for the purpose to which they are put here: the estimation of a mean density figure applying to the whole lake. The calculations are intended to illustrate the estimation method, not to lead to a particular numerical answer.

seen that the mean cell counts within each stratum are given by

$$\bar{y}_1 = \frac{1}{n_1}\Sigma y_1 = \frac{6754}{15} = 450.267 \text{ cells}$$

and

$$\bar{y}_2 = \frac{1}{n_2}\Sigma y_2 = \frac{1927}{8} = 240.875 \text{ cells.}$$

TABLE 6.2   Data for estimating phytoplankton density

Cell counts of green algae	
Inshore waters (Stratum 1) y_1	Offshore waters (Stratum 2) y_2
344	163
1400	340
674	273
511	116
482	166
234	188
162	410
281	271
309	$\Sigma y_2 = 1927$
312	
573	
528	
339	
399	
206	
$\Sigma y_1 = 6754$	

Recall that we wish to estimate $\bar{Y}$, the overall mean cell count for the surface waters of the whole lake. To do this it is necessary to take account of the relative sizes of the different strata; (their absolute sizes would be required only if we wished to estimate Y, as opposed to $\bar{Y}$). The volume of water under consideration is a thin surface layer and thus the relative sizes of the strata are given by their relative areas. The areas are, approximately,

$$N = 19,500 \text{ sq. km.} \text{ for the whole lake,}$$

$$N_1 = 6,500 \text{ sq. km.} \text{ for the inshore waters, and}$$

$$N_2 = 13,000 \text{ sq. km.} \text{ for the offshore waters}$$

whence

$$\frac{N_1}{N} = \frac{1}{3} \quad \text{and} \quad \frac{N_2}{N} = \frac{2}{3}.$$

It is intuitively clear that an estimate, $\hat{Y}$, of $\bar{Y}$ is given by the weighted mean of the within-stratum means, i.e. by

$$\bar{y} = \sum_{i=1}^{2} \frac{N_i}{N} \bar{y}_i = 310.67.$$

The results thus far are brought together in Table 6.3a.

TABLE 6.3   Stages in estimating population size from a stratified sample

a)   Calculating $\bar{y}$

	Relative sizes of strata	Sample size, n_i	Stratum totals Σy_i	Stratum means $\bar{y}_i = \dfrac{1}{n_i} \Sigma y_i$
Stratum 1	$\dfrac{1}{3}$	15	6754	450.267
Stratum 2	$\dfrac{2}{3}$	8	1927	240.875
		$n = 23$		

$$\bar{y} = \frac{1}{3} \times 450.267 + \frac{2}{3} \times 240.875 = 310.67$$

b)   Calculating var $(\bar{y})$

	Within-stratum sums of squares Σy_i^2	Sampling variances of stratum means var $(\bar{y}_i)$	P_i	$P_i^2/(n_i - 1)$
Stratum 1	4,302,514	6006.728	0.5442	0.02115
Stratum 2	534,595	1257.659	0.4558	0.02968
				0.05083

$$\text{var}(\bar{y}) = \frac{1}{9} \times 6006.728 + \frac{4}{9} \times 1257.659 = 1226.37$$

Next, so that confidence limits for $\hat{Y}$ may be calculated, it is necessary to obtain var $(\bar{y})$ (in other words var $(\hat{Y})$] and hence $s_{\bar{y}}$, the standard error of $\bar{y} = \hat{Y}$. The results of the various steps in the calculation are tabulated in Table 6.3b.

The first column shows the sum of squares of the observations in each stratum.

The second column shows var $(\bar{y}_i)$, the variance of the ith stratum mean, which is given by

$$\text{var}(\bar{y}_i) = \frac{1}{n_i(n_i - 1)}\left\{\Sigma y_i^2 - \frac{(\Sigma y_i)^2}{n_i}\right\}. \qquad (6.6)$$

From these var $(\bar{y})$ can be calculated thus:

$$\text{var}(\bar{y}) = \sum_{i=1}^{2}\left(\frac{N_i}{N}\right)^2 \text{var}(\bar{y}_i) = 1226.37.$$

Hence $s_{\bar{y}} = \sqrt{1226.37} = 35$.

To find the number of degrees of freedom of $s_{\bar{y}}$ it is necessary to combine the numbers appropriate to each stratum considered alone and so obtain the so-called "effective degrees of freedom". This is done as follows: for each stratum, calculate P_i (shown in the third column) from

$$P_i = \left(\frac{N_i}{N}\right)^2 \frac{\text{var}(\bar{y}_i)}{\text{var}(\bar{y})};$$

next obtain the values of $P_i^2/(n_i - 1)$ as shown in the last column. The effective degrees of freedom, ν', is the reciprocal of the sum of this column.

That is

$$\nu' = \frac{1}{\Sigma P_i^2/(n_i - 1)}.$$

In the present case $\nu' = 19.7$ or approximately 20; (note, as a check, that ν' is always less than $n - h$ where h is the number of strata, in this case 2). From the table in Appendix 6.1 the required value of $t_{0.025}$, for 20 degrees of freedom, will be found to be 2.086.

Finally, therefore, the confidence limits of $\bar{Y}$ can be found. They are

$$310.67 \pm 2.086 \times 35.$$

That is, the probability is 95% that the overall mean cell count lies between the limits 237 and 384.

In the example just described the space containing the population was divided into only two parts (i.e. there were only two strata). Generalization to any number of strata is perfectly straightforward.

Also, the number estimated was the mean number of organisms per sampling unit, $\bar{Y}$, rather than the total size of a population, Y. When Y is to be estimated by stratified sampling, it should be recalled that [see equation (6.5)]

$$\text{var}\,(\hat{Y}) = \text{var}\,(N\hat{\bar{Y}}) = N^2\,\text{var}\,(\hat{\bar{Y}}).$$

Another difference between the present example and that of Section 3 (the sunflowers) is that for the phytoplankton population the sampling fraction is so small (less than 20 liters from the surface water of lake Ontario) that the finite population correction can obviously be neglected; removal of the sample makes no perceptible difference to the population, which can consequently be treated as infinite. In fact it is customary to ignore the finite population correction [the term $(1 - n/N)$ in equation (6.4)] whenever the sampling fraction is less than 0.05. Cases often arise, of course, when the correction should be made in estimating the variance of a stratum mean. Thus if n_i/N_i, the sampling fraction within the ith stratum, exceeds 0.05, equation (6.6) should be replaced by the following

$$\text{var}\,(\bar{y}_i) = \frac{1}{n_i\,(n_i - 1)}\left\{\Sigma y_i^2 - \frac{(\Sigma y_i)^2}{n_i}\right\}\left(1 - \frac{n_i}{N_i}\right).$$

The foregoing discussion has concentrated on the computational details of stratified sampling, or how to treat the observations after they have been made. Quite as important is careful advance planning of the sampling. Any ecologist can visualize populations that obviously lend themselves to stratification. But often there is room for choice as to whether there should be many narrowly-defined strata that are very homogeneous, or fewer, more broadly-defined strata that are somewhat inhomogeneous.

Another decision that has to be made before field work can begin is on the values of the n_i, that is, the numbers of sampling units to be drawn from each stratum. The straight-forward approach is to make the sampling fraction the same in every stratum, so that

$$\frac{n_i}{N_i} = \frac{n}{N}\text{for all } i.$$

This is called *proportional allocation* of sampling units to strata and will give the most precise estimates if the variances within the several strata are all equal. It is the best way to allocate the sampling units if one has no prior information on the within-stratum variances. However, if the within-stratum variances are unequal, as was true of the lake phytoplankton, for example [compare var $(\bar{y}_1)$ and var $(\bar{y}_2)$ in Table 6.3], it is clearly desirable to sample more intensively within the more variable strata. Methods of determining optimum sampling fractions for the strata will be found in any of the books referred to earlier (page 104).

5 TWO-PHASE OR DOUBLE SAMPLING

Two-phase sampling was briefly mentioned on page 101. The need for it is common in the study of natural populations. Whenever the individuals of the population under investigation are found only in separate, scattered sites and the number of sites in the area concerned is unknown, two-phase sampling is necessary. Entomologists, in particular, have many occasions for using the method since so many insect species are found only at specialized sites; for instance, to sample populations of fruit-eating insects, pollen-eating insects, leaf miners, fungus insects, gall insects or insects parasitic on small mammals, one might begin by collecting a sample of fruits, flowers, leaves, fungi, galls or small mammals, and estimating the numbers of individuals per unit. But to infer from this estimate the number of individuals of the insect species within a specified area one would need to know, also, the total number of units in the area. It is seldom feasible to determine this number by a complete census, so it too must be estimated.

It is important to make clear the distinction between *two-phase* and *two-stage* sampling. In two-phase sampling, as already explained, the size of the population of sampling units, N, is estimated by sampling. In two-stage sampling, on the other hand, N is exactly known. The whole area is subdivided into "clusters" (or blocks) of sampling units; a sample of clusters is chosen at random, and then a sub-sample of individual sampling units is chosen at random from each of the chosen clusters.

Thus in two-phase sampling it is the estimation procedure that involves two phases. In two-stage sampling it is the choice of sampling units that involves two stages. *Multi-phase* and *multi-stage* sampling are analogously defined. Two further terms are worth noting in connexion with multi-stage sampling. It is also known as *nested sampling*. And sometimes all the units in a cluster (instead of a subsample of them) are observed, in which case the process is called *cluster sampling*.

It is often desirable to stratify a population before carrying out either multi-phase or multi-stage sampling in each stratum separately. Sampling plans can, indeed, become extraordinarily intricate. A good brief discussion of the general principles of sampling insect populations, and many references, will be found in Harcourt (1969).

We now consider a numerical example of two-phase estimation. The data to be used are artificial but for concreteness we shall suppose that the population whose size is to be estimated consists of all the adult individuals of a species of pollen-eating insect inhabiting the flowers of a particular plant species in a given area of $N = 400$ square meters.

A first-phase sample of $n_1 = 12$ meter-square quadrats is chosen at random and the number of flowers, x, counted in each of them. Denote the mean of these x's by $\bar{x}_1$; the subscript 1 is used to show that this mean is based on n_1 observations. Clearly $N\bar{x}_1$ is an estimate of X, the total number of flowers in the area.

The second-phase sample is next obtained by choosing, at random, $n_2 = 8$ of the quadrats belonging to the first-phase sample. In each of these quadrats the number of insects, y, in *all* the flowers is counted. Let $\bar{y}_2$ be the observed mean; the subscript 2 shows that this mean is based on n_2 observations.

Now consider Table 6.4 in which the observations (artificial) are tabulated. (Since this is an imaginary example n_1 and n_2 have been kept small; in practice n_2 should not be less than 30). It should be realised that each sampling unit at the second phase is a *quadrat* as at the first phase, not a flower. However, because counting all the insects in a quadrat is laborious and slow, counts of insects have been made in only n_2 of the quadrats; these n_2 quadrats are a subset of the n_1 quadrats on which flowers were counted, a quick and easy task.

The desired estimate, $\hat{Y}$ of the total number of insects in the area is easily obtained from the data as follows.

TABLE 6.4   Data (artificial) for two-phase estimation of a population of pollen-feeding insects

Quadrat	Flowers per quadrat x	Insects per quadrat y
1	12	57
2	13	49
3	8	29
4	10	60
5	8	24
6	5	15
7	11	41
8	3	5
9	3	
10	10	
11	9	
12	7	

$$n_1 = 12 \quad \sum_1^{12} x = 99$$

$$n_2 = 8 \quad \sum_1^8 x = 70; \quad \sum_1^8 x^2 = 696$$

$$\Sigma y = 280; \quad \Sigma y^2 = 12{,}598$$

$$\Sigma xy = 2886$$

Let $R = Y/X$ be the ratio of insects to flowers in the whole population. Then

$$Y = RX. \tag{6.7}$$

The observations on the n_2 quadrats of the second-phase sample allow R to be estimated. The estimate is

$$\hat{R} = \frac{\sum_1^{n_2} y}{\sum_1^{n_2} x} = \frac{280}{70} \quad \text{or} \quad 4.0 \text{ in the example.}$$

The observations on the n_1 quadrats of the first-phase sample permit calculation of $\hat{X}$, an estimate of X, from

$$\hat{X}_1 = N\bar{x}_1 = \frac{N}{n_1} \sum_1^{n_1} x = \frac{400}{12} \times 99 = 3300.$$

Obviously this is a more precise estimate of X than could have been obtained from the smaller, second-phase sample.

Finally,

$$\hat{Y} = \hat{R}\hat{X} = 4 \times 3300 \text{ or } 13,200$$

and thus a point estimate of Y is obtained.

The sampling variance of $\hat{Y}$ is calculated as follows:
First, obtain

$$s_y^2 = \frac{1}{n_2 - 1} \left\{ \sum_1^{n_2} y^2 - \frac{\left(\sum_1^{n_2} y\right)^2}{n_2} \right\}$$

$$= \frac{1}{7} \left\{ 12,598 - \frac{280^2}{8} \right\} \quad \text{or} \quad 399.71.$$

Next obtain

$$s_{y.x}^2 = \frac{1}{n_2 - 1} \sum_1^{n_2} (y - \hat{R}x)^2$$

$$= \frac{1}{n_2 - 1} \{ \Sigma y^2 - 2\hat{R} \Sigma xy + \hat{R}^2 \Sigma x^2 \}$$

$$= \frac{1}{7} \{ 12,598 - 2 \times 4 \times 2886 + 16 \times 696 \}$$

$$= 92.29.$$

Now

$$\text{var}\,(\hat{Y}) = N^2 \left\{ \frac{s_{y.x}^2}{n_2} \left(\frac{n_1 - n_2}{n_1} \right) + \frac{s_y^2}{n_1} \left(\frac{N - n_1}{N} \right) \right\}$$

$$= 400^2 \left\{ \frac{92.29}{8} \times \frac{4}{12} + \frac{399.71}{12} \times \frac{388}{400} \right\}$$

$$= 5{,}784{,}849. \tag{6.8}$$

(Notice that in equation (6.8) the term $(N - n_1)/N$ in the second pair of parentheses is the finite population correction mentioned on page 106. It is usually replaced by 1 when $n_1/N < 0.05$). The standard error of $\hat{Y}$ is

$$s_{\hat{Y}} = \sqrt{\text{var}\,(\hat{Y})} = 2405;$$

finally, a 95% confidence interval for Y is given by

$$Y \pm t_{0.025} s_{\hat{Y}} \quad \text{or} \quad 13{,}200 \pm (t_{0.025} \times 2405).$$

The number of degrees of freedom of the standard error, and hence an appropriate value for $t_{0.025}$ are hard to arrive at. This is immaterial in practice since two-phase estimation should not be attempted unless $n_2 \geqq 30$ and then it is satisfactory to put $t_{0.025} = 2$. (In the artificial illustrative example, n_1 and n_2 were kept small to simplify demonstration of the method).

Two-phase sampling is useful in many ecological investigations. Continuing with the example dealt with above, we might wish to know whether the size of the insect population changed from one year to the next. Merely to estimate the number of insects per flower would not provide the answer since the numbers of flowers might be different in the two years. Probably we should wish to obtain two results, the total number of insects and also the mean number of insects per flower, and should want to discover how each changed and how they were interrelated.

It should be observed that in estimating Y from $\hat{Y} = \hat{R}\hat{X}$ [see equation (6.7)], we tacitly assumed that y (the number of insects in a quadrat) was directly proportional to x (the number of flowers in the quadrat); or, equivalently, that the number of insects per flower is roughly constant. The procedure would therefore lead to erroneous conclusions if the number of insects per flower tended to be more (or less) numerous where the flowers themselves were abundant. Complications such as this have to be dealt with as they arise; the important thing is for the ecologist never to forget the assumptions that underlie his estimation methods and the need to consider whether they hold; modified methods must be improvised when they do not. Thus, if, in the example above, it were believed that R depended on the number of flowers per quadrat, one could stratify the population so that within each stratum the density of the flowers was fairly constant.

6 ESTIMATION BY THE REGRESSION METHOD

It sometimes happens that the number of individuals per sampling unit is difficult to determine but that some other, "indicator" quantity, related to the number of individuals present, can be quickly and easily measured. When this is so, population size may be estimated indirectly. The relation between the number of individuals and the magnitude of the indicator can be estimated by determining both of them for a sample of n sampling units. The indicator can be measured on all N sampling units (containing the whole population). And hence the number of individuals in the whole population can be inferred.

For example, suppose the number of beavers in a tract of forest is to be estimated. Determining the number of resident beavers in a unit of area (a sampling unit) may require prolonged observation, and is only practicable for comparatively few units. But counting the number of occupied beaver lodges per unit is straightforward and this number could be used as indicator of the number of beavers. Any naturalist can think of similar examples, in which evidence for the presence of a species comes more often from indirect traces than from sightings of individuals. This is especially true of nocturnal animals or those that are very retiring. There are many kinds of traces: nests, shed antlers, droppings, tracks, and owl pellets; and feeding signs such as clam shells left by mink, chewed pine cones dropped by squirrels, and girdled trees where rodents have fed. Numbers of insects, also, can sometimes be judged indirectly, for example by the amount of frass (feces), by numbers of empty cocoons, or by the degree of defoliation in the case of leaf-eating species. In general, indirect estimation of population size is worth considering whenever the appropriate indicator, whatever it may be, can be observed much more quickly and cheaply than can numbers of individuals.

We now consider a numerical example. It was desired to estimate the number of individuals of a species of beetle (*Diaperus maculatus*) within a collection of fruiting bodies ("brackets") of the birch bracket fungus *Polyporus betulinus*; (this is a bracket fungus, or shelf fungus, that grows on the trunks of standing or fallen dead birch trees). When the brackets are kept in cages in the laboratory, the beetle larvae within them mature over a period of weeks and the adults emerge and can be removed. It was known that the larger the bracket the greater the number of beetles it was likely to yield, and it therefore seemed reasonable to use the weight of a bracket as an indicator of the number of beetles it contained.

The beetle population whose size was to be estimated consisted of all the individuals in a collection of $N = 60$ fungus brackets. The total weight of these brackets was $X = 7620$ grams. A sample of $n = 25$ brackets was picked at random from the collection and each was weighed and then caged (in a separate cage) until all the beetles it contained had emerged and been counted. The observations on these 25 brackets are shown in Table 6.5 and as a scatter diagram in Figure 6.1. Each of the 25 points shows the weight of a bracket, x, and the number of beetles that emerged from it, y. We now require to find the equation of the straight line

$$y = \hat{a} + \hat{b}x$$

which best fits these sample points in the sense that the sum of the squared distances (parallel with the y-axis) of the points from the line is a minimum. The data in Table 6.5 are to be used to determine the constants $\hat{a}$ and $\hat{b}$. They have "hats" to show that they will be regarded as estimates of two unknown numbers A and B which are such that

$$Y = A + BX. \tag{6.9}$$

Here Y is, as always, the true size of the population under study, which is to be estimated. And X is the total value, over all N sampling units, of the indicator variable, in this case the weight of all 60 fungus brackets. Thus $Y = A + BX$ specifies the relation between Y and X.

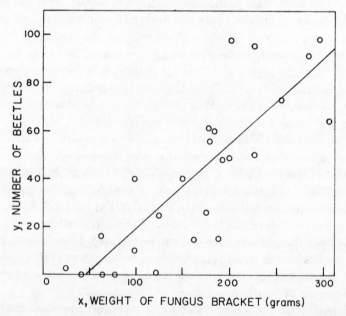

FIGURE 6.1   The scatter diagram, and fitted straight line, show the relationship between the number of beetles contained in a fungus bracket and the weight of the bracket. The use of these data in estimating the number of beetles in a large collection of brackets is explained in the text

In what follows, the method of calculation is set down without explanation. The formulas for $\hat{a}$ and $\hat{b}$ are derived in Appendix 6.2.

TABLE 6.5   Data for estimating the size of a population of beetles by the regression method

Weight of fungus (grams) x	Number of beetles y	Weight of fungus (grams) x	Number of beetles y
62	16	125	25
226	50	177	61
175	26	307	64
255	73	42	0
226	95	99	10
99	40	122	1
150	40	88	15
25	3	201	98
200	49	183	60
77	0	63	0
178	56	296	98
283	91	162	15
192	48		

$$N = 60 \qquad n = 25$$

$$X = \sum_1^{60} x = 7620$$

$$\sum_1^{25} x = 4013; \quad \bar{x} = 160.52; \quad \sum_1^{25} x^2 = 796{,}253$$

$$\sum_1^{25} y = 1034; \quad \bar{y} = 41.36; \quad \sum_1^{25} y^2 = 68{,}918$$

$$\Sigma xy = 219{,}817$$

Table 6.5 shows the sums, sums of squares, and sum of cross-products that are required in the calculations. Values of $\hat{b}$, and then $\hat{a}$, are obtained as follows:

$$\hat{b} = \frac{n \, \Sigma xy - (\Sigma x)(\Sigma y)}{n \, \Sigma x^2 - (\Sigma x)^2} \tag{6.10}$$

$$= 0.354 \quad \text{in the example;}$$

$$\hat{a} = \bar{y} - \hat{b}\bar{x}$$

$$= -15.464 \quad \text{in the example.}$$

Therefore the equation of the straight line that best represents y as a function of x is, on the evidence available from the sample,

$$y = -15.464 + 0.354x.$$

This line is shown in Figure 6.1. A point estimate, $\hat{Y}$, of Y can now be calculated from

$$\hat{Y} = \hat{a} + \hat{b}X$$

$$= -15.464 + 0.354 \times 7620 = 2682.$$

To find a confidence interval for Y we require var $(\hat{Y})$. This is obtained by first calculating

$$s_e^2 = \frac{1}{n-2}\left\{\Sigma y^2 - \frac{(\Sigma y)^2}{n} - \frac{b}{n}[n\,\Sigma xy - (\Sigma x)(\Sigma y)]\right\} \qquad (6.11)$$

$$= 308.36.$$

Then

$$\text{var}(\hat{Y}) = s_e^2\,\frac{N^2}{n}\left(1 + \frac{1}{n}\right)\left(1 - \frac{n}{N}\right) \qquad (6.12)$$

$$= 26{,}938.33.$$

The standard error of $\hat{Y}$ is therefore

$$s_{\hat{Y}} = \sqrt{\text{var}(\hat{Y})} = 164.13.$$

The number of degrees of freedom of $s_{\hat{Y}}$ is $v = n - 2$ or 23 in this example. Then, from Table 6.6 in Appendix 6.1, $t_{0.025} = 2.07$. Finally, the desired 95% confidence interval for Y, the total number of beetles in all 60 fungus brackets is

$$2682 \pm 2.07 \times 164.13 \quad \text{or} \quad 2342 \text{ to } 3022.$$

A few more points to notice are the following. If it had been assumed in advance that the relationship between Y and X was given by equation (6.7), i.e. by $Y = RX$, instead of by equation (6.9), i.e. by $Y = A + BX$, (or equivalently, that Y was directly proportional to X so that the regression line in Figure 6.1 passed through the origin), then the "ratio method" of estimation could have been used instead of the regression method. The calculations are somewhat simpler, but errors are introduced if the assumption is incorrect.

The assumption that $Y = RX$ underlies the method of two-phase estimation described in Section 5 but clearly two-phase estimation could also be based on the assumption that $Y = A + BX$. Thus four different procedures are possible and can be summarized as follows:

1   Assume $Y = RX$ and determine the true population value of X. Hence estimate Y by the ratio method.

2 Assume $Y = A + BX$ and determine the true population value of X. Hence estimate Y by the regression method.

3 Assume $Y = RX$ and estimate X from a sample. Hence estimate Y by two-phase ratio estimation.

4 Assume $Y = A + BX$ and estimate X from a sample. Hence estimate Y by two-phase regression estimation.

The examples described in detail above deal with method 3 (the pollen-eating insects example; see page 114) and method 2 (the fungus beetles example; see page 119). Full descriptions of methods 1 and 4 may be found in any of the books on sampling listed on page 104, which also describe the more elaborate methods that may be required if the relationship between X and Y is non-linear.

7 THE CAPTURE-RECAPTURE METHOD OF POPULATION ESTIMATION

The estimation methods described thus far in this chapter require that the number of population members within each of several sampling units be observed and recorded. There are many kinds of animals for which such a procedure is impossible. Given a population of animals that are active and fairly well concealed (e.g. the fish in a lake, the voles in an old field) there is obviously no way in which one can delimit a sampling unit and count the individuals it contains. Populations such as these are best estimated by the "capture-recapture" method, which, of all the many ways in which a population's size can be estimated, is the most distinctively ecological.

The general notion underlying the method is well known and is as follows. Suppose the population is of size N, so that N is the number we wish to estimate. Let M of the population members be captured, tagged in some recognizable way and released to mingle with the rest of the population. Some time later, let another batch of animals, say n in number, be caught and suppose it is found that m of these n are tagged. Assume that the proportion of tagged animals in the second catch is the same as the proportion present before the catch was taken, or equivalently, that

$$\frac{m}{n} = \frac{M}{N};$$

then, since m, n and M are known, N can apparently be found without difficulty. Unfortunately, the "naive" estimate given by $N = nM/m$ has mathematical imperfections and is also, of course, subject to variation due to chance. We return to the mathematical details below. First, some other comments on the method should be made.

As to nomenclature: the method is variously called the "capture-recapture method", the "mark-recapture method", the "tag-recapture method", Petersen's method (by fishery biologists) and the Lincoln Index method (by wildlife biologists).

A tremendous assortment of techniques exist for tagging or marking captured animals so that they may be easily recognized on recapture. Birds can be banded or dyed; mammals, also, can be dyed. Small rodents such as squirrels, chipmunks, ground squirrels, mice and voles (and also lizards) can be "tagged" by the amputation of one or more toe joints. Flyger (1959) described experiments in which squirrels were dyed purple and had the hairs of their tails clipped. Because capture-recapture is so indispensible to fishery biologists, numerous ways of tagging fish and marine invertebrates have been devised. Many of these are used to enable migrations to be followed rather than to allow numbers to be estimated, but they are worth mentioning her nonetheless. Thus, in a symposium on fish marking published by the International Commission on Northwest Atlantic Fisheries (1963), procedures for the following animals (among others) were given; cod, char, haddock, salmon, sole, plaice, hake, whiting, redfish, tuna, herring, menhaden, shark, crabs, shrimps, lobsters, whelks and scallops.

The commonest ways of marking fishes are to wire colored plastic tags to their fins, or to clip the fins. Or fishes can be sprayed with, or immersed in, coloring fluids such as biological stains or solutions of fluorescent pigments. Fish marking has a long history: in *The Compleat Angler*, published in 1653, Izaak Walton described how Sir Francis Bacon marked "young salmons" by tying a riband in their tails (Cormack, 1968).

Shellfish such as whelks are easily marked with a dab of paint on the shell. Insects, also, can be marked with paint. Such insects as beetles, grasshoppers and tsetse flies, for example, have enough space on the dorsal surface of the thorax to accomodate several distinct spots of oil paint, making it possible to devise elaborate color codes for labelling individual insects (Swynnerton 1936; Southwood, 1961). Butterflies have been marked by rubbing the scales off a small patch on the hindwing and painting a number on the wing membrane. The list of marking methods could be continued almost indefinitely.

Estimation of a population's size by capture-recapture is entirely straightforward *provided* all the following seven conditions are met.

1 The population must be "closed". That is, there must be no immigration or emigration, and likewise no births or deaths, between release of the first,

tagged catch of M individuals and the capture for inspection of the second catch, of n individuals. Clearly emigration and deaths would remove some tagged individuals and immigration and births would bring in animals that had had no chance to be tagged. Consequently the tagged part of the population would be steadily diluted.

2   The probability of being caught must be the same for all animals in the population.

3   It is obviously essential that tags not become lost or unrecognizable.

4   An animal caught and tagged in the first catch must be neither more nor less likely than other population members to be caught on the second occasion. Thus the animals must not become trap-happy (usually in response to tempting bait) or trap-shy (in response to a painful trapping method).

5   Tagging must not alter an individual's probability of death. For example, attaching a tag must not leave a wound that is likely to become infected. An array of paint dots on an insect's thorax may change its chance of survival though it is difficult to foresee in what way. The dots make the insect more conspicuous to its predators but this will not necessarily make it more vulnerable as the pattern might simulate warning coloration.

6   When tagged animals are released, they must mingle freely with the rest of the population.

7   Sufficient time must elapse between the first and second catches for tagged animals to disperse through the whole area.

There is often reason to suspect that one or more of these conditions is not met. In that case tests must be carried out to judge whether the assumptions do, indeed, hold; and in some cases it is possible to make appropriate allowances if they do not. There is an enormous literature on this and related topics and Cormack (1968) has given a comprehensive review with a large bibliography.

We now consider a fairly straightfoward application of the method, using data from Buck and Thoits (1965). The population whose size was to be estimated consisted of all the Yellow Perch (*Perca flavescens*) having lengths in the range four to seven inches, in a one-acre pond. The greatest depth of the pond was eight feet. It seemed reasonable to suppose that all the conditions listed above were fulfilled. The first capture was made with a single haul of a seine and netted 284 fish in the specified size range. Each fish was marked by clipping its soft dorsal fin. All the marked fish were then returned to the pond, on the same day, by releasing them in batches at different points around the shore; this was done to hasten their diffusion through the pond, a process that was allowed to continue for 48 hours before the second catch was seined. This time 1392 fish were caught and 86 of them were found to be marked.

Thus, using the symbols defined earlier, we have

$$M = 284, \quad n = 1392 \quad \text{and} \quad m = 86.$$

As already explained, a naive estimate of N, the unknown number of perch in the pond, would be given by $Mn/m = 4597$. However, Mn/m is a *biased* estimator*. That is to say, if this estimator were used with replicate data obtained in an identical manner, many times, from the same population, it would be found that the mean of the estimates does *not* tend to get steadily closer to the true value of N as the number of replicates (repetitions) is increased. Instead, the mean would converge to some other number, say $N - b$, where b denotes the size of the bias. This somewhat surprising result is because m is not independent of n, and the expected value of their ratio is not equal to the ratio of their expected values. In symbols,

$$\mathscr{E}\left(\frac{n}{m}\right) \neq \frac{\mathscr{E}(n)}{\mathscr{E}(m)}.$$

It seems impossible to find a truly unbiased estimator (i.e. one for which $b = 0$) but it has been shown (see Seber, 1970) that the formula

$$\hat{N} = \frac{(M + 1)(n + 1)}{(m + 1)} - 1$$

has negligibly small bias provided $m \geq 7$. For the perch in the pond, therefore,

$$\hat{N} = \frac{285 \times 1393}{87} - 1 = 4562 \quad \text{is a point estimate of } N.$$

The variance of the estimate is given by

$$\text{var}(\hat{N}) = \frac{(M + 1)(n + 1)(M - m)(n - m)}{(m + 1)^2(m + 2)}$$

$$= 154{,}128 \quad \text{in the present example.}$$

---

* The words *estimator* and *estimate* should be kept distinct. An *estimator* is a formula, and an *estimate* is some particular numerical result obtained by substituting observed values for the symbols in the formula. Thus Mn/m is an estimator, and 4597 an estimate, of the number of perch in the pond.

We may assume $\hat{N}$ to be normally distributed and therefore the 95% confidence interval for N is $\hat{N} \pm 2\sqrt{\text{var}(\hat{N})}$. Thus the 95% confidence interval for the number of perch in the pond is 3776 to 5348. The true value, which was subsequently found by draining the pond, was 4123.

The classic capture-recapture method described above has been elaborated in many ways. One can, for instance, tag all the animals caught in the second catch in some distinctive manner and release them again. The process can be repeated many times; with this "multiple recapture" technique, an animal captured in the last catch could bear any or all of the several dinstictive markings used at successive captures.

Another development is so-called "optional stopping" at the recapture stage. If, when animals are recaptured, they are caught one at a time, one can keep on enlarging the recaptured sample for as long as desired. Thus, one might continue until n, or m, or the ratio m/n, had increased to some specified value. It would therefore be necessary to decide on a stopping rule beforehand; also, with optional stopping, the way in which the confidence interval for N should be calculated depends on the stopping rule chosen. The mathematics of the problem have been discussed by Samuel (1969).

8 THE REMOVAL METHOD OF POPULATION ESTIMATION

The last of the methods of population estimation we shall consider in this chapter is usually applicable only when we are willing to deplete the population in the process of estimating its original size. This is true of populations of vermin (rats in houses, for instance) when we want to reduce the population as much as possible and estimate its initial size in so doing; it is assumed that total destruction of the population cannot be achieved.

A second example: we might wish to remove as many as possible of the fish in a small body of water and at the same time estimate the number present before removal began; it is assumed that the population cannot be entirely removed, for if it could be a complete census would be obtained in the process.

A third example: suppose we want to estimate the number of mites in a soil sample of given size and can extract from the sample a proportion (not all) of the mites it contains. To do this one might use a Berlese funnel in which animals in the soil sample are driven out by mild heat and can then be collected and counted. In this example, as in the other two, the sampling is destructive in the sense that the individuals counted are removed from the population and not put back. Obviously, when the whole population consists of the organisms in a single soil sample, its destruction is immaterial.

The data required for the method are obtained as follows. A "catch" is taken from the population by some prescribed, exactly repeatable method; let the number trapped (the size of the catch) be C_1. A second catch is taken by a repetition of the method; denote by C_2 the size of the second catch. We shall see that if the trapping procedure depletes the population by a constant proportion on both occasions, its initial size, N, is expressible as a function of C_1 and C_2.

Before deriving an estimator for N, it is necessary to list the assumptions that underlie the estimation method. There are three.

1   It is assumed that the probability that an animal will be trapped is the same for all animals and remains unchanged from one catch to the next. It is also necessary that the probability of being trapped should be the same for the last animal taken on a particular occasion as for the first. Thus difficulties arise when traps are used that can capture only one animal; once sprung, they cannot be sprung again. Traps of this kind must be numerous enough, or inefficient enough, for there to be a plentiful oversupply of them all the time one catch is being taken, otherwise the probability of capture will steadily decrease as more traps become occupied and fewer remain empty. Often this problem does not arise; for example, when fish are caught by first stunning them with electric shocks, or when soil animals are driven from soil samples by heat.

2   It is assumed that the proportion trapped each time is large enough to bring about an appreciable reduction in population size.

3   It is assumed that the population is stationary in the sense that gains (from births and immigration) exactly balance losses (from deaths and emigration). Removal of the catches is the *only* cause of population change. This assumption is less restrictive than the corresponding one (see 1 on page 122) for capture-recapture estimation. Since the removal method does not entail tagging and subsequently recognizing particular individuals, we require only that gains and losses should balance during the observation period.

Now denote by p the probability that an animal is caught. As already explained, p is constant for all animals and at all times. Then the expected size, $\mathscr{E}(C_1)$, of the first catch is

$$\mathscr{E}(C_1) = Np;$$

consequently the expected size of the population remaining is $N - Np$ $= N(1 - p)$ and this remaining population is now subjected to the second round of trapping. Thus $\mathscr{E}(C_2)$, the expected size of the second catch, is

$$\mathscr{E}(C_2) = N(1 - p)p.$$

To estimate N, let us assume that the sizes of the catches actually obtained, C_1 and C_2, have their theoretically expected values, that is that

$$C_1 = \mathscr{E}(C_1) \quad \text{and} \quad C_2 = \mathscr{E}(C_2).$$

Then

$$C_1 = Np$$

and

$$C_2 = Np(1 - p).$$

Solving these equations gives

$$\hat{p} = \frac{C_1 - C_2}{C_1}$$

and

$$\hat{N} = \frac{C_1^2}{C_1 - C_2} \tag{6.13}$$

the required estimator of N.

It has been shown (see Seber and Le Cren, 1967) that

$$\text{var}(\hat{N}) = \frac{(C_1 C_2)^2 (C_1 + C_2)}{(C_1 - C_2)^4}. \tag{6.14}$$

An example is given below after an extension of the method has been discussed.

Obviously, if data from two catches enable N to be estimated, data from three or more should lead to estimates of greater precision. Zippin (1956, 1958) has presented graphs for deriving estimates of N when several (up to seven) catches have been taken. Here we develop the theory applicable to data on three catches, of sizes C_1, C_2, and C_3.

Define N and p as before, and put $q = 1 - p$.

For simplicity we write C_j in place of its expected value $\mathscr{E}(C_j)$, for $j = 1, 2, 3$.

As before, $Np = C_1$, $Npq = C_2$ and, analogously, $Npq^2 = C_3$. Thus

$$C_1 + C_2 + C_3 = Np(1 + q + q^2) = N(1 - q)(1 + q + q^2)$$

$$= N(1 - q^3),$$

whence

$$N = \frac{C_1 + C_2 + C_3}{1 - q^3}. \tag{6.15}$$

Now put

$$T = C_1 + C_2 + C_3,$$

and also put

$$Q = \frac{C_2 + 2C_3}{T}.$$

Observe that Q is a function of observed quantities; its numerical value can be calculated from the data.

Then

$$Q = \frac{Npq + 2Npq^2}{N(1 - q^3)}$$

$$= \frac{Np(q + 2q^2)}{N(1 - q)(1 + q + q^2)}$$

$$= \frac{q(1 + 2q)}{1 + q + q^2}.$$

Now a quadratic equation in q is obtainable, namely

$$(2 - Q)q^2 + (1 - Q)q - Q = 0. \tag{6.16}$$

An estimate of N is arrived at as follows. Substitute the numerical value of Q in equation (6.16); solve this equation to yield $\hat{q}$, an estimate of q; substitute this in equation (6.15) and hence obtain $\hat{N}$.

Zippin (1958) gives the sampling variance of $\hat{N}$. It is

$$\text{var}(\hat{N}) = \frac{\hat{N}T(\hat{N} - T)}{T^2 - 9\hat{N}(\hat{N} - T)(1 - \hat{q})^2/\hat{q}}. \tag{6.17}$$

We shall not here derive estimators of N based on data from more than three catches, but it should be noticed that the expected sizes of a succession of r, say, catches are

$$Np, Npq, Npq^2, \ldots, Npq^{r-1}.$$

They form a geometric series.

Also, since

$$\mathscr{E}(C_r) = Npq^{r-1} = \frac{Np}{q} q^r$$

we have

$$\log \mathscr{E}(C_r) = \log \frac{Np}{q} + r \log q = \log \frac{Np}{q} - r \log \frac{1}{q}$$

and it is seen that $\log \mathscr{E}(C_r)$ is proportional to $-r$. Thus when several catches are taken from a population, one can plot the observed values of $\log C_r$ against r for $r = 1, 2, 3, \ldots$ If the plotted points fall on a straight line with negative slope as we expect them to, this suggests that the assumptions underlying the estimation process are justified.

Now consider an actual example of estimation by the removal method. The data, which are from Leslie and Davis (1939), give the number of rats (*Rattus rattus*) caught on successive occasions over a six-week period in 1937 in a 22.5 acre block of houses in Freetown, Sierra Leone.

To illustrate the estimation method we here use the numbers of rats caught in the first, second and third fortnights of the observation period. Other evidence (not given here) suggests that the population was stationary (see page 28) so that the results are unaffected by the length of the observation period.

The data are:
$$C_1 = 195, \quad C_2 = 119, \quad C_3 = 80.$$

Let us first estimate N from C_1 and C_2 only, using equations (6.13) and (6.14). It is found that
$$\hat{N} = 500 \quad \text{and} \quad \sqrt{\text{var}\,(\hat{N})} = 71;$$

therefore the 95% confidence interval for N is
$$500 \pm 2 \times 71 \quad \text{or} \quad 358 \text{ to } 642.$$

Now use all three catch sizes and equations (6.15), (6.16) and (6.17).

We have $Q = 0.7081$ and if this value is substituted in (6.16) it is found that
$$\hat{q} = 0.6359.$$

(Notice that though (6.16) has two roots, the only admissible one is that for which $0 < \hat{q} < 1$).

Then, from (6.15),
$$\hat{N} = 530.$$
Also, from (6.17),
$$\sqrt{\text{var}\,(\hat{N})} = 38$$

so that the 95% confidence interval for N is
$$530 \pm 2 \times 38 \quad \text{or} \quad 454 \text{ to } 606.$$

It is seen that the confidence interval is narrower than that calculated from C_1 and C_2 only. Taking account of three catches instead of only two permits greater precision in the estimation of N.

In this chapter we have briefly considered the chief methods for estimating the numbers of individuals in natural populations. This is only one of many estimation problems that confront ecologists. The estimation of "cover" (by students of vegetation) is described in Chapter 8, and the estimation of "diversity" (a property of many-species communities) is described in Chapter 12.

CHAPTER 7

Spatial Patterns I
The Pattern of a Collection of Individuals

THROUGHOUT THE FIRST five chapters of this book the underlying topic was the temporal variation of natural plant and animal populations, both in size and in age composition. It was repeatedly stressed that a population's size rarely, if ever, remains constant for long. Thus an understanding of the way in which populations vary, of the rate, magnitude and regularity of changes, and of their causes and consequences, are basic objectives of a great many ecological studies.

We come now to a discussion of the spatial, as opposed to the temporal, variations of a population. The fact of spatial variation is so obvious to all naturalists and field ecologists that the interesting problems it poses are easily overlooked. Even in quite small areas, the density of a species (the number of individuals per unit area or volume) is apt to vary widely. Rarely does one find the individuals of a population spread uniformly and evenly throughout the available space. Whether the population being investigated consists of the trees in a forest, the snails in a marsh or the barnacles on a rocky shore, it can usually be taken for granted that the density will be highly variable over the region delimited for study.

Indeed, it is this high degree of spatial variability that often makes estimates of population size so imprecise. The examples in Chapter 6 are typical; thus, to mention two of them, it will be recalled that the numbers of green algae per sample of lake-water (Table 6.2 on page 109), and of beetles per fungus bracket (Table 6.5 on page 119) were both highly variable. This variability may make it difficult to obtain anything more than a very crude estimate of population size unless extremely large samples can be taken. The desired precision is often unattainable because a shortage of time, or of funds, makes it impossible to take a sample of the necessary size.

Spatial variation is, however, as much a property of a natural population as temporal variation, and is as worthy of study. The reasons why a population's members should be crowded in some places and sparse in others may

9*
131

not be apparent, especially in a seemingly homogeneous habitat. The first point to consider is: how much variation should be accepted as, so to speak, natural and inevitable and thus in no need of explanation? (Compare page 73 for the analogous question on temporal change). If we regard both "excessive" uniformity and "excessive" variability as puzzling departures from the ordinary that demand explanation, then what are the limits of ordinary variability? Before we consider this question, two preliminaries must be dealt with.

First, it should be remarked that we are mainly concerned with the population patterns of sessile, sedentary, or very small organisms; for example with plants, and with invertebrate animals such as mollusks, immature or wingless insects, and soil and litter animals (cryptozoa). Concerning such populations, it is reasonable to inquire how the individuals established themselves and survive at the locations where we find them.

Secondly, it is necessary to point out a fundamental distinction between two wholly different kinds of organisms: those that occur as individuals and those that occur as colonies. The contrast is between separate "bodies" that can be counted, on the one hand, and extensive interconnected structures such as vegetatively reproducing plants and colonial coelenterates, sponges and bryozoa, on the other. It should be noticed that the distinction is *not* between genetically distinct organisms and clones, but between things that can be counted and things that cannot. We shall use the word "individuals" to connote distinct, indivisible organisms that can be counted, regardless of whether they differ from one another genetically; thus clones resulting from agamospermy (asexual seed production) in plants or parthenogenesis in animals are made up of separate individuals even though all have the same genotype. Conversely, the word "colony" can be used to connote an extended organism (composed of one or several genotypes) whose parts are connected by living or skeletal links; for example, a group of trees interconnected by natural root grafts forms a single "colony" with several genotypes. Such supra-organisms are usually not amenable to counting since they often have indefinite boundaries. For students of ecological patterns, by far the most frequently studied of these colonial organisms are plant clones formed by vegetative reproduction. These have, of course, a common genotype throughout, but it is the indefinite boundary and not the genetic constitution that concerns us here. In this chapter we shall consider populations of countable individuals; a discussion of pattern in colonial species is deferred to Chapter 8.

1 RANDOM PATTERNS

One way of studying the spatial variability, or pattern, of a population of separate individuals is to count the number of them in each of many sampling units; thus each sampling unit yields a variable, or *variate*, namely the number of individuals it contains. The data to be examined and interpreted can then be tabulated in the form of a *frequency distribution* (or simply a *distribution*) using the word in its statistical sense to mean a list of the frequencies of occurrence of different values of the variate. Table 6.1 on page 105 (which lists the frequencies, in a sample, of quadrats containing different numbers of sunflower plants) is an example.

The use of statistical frequency distributions in investigations of spatial patterns has unfortunately led to some appalling terminological muddles. These have arisen because the word "distribution" has been indiscriminately used in its technical statistical sense, to mean a list of frequencies, and also in its colloquial sense to mean "arrangement" or "pattern". We shall here use the word in its statistical sense only; it should be understood that whenever one lists the frequencies of the different variate values obtained from a collection of sampling units, the result is a "distribution" by definition, even when the spatial pattern of the observed objects is irrelevant. In the present discussion, of course, spatial pattern is relevant: we shall be discussing statistical distributions that summarize observations on spatial patterns.

When patterns are studied by recording the distributions of the numbers of individuals in sampling units, the units may be natural or arbitrary. Natural units exist for animals that live in separate, discrete pieces of habitat; some examples have already been given (see page 113) and many more could be listed, for instance tubers, roots, carrion, dung, decaying logs, and birds' nests. For nearly all plants, and also for animals that occupy continuous habitats such as soil or water, arbitrary sampling units have to be defined. Choice of the size of sampling unit to use entails an arbitrary decision on the part of the ecologist, and difficulties arise if the results obtained with units of different sizes lead to different conclusions. Confusion from this source does not occur, however, when the simplest of all patterns, a *random pattern*, is investigated. This is the pattern we shall describe first.

Imagine that every sampling unit can be thought of as consisting of a very large number, say z, of "elementary habitable sites", each capable of containing only one individual. The probability that any one elementary site is occupied by an individual is the same for all sites in all sampling units and is assumed to be extremely small; it will be denoted by m/z. We are, in fact, supposing that the probability of occupancy of a site is inversely proportional to the number of sites in a sampling unit, or equivalently, directly proportional to the size of the site; m is the constant of proportionality. It is assumed that an elementary site is too small to contain more than one individual; it must be either occupied or unoccupied. Consequently, the probability that it is not occupied is $\left(1 - \dfrac{m}{z}\right)$.

Now consider a single sampling unit containing z elementary sites. Denote by p_0 the probability that it contains 0 individuals, i.e. that all z of its elementary sites are unoccupied.

Then

$$p_0 = \left(1 - \frac{m}{z}\right)^z.$$

If we now let z be very large indeed, p_0 can be approximated by the limiting value of $\left(1 - \dfrac{m}{z}\right)^z$ as z tends to infinity. That is, we can put

$$p_0 = \lim_{z \to \infty} \left(1 - \frac{m}{z}\right)^z$$
$$= e^{-m}.$$

(See Appendix 1.1. Here $e = 2.718\ldots$ is the base of the natural logarithms).

Next, we find p_1, the probability that 1 elementary site is occupied and $(n - 1)$ are empty; (in other words, that a sampling unit contains exactly one individual). Clearly p_1 is given by the binomial probability*

$$p_1 = z \frac{m}{z} \left(1 - \frac{m}{z}\right)^{z-1} = m \left(1 - \frac{m}{z}\right)^{z-1}.$$

Allowing z to become indefinitely large as before,

$$p_1 = \lim_{z \to \infty} m \left(1 - \frac{m}{z}\right)^{z-1}.$$

But clearly

$$\lim_{z \to \infty} \left(1 - \frac{m}{z}\right)^{z-1} = \lim_{z \to \infty} \left(1 - \frac{m}{z}\right)^z$$

---

* It should be recalled that the probability of j successes in z independent trials, when the probability of success at a single trial is p and of failure is $1 - p = q$, is given by

$$\binom{z}{j} p^j q^{z-j}, \quad \text{or} \quad {}^z C_j p^j q^{z-j}, \quad \text{or} \quad \frac{z!}{j!\,(z-j)!} p^j q^{z-j}.$$

In the case discussed in the text, $p = m/z$.

and therefore
$$p_1 = me^{-m}.$$

Analogous arguments yield the probability that a sampling unit contains 2, 3, … individuals.
Thus
$$p_2 = \lim_{z \to \infty} \frac{z(z-1)}{2!} \left(\frac{m}{z}\right)^2 \left(1 - \frac{m}{z}\right)^{z-2}.$$

As before,
$$\lim_{z \to \infty} \left(1 - \frac{m}{z}\right)^{z-2} = \lim_{z \to \infty} \left(1 - \frac{m}{z}\right)^z = e^{-m};$$

also
$$\lim_{z \to \infty} z(z-1) = z^2.$$

Then we have
$$p_2 = \frac{z^2}{2!} \frac{m^2}{z^2} e^{-m} \quad \text{or} \quad \frac{m^2}{2!} e^{-m}.$$

Similarly
$$p_3 = \lim_{z \to \infty} \frac{z(z-1)(z-2)}{3!} \left(\frac{m}{z}\right)^3 \left(1 - \frac{m}{z}\right)^{z-3}$$
$$= \frac{z^3}{3!} \frac{m^3}{z^3} e^{-m} \quad \text{or} \quad \frac{m^3}{3!} e^{-m}.$$

In general, the probability that a sampling unit contains r individuals is

$$p_r = \frac{m^r}{r!} e^{-m}.$$

It might be objected that we ought not to assume that

$$z(z-1)(z-2) \dots (z-r+1) \to z^r$$
and
$$\left(1 - \frac{m}{z}\right)^{z-r} \to \left(1 - \frac{m}{z}\right)^z$$

when r becomes large. The difficulty is only apparent, however. Successive values of p_r quickly decrease to vanishingly small magnitudes as soon as r becomes large; in concrete terms, the probability is virtually zero that more than a negligibly small proportion of the elementary sites in any one sampling unit will be occupied simultaneously.

Summarizing, we see that the probabilities of finding 0, 1, 2, 3, … individuals in a sampling unit are

$$e^{-m}, \quad me^{-m}, \quad \frac{m^2}{2!} e^{-m}, \quad \frac{m^3}{3!} e^{-m}, \quad \dots$$

Equivalently, the expected frequencies of sampling units containing 0, 1, 2, 3, ... individuals, in a sample of size n, are

$$ne^{-m}, \quad nme^{-m}, \quad n\frac{m^2}{2!}e^{-m}, \quad n\frac{m^3}{3!}e^{-m}, \quad ...$$

This theoretical frequency distribution is known as the Poisson distribution.

Now let us put x for the variate value, i.e., the number of individuals in a sampling unit. Its expected value, i.e. the mean of the theoretical, or expected, distribution can be shown to be

$$\mathscr{E}(x) = m.$$

Likewise, the variance* of the theoretical distribution is

$$\text{Var}(x) = m.$$

A proof that $\mathscr{E}(x) = \text{Var}(x) = m$ for the Poisson distribution is given in Appendix 7.1.

Let us now consider the relevance of the foregoing theoretical arguments to actual ecological problems. We must first explore the implications of the assumptions on which the arguments were based; i.e. we must reformulate them in terms applicable to real-life situations. Recall that the pattern of a population was said to be random if all the sampling units contained an equal number, z, of "elementary habitable sites" each with equal probability of containing exactly one individual. Translating from the abstract to the concrete, the crucial assumption is that the expected number of occupants is the same for all sampling units.

From this assumption, three others follow:

First, it is assumed that all the individuals are independent of one another. They do not attract or repel one another; nor do they compete for space since their density is everywhere too low for crowding; and if they originated in groups (as several offspring of one parent) they have since dispersed sufficiently, or been so depleted by deaths, that no evidence of their group origin remains.

Second: it is also assumed that every sampling unit is equally hospitable to the species. There are no "good locations" or "bad locations" since all are equal.

---

* It is customary to write Var (x) (with capital V) for the variance of a theoretical, or expected, distribution; and var (x) (with lower case v) for the variance of an observed or empirical distribution.

Third: the expected number of individuals per sampling unit is the same for all individuals and all sampling units regardless of any size differences among individuals or among sampling units. This implies that the expected number is unaffected by any variation in size of natural sampling units (flowers, fruits, cones and so on). However, if the sampling units have arbitrary boundaries (as with water samples, soil samples, or the quadrats used to sample vegetation), the assumption is that the units are all identical in quality and thus that the expected number of occupants is exactly proportional to the size of the unit.

Now suppose we wish to test the hypothesis that a particular pattern is random. This may be done by counting the number of individuals in each of many sampling units and comparing the observed distribution with the expected Poisson distribution. An example is given below. First it is necessary to emphasize a point that can scarcely be overemphasized and is too often ignored, especially by those who appear to believe that statistical argument obviates the need for common sense.

It is this: there is never any justification for testing an hypothesis unless there were good reasons for entertaining it in the first place. It is not at all unusual for a theoretical frequency distribution to resemble an empirical distribution very closely, even when it is known beyond doubt that the assumptions on which the theoretical distribution is based cannot possibly be true of the actual situation. The ecologist must avoid being lulled into accepting an intrinsically unreasonable hypothesis merely because there is a correspondence, in some aspects at least, between the consequences of the hypothesis and his field observations. To repeat: hypotheses are only worth testing if they seem intuitively reasonable in the light of existing biological knowledge. And, whenever possible, several (not just one) of the consequences of a hypothesis should be compared with observation before the hypothesis is accepted with any confidence.

As an example, we consider the pattern of the trees in an even-aged second growth forest of lodgepole pine (*Pinus contorta*). The forest concerned is in the sub-alpine region of interior British Columbia and is typical of many such forests that have grown up following fire. A fire destroys all existing vegetation and the heat causes the opening of scattered lodgepole pine cones that, in the absence of fire, remain closed with viable seeds inside for many years. The liberated seeds germinate in millions in the nutrient-rich ashes and there are no other survivors of the fire to compete with them. The resultant even-aged forest may be extraordinarily dense; in some 40-year old forests there are more than 10,000 trees per acre. The forest here considered was between 40 and 50 years old, and was far less dense, having about 1600 trees per acre. It was therefore reasonable to suppose that the trees were not so crowded that they interfered with one another. Also, even if the seeds had originally germinated in groups around fallen cones, so few of the original seedlings had survived until the time of the investigation that no trace of such grouping would remain. The habitat throughout the area delimited for study seemed uniform. Having regard to all these facts, therefore, it was reasonable to hypothesize that the pattern of the trees was random, and to test the hypothesis.

Observations were made by counting the number of trees in a sample of 100 quadrats, each of area 7.3 square meters, that were placed at random (but without overlaps) in the study area. The observed results are shown in Table 7.1 and we now wish to judge how closely they resemble the frequencies to be expected on the hypothesis that the tree pattern is random.

TABLE 7.1   The frequency distribution of lodgepole pines in quadrats placed at random

Number of trees per quadrat x	Frequency Observed f_x	Expected np_x
0	7	5.7
1	16	16.4
2	20	23.4
3	24	22.3
4	17	16.0
5	9	9.1
6	5 ⎫	
7	1 ⎪	7.1
8	1 ⎬	
9 and over	0 ⎭	
	100	100.0

$$\Sigma x = 286 \quad \bar{x} = \frac{1}{n}\Sigma x = 2.86 \quad \Sigma x^2 = 1102$$

$$X^2 = \Sigma \frac{(f_x - np_x)^2}{np_x^2} = 0.995 \quad \text{(see Appendix 7.2)}$$

With 5 degrees of freedom, $P(\chi^2) > 0.9$

As shown in the table, the mean number of trees per quadrat is $\bar{x} = 2.86$. To test the hypothesis, we therefore compare the observed frequency distribution with the expected Poisson distribution of mean $m = 2.86$. The expected frequencies may be obtained by looking up Poisson probabilities in published tables (e.g. Table X in Rohlf and Sokal, 1969) and multiplying these probabilities by n, the sample size. Alternatively, they may be calculated from the formulae on page 136. This is conveniently done by first obtaining p_0 and then deriving successive probabilities using the fact that

$$p_r = \frac{m^r}{r!} e^{-m} = \frac{m}{r} \frac{m^{r-1}}{(r-1)!} e^{-m} = \frac{m}{r} p_{r-1}.$$

Thus the expected frequencies in Table 7.1 are obtained as follows:

$$np_0 = ne^{-m} = 100e^{-2.86} = 5.73$$

$$np_1 = m\,(np_0) = 2.86 \times 5.73 = 16.39$$

$$np_2 = \frac{m}{2}\,(np_1) = 23.44$$

and so on.

When $r > m$, successive values of p_r become steadily smaller, but never (in theory) decrease to zero. Therefore it is necessary to pool the frequencies in the "tail" of the distribution. Thus in the example the expected frequencies of quadrats with six *or more* trees is obtained by subtraction as

$$n - n\sum_{r=0}^{5} p_r = 100 - 92.9 = 7.1.$$

The close correspondence between observed and expected frequencies in Table 7.1 is obvious. Usually the discrepancies between observation and theory are much larger. In any case, it is never possible to judge with certainy whether the hypothesis is true and the discrepancies are simply due to chance, or whether the hypothesis is wrong. Complete certainty on this point is intrinsically unobtainable, and performance of a statistical "goodness of fit" test is the only way to resolve the problem. The test consists in finding the probability that, *if* the hypothesis were true, the observed results would depart from expectation by as much as they are found to do. If this probability is small, that is, if the observations are highly improbable given the hypothesis, then the best course is to reject the hypothesis as untenable. Conversely, if the probability is high, that is if, given the hypothesis, the observations are unsurprising, it is reasonable to accept the hypothesis provided (as was emphasized earlier) it was a reasonable hypothesis in the first place. It is up to the investigator to choose the dividing point between "high" and "low" probabilities. The chosen value is called the "significance level" of the test. A common convention is to reject the hypothesis if the discrepancies between observation and expectation have a probability of less than 0.05 (or 5%). This is only a convention and there is nothing sacrosanct about it.

The appropriate test for comparing observed and expected frequency distributions is the χ^2-test for goodness of fit. The values in the present example of the test statistic, X^2, and the probability associated with it, $P(\chi^2)$, are given in Table 7.1. The method of calculation is fully described in Appendix 7.2. The foregoing discussion should have made it clear, however, that carrying out prescribed calculations in obedience to a set of rules does not absolve the ecologist from devising an ecologically realistic hypothesis and, after calculating the relevant probability, deciding what conclusion is justified. It is seen that the final step is to choose a probability level (a significance level) to divide the "acceptable" from the "unacceptable"; this choice is inescapably subjective.

Table 7.1 permits another comparison between observation and expectation. As remarked earlier, for a theoretical Poisson distribution, the variance is equal to the mean. In the lodgepole pine example, the sample values of the mean and variance* of the number of trees per quadrat are

$$\bar{x} = 2.86$$

and

$$\text{var}(x) = \frac{1}{n} \sum (x - \bar{x})^2 = \frac{1}{n} \left\{ \sum x^2 - \frac{(\sum x)^2}{n} \right\} = 2.84.$$

Thus the observed $\bar{x}$ and var (x) are nearly equal, as we should expect if the pattern is random.

A test based on this comparison may be done as follows. Calculate the test statistic I defined as

$$I = \frac{\sum (x - \bar{x})^2}{\bar{x}} = \frac{n \sum x^2 - (\sum x)^2}{\sum x}$$

$$= \frac{100 \times 1102 - 286^2}{286}$$

or 99.3 in the present example. I is known as the *index of dispersion*. If the pattern is random (and assuming $n \geq 30$)

$$\Pr \left\{ -1.96 \leq \sqrt{2I} - \sqrt{2n - 1} \leq +1.96 \right\} = 0.95.$$

---

* Observe that $\dfrac{1}{n} \sum (x - \bar{x})^2$ is the variance of a sample of size n; that is it gives, exactly, the variance of the sample itself. Whereas $\dfrac{1}{n-1} \sum (x - \bar{x})^2$ is the formula for estimating, without bias, the unknown variance of a large population from which a sample of size n has been taken. This unknown population variance has the true value $\dfrac{1}{N} \sum (x - \bar{x})^2$ where N is the size of the whole population. It could be calculated exactly if all N variate values were known.

Therefore, if $\sqrt{2I} - \sqrt{2n-1}$ lies outside the range -1.96 to $+1.96$ we can say that, given the hypothesis of randomness, an improbable set of observations has been obtained; the probability is 5% or lower. Then, if we have chosen to use 5% as the significance level for the test, we would reject the hypothesis of randomness. For the lodgepole pine population

$$\sqrt{2I} - \sqrt{2n-1} = -0.013$$

and the result accords with what we should expect given a random pattern.

The two tests just described (the χ^2 goodness of fit test and the index of dispersion test) both use the same data and are therefore not independent. Although these data constitute evidence to support the belief that the pattern of the trees is random, they are only a single set of observations and confirmation from *independent* observations is desirable. One way of confirming the conclusion would be to sample the area a second time, with quadrats of a different size, say C times as large as those used the first time; (it is immaterial whether $C < 1$ or $C > 1$). Then if the observed frequency distribution of the number of trees per quadrat were found to be a Poisson distribution again, this time with a mean of $2.86C$, the evidence for accepting the hypothesis that the pattern is random would be very strong indeed. A confirmatory test based on data of an entirely different kind is described in Section 4.

2 NON-RANDOM PATTERNS

Now let us consider non-random patterns and the ways in which they contrast with random ones. There are always two different types of pattern to envisage: those possible for species that occur only in specialized pieces of habitat which serve as natural sampling units (fruits, funguses, carcasses etc.); and those possible for species whose individuals may be scattered anywhere throughout a continuous space. Figure 7.1 shows examples and the discussion that follows uses the figure for illustration.

Consider first the three populations on the left in the figure, labelled A, C and E. In all of them there are 57 individuals distributed among 20 natural units and the density of the individuals, or mean number per unit, is thus $57/20 = 2.85$. It is obvious at a glance, however, that the variance of the number of individuals per unit has different values in the three cases.

For population A the variance is 3.17, which is not very different from the mean. The pattern is, in fact, random: it was constructed by selecting at random, for each of the 57 individuals in turn, one of the 20 units for it to occupy. Thus every unit was a likely as every other to receive each individual and the individuals are independent. The number of individuals per unit has a Poisson distribution.

Next consider population *C*. In this case the variance of the number of individuals per unit is 15.43 which greatly exceeds the mean (2.85, as before). Clearly the high variance is due to the fact that the individuals are concentrated in large numbers in only a few of the units, leaving many units empty. Several terms have been coined to describe this type of pattern: *aggregated*, *clumped*, *clustered* and *contagious* are all frequently used and as all are descriptive no misunderstandings are likely because of the profusion of names. The crucial diagnostic character is that the variance exceeds the mean. There are two obvious causes for the occurrence of these clumped patterns and, indeed, both causes can operate at the same time. Sometimes the clumping stems from the fact that the units are not all equal to one another; there may be qualitative differences among them so that some can attract, contain, or nourish more individuals than others. Alternatively,

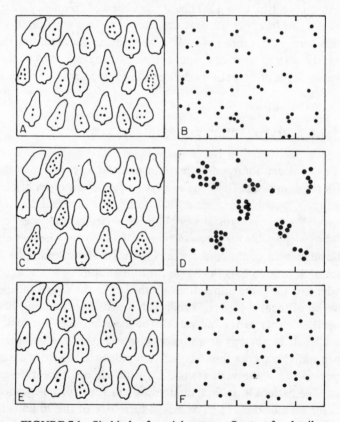

FIGURE 7.1   Six kinds of spatial patterns. See text for details

even if the units are identical in all respects, the organisms may occur as "familial" clumps because they originated from clumps of eggs laid by a single mother; or because several generations are present and small inbreeding populations, each restricted to a unit, have flourished in some units and died out in others.

The pattern of population E contrasts with the random pattern of population A in the opposite sense. Its variance is 1.83 which is much less than the mean, 2.85, and it is apparent that the cause of this lies in the very even manner in which the individuals are distributed among the units. None of the units is empty and none contains more than five individuals. This sort of pattern, which is known as *regular*, is found when there is a fairly low upper limit to the number of individuals a unit can accommodate; or when individuals exclude one another from the units by driving off late arrivals or making the already occupied units unattractive to them. And sometimes a population which has an aggregated or random pattern while its members are young may subsequently come to have a regular pattern owing to the dying off of some surplus individuals in the overcrowded units.

The patterns on the right in Figure 7.1, labelled B, D and F, are obviously analogous to patterns A, C and E respectively. The same descriptive names (random, clumped and regular) are applied to them and similar causes give rise to them, except that now the sampling units are no longer natural entities but merely arbitrarily delimited subareas of a continuous habitable region. There is not, however, a complete parallel between patterns A, C and E on the one hand and B, D and F on the other. The continuous patterns D and F differ in a very important respect from the random pattern B and there is no analogous separation of A from C and E.

To see this we must notice a special characteristic of population B. In the following discussion we shall assume for simplicity that the individuals are plants and the sampling units quadrats. When population B is sampled with quadrats then, the expected frequency distribution is a Poisson distribution in all circumstances. The quadrats may be placed at random, or in a single block, or in groups or rows, or at the corner points of a lattice. Whatever arrangement is used, the result (except, as always, for inevitable sampling variations due to chance) will be a Poisson distribution. Similarly, the shapes of the quadrats can be allowed to vary in any way whatever. So long as they are all of the same size they may be squares, or circles, or long, narrow rectangles or any shape, regular or irregular, that one cares to choose; indeed, a mixture of different shapes could be used to obtain a single sample. Provided only that their sizes are constant, the observed frequency distribution will be a Poisson distribution. If a sample is taken using quadrats of a different size, yielding a different mean number of individuals per quadrat, the result will again be a Poisson distribution, but with the new mean. These remarks can be summed up by saying that the salient characteristic of a random pattern is that the *expected* number of individuals per quadrat of any given size is the same everywhere.

With non-random patterns, the results of quadrat sampling will obviously be affected by the shape, size and arrangement of the quadrats. This is most apparent for population D with its fairly compact clumps. Clearly, long, narrow quadrats would yield a frequency distribution with lower variance than would square quadrats of the same size. Small quadrats that would often lie completely inside or completely outside a clump would yield a distribution with proportionately greater variance (relative to the mean) than would large quadrats. And if quadrats were located at random, they might all chance to contain the central parts of dense clumps. In brief, it is not true for population D that the expected number of individuals per quadrat is the same throughout the area.

The dependence of the results on the accidents of sampling is less obvious for population F with its very regular pattern. Notice, however, that although the exected number of individuals per quadrat may be everywhere constant for large quadrats, the same is not true for small quadrats. If sampling is done with quadrats so small that none contains more than one individual, then it is clear that a quadrat next to an empty quadrat has a higher probability of being occupied than one next to an occupied quadrat. In this case also, therefore, the essential condition for true randomness, namely that the number of individuals in a quadrat should be independent of its location, is not met.

The foregoing arguments justify the following statement: the frequency distribution of the number of individuals per sampling unit provides a compact and unambiguous description of a spatial pattern whenever the sampling units are natural entities*; and also when they are arbitrarily delimited parts of a continuous area *if* the pattern is random. But non-random continuous patterns cannot be summarily described in this way; the observed frequency distributions are inevitably influenced by the chosen sampling procedure.

If the population whose pattern we are investigating consists of motionless, plainly visible individuals (such as barnacles on a rock face, or tree seedlings in a woodlot) it is, indeed, wasteful of effort to list the numbers of

---

* Ideally one would also list the original locations of the natural sampling units but it is seldom feasible to do this at all exactly. However, if the units are obtained by stratified sampling as if often the case, at least a crude classification of their regions of origin may be possible.

individuals in the sampling quadrats without keeping a record of the quadrats' locations. When a *spatial* pattern is being studied, the spatial relations of the various observations are obviously relevant and ought to be taken into account. This is especially true of vegetation sampling; we shall discuss in Chapter 8 some of the many ways of studying patterns when the locations as well as the contents of sampling quadrats can be recorded.

Plant populations, however, are among the easiest to deal with. The population patterns of small, active, concealed animals are much harder to study. The cryptozoa of soil and litter, and the plankton of marine and inland waters are the two chief types of community for which the best and sometimes the only evidence on pattern consists of frequency distributions of the numbers of individuals in sampling units. The species populations making up these communities are usually found to be clumped or aggregated. Aggregated patterns are commoner than random ones and regular patterns are very unusual.

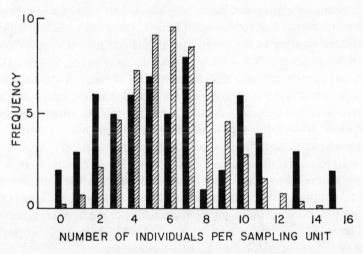

FIGURE 7.2 The frequency distribution of copepod nauplii (*Microcalanus pygmaeus*) in sixty small sampling units of sea water. Black bars: observed frequencies. Hatched bars: expected frequencies for a Poisson distribution with the same mean. (Data from Barnes and Marshall, 1951)

Figure 7.2 shows, as a bar graph, a typical one of these frequency distributions. The data on which the figure is based are from Barnes and Marshall (1951). The organisms concerned are nauplii of an abundant copepod of the marine plankton, *Microcalanus pygmaeus*, and the observations were made in coastal waters, 60 meters deep, off western Scotland. Sixty small sampling units were collected in quick succession from water 10 meters deep; the whole sample was taken within $1\frac{1}{4}$ hours, a period short enough for there to be no appreciable change in the population being sampled owing to diurnal vertical migrations of the plankton. In the figure, the black bars show the observed frequencies of the numbers of nauplii per sampling unit and the hatched bars the frequencies of a Poisson distribution with the same mean. The discrepancies are obvious. As always happens with aggregated populations, a large proportion of the units had very many or very few individuals and only a comparatively small proportion had intermediate numbers.

For these data, $n = 60$; the total number of nauplii in the whole sample was $\Sigma x = 374$; and $\Sigma x^2 = 3182$. The index of dispersion (see page 140) is thus $I = 136.5$ and the probability of obtaining as high a value as this if the pattern of the nauplii were random is less than 1 in 10,000.

In the past a great deal of effort has gone into attempts to analyse empirical frequency distributions such as this one and to try to wring from them, by mathematical reasoning, clues as to the causes of aggregation in particular cases. Mathematical models have been devised to predict what forms the observed frequency distributions will have if (i) the individuals of a population remain together in family groups; or (ii) densities are patchy because the habitat is patchy. It has been known for many years, however (Feller, 1943), that entirely different underlying models can lead to identical observable outcomes and consequently an attempt to deduce the underlying causes of a pattern using a frequency distribution as the only evidence is foredoomed to failure. This is hardly surprising; a single empirical frequency distribution is a somewhat sketchy foundation for an elaborate theory. The intuitive explanations of clumping offered by observers familiar with the biology of the species they are studying are far more believable. The clumping of the copepod nauplii appeared to Barnes and Marshall to be due to the presence of a number of distinct subpopulations, each associated with a different water mass; these subpopulations had not mingled but had retained their separate identities long enough for their numbers to build up.

The causes of clumping in plankton and other mobile organisms would be easier to discover if one could tell whether clump boundaries were clearcut or indefinite; i.e. whether the density variations were abrupt or gradual. This cannot be judged from a frequency distribution. Cassie (1962) made underwater observations on plankton and found that distinct swarms, though rare, were occasionally encountered.

It should now be clear that arriving at a complete explanation of the origin of an observed spatial pattern is not at all easy even if it can be done at all. It is a delusion to suppose that mysterious mathematical manipulations will reveal a full-blown theory lurking unsuspected in a modest column of numbers. (Cf. page 52 for analogous remarks on modes of population growth).

An inquiry into the causes of an observed pattern is not, in any case, an end in itself. The true objective of pattern studies is to discover all that one can about the biology of a population using its spatial pattern as evidence. Comparative studies are therefore likely to be especially rewarding. For a given species, it would be interesting to know whether its different populations all have similar patterns, or whether the patterns vary. Population patterns may differ in different parts of a species' geographical range, or because of differences in overall density or in various environmental factors. By observing the changes in pattern that accompany a reduction in size of an animal population it may be possible to judge whether the reduction is density-dependent; this is discussed in Section 3 (see page 152).

Investigations of the way in which a population's pattern alters with time are undoubtedly of great interest. There are seasonal changes, and also long-term changes, to consider. It may sometimes happen that density, pattern and the behavior of the organisms undergo change simultaneously and interdependently. This has been suggested by some remarkable work by Wellington (1957, 1960) on the Western Tent Caterpillar (*Malacosoma pluviale*) a species of moth whose caterpillars live together in dense swarms in tentlike webs that they spin in the branches of trees, chiefly alders, willows and orchard trees. Wellington observed the changes that took place, during the late 1950's, in colonies of this species in Vancouver Island. He found that the caterpillars' behavior in the early stages of an infestation, while populations were building up, was entirely different from what it became later when the infestation peaked and subsequently declined.

Early in the infestation the colonies contained a high proportion of very active caterpillars; they moved quickly, responded readily to outside stimuli and foraged independently without following pre-existing silk trails. Colonies dominated by these active caterpillars tended to build several tents, of elongate shape, and to forage over a wide area. As the infestation progressed, however, the new generation of caterpillars contained increasing proportions of sluggish individuals. These were caterpillars that failed to move far from their birthplaces and remained in very dense swarms; they would travel only along silk trails left by active caterpillars; a colony containing a high proportion of them would build only one or two compact, pyramidal tents and feed nearby. They exhausted the available food in the vicinity of their tents and then, being too sluggish to move, starved. Also, their high density promoted the rapid spread of disease. Adult behavior reinforced the effects of juvenile behavior; sluggish adult moths (from sluggish caterpillars) did not fly far to lay their eggs, whereas active moths did. Consequently sluggish

10*

colonies became numerous in restricted areas. As the environment deteriorated from their depredations, so did the quality of the colonies, until finally they died out causing an abrupt decrease in density and an end to the infestation in that locality.

It was particularly noteworthy that the behavior, and hence success, of a colony depended on the proportions of active and sluggish caterpillars comprising it. In Wellington's words, in a "good" colony sluggish individuals are insulated from their own ineptitude. A good colony is controlled by its active members, and travels far enough and disperses widely enough to avoid destroying its own environment; the weak as well as the strong members of the colony benefit.

The whole phenomenon, a density cycle going from sparse to dense to sparse again, accompanied by qualitative changes in the individuals constituting the population, resembles the vole cycles described by Chitty (see page 96). In the tent caterpillars however, Wellington (1965) found that the cyclical changes in population quality result from a vicious circle. If female caterpillars are ill-nourished, the eggs they lay when they become adult are deficient in yolk and yield sluggish caterpillars; these, because of their sluggishness, consume even less food and yield offspring of still lower quality. The deterioration is cumulative and every caterpillar "inherits" its vigor from its mother, not in the genetic sense but through the supply of yolk available to it in the egg. The cycle starts because the eggs laid by a female moth are not of uniform quality; her food reserves are unequally partitioned among them. The eggs laid first are rich in food and yield agile progeny whereas eggs deposited later, and the caterpillars that emerge from them, are progressively poorer in quality. As remarked earlier, concomitantly with these changes, a tent caterpillar population undergoes changes in spatial pattern. It would be interesting to know if other gregarious animals exhibit cyclical changes in pattern brought about by similar mechanisms.

3   MEASURING AND COMPARING DIFFERENT DEGREES OF AGGREGATION

Suppose we accept that if pattern studies are to yield appreciable gains in biological knowledge, it will be necessary to compare the patterns of many populations living in different conditions. That is, instead of indulging in untestable speculations on the causes of particular patterns, we shall conduct a more general search for relations between patterns and the many other properties of ecological populations. It then becomes necessary to devise ways of measuring pattern, to quantify this particular property of a population and so enable comparisons to be made among different populations.

Let us now consider what it is that we wish to measure. Because of their commonness in nature, we shall be concerned in this section with aggregated patterns.

The pattern of a population spread over a continuum (trees in a forest, for example) has two obvious properties which are candidates for measurement, and they are independent of each other. These properties may be called *intensity* and *grain* and can be defined as follows. The intensity of a pattern is high if a wide range of densities is present; conversely, it is low if the density contrasts are slight. The grain of a pattern is coarse (or the pattern is coarse-grained) if its clumps and the gaps among them are large; in the converse case the pattern is fine-grained. Figure 7.3 shows examples. It is seen that the intensity and grain of a pattern are both independent of its density; it is the relative densities, not their absolute values, that determine intensity. It should also be observed that the patterns of organisms living in discrete

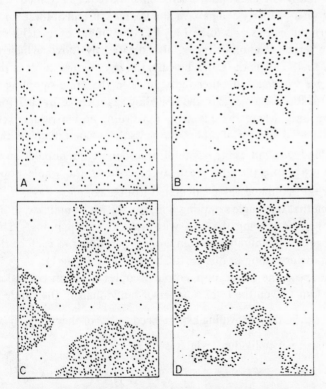

FIGURE 7.3 Patterns of different intensities and grains. The intensity of the patterns is low in maps A and B; and high in maps C and D. The grain of the patterns is coarse in maps A and C; and fine in maps B and D

natural habitable units can properly be said to have intensity but not grain. Thus in Figure 7.1 (page 142) the intensity of pattern C is greater than, and of E less than, that of the random pattern A. It is clear that measurement of the grain of a continuous pattern cannot be made unless the locations as well as the contents of sampling units (usually quadrats, soil cores, or water samples) are taken into consideration. This assertion is obviously true even though we have not yet decided on a method for measuring grain.

We shall now concentrate on populations that either (i) live in discrete habitable units, or (ii) though living in continuous regions, are hard to sample in a way that permits exact mapping of the locations of sampling units or individuals. The data obtainable for analysis thus consist of frequency distributions of the numbers of individuals per sampling unit and inferences about grain (relevant only in case (ii)) cannot be made. Such data do lend themselves to the measurement of pattern intensity, however, and the question is how best to do it.

A great many "indexes" and "coefficients" have been invented to measure one or another property of a pattern and the most useful of these for measuring pattern intensity seems to be Lloyd's (1967) index of "patchiness". It is defined in terms of another index, also devised by Lloyd, which must be described first. This is the index he calls the *mean crowding*. It is the mean number, *per individual* of other individuals in the same sampling unit. In other words it is the average of the crowding experienced by each individual in its own unit, where the quantity of crowding an individual experiences is equated to the number of other occupants that share the unit with it. Following Lloyd we shall use the symbol $\overset{*}{m}$ to denote the population value of the mean crowding (i.e. the mean averaged over the whole population) and $\overset{*}{x}$ for the estimated value calculated from a sample.

It is seen that if a single sampling unit, say the jth, contains x_j individuals, each of them is sharing the unit with $x_j - 1$ "companions". Therefore, adding together the number of companions pertaining to each of the x_j individuals, the crowding in this particular unit is $x_j (x_j - 1)$.

Now suppose the whole population occupies a region equivalent to Q sampling units. Then the total number of individuals in the whole population is $\sum_{j=1}^{Q} x_j$; the total crowding experienced by all of them is $\sum_{j=1}^{Q} x_j (x_j - 1)$; and the mean per individual, which is $\overset{*}{m}$, is

$$\overset{*}{m} = \frac{\sum_{j=1}^{Q} x_j (x_j - 1)}{\sum_{j=1}^{Q} x_j}. \tag{7.1}$$

Next, let a sample of q sampling units be selected at random. As an estimator of $\overset{*}{m}$ we could use formula (7.1) taking the sums over the q units in the sample instead of over the Q units that contain the whole population. This estimator is biased, however. An unbiased estimator, $\overset{*}{x}$, is obtained as follows.* Let $\bar{x}$ and s^2 be the estimates of the mean and variance of x calculated from the sample data.

That is

$$\bar{x} = \frac{1}{q} \sum_{j=1}^{q} x_j \quad \text{and} \quad s^2 = \frac{1}{q-1} \left\{ \sum_{j=1}^{q} x_j^2 - q\bar{x}^2 \right\}.$$

Then

$$\overset{*}{x} = \bar{x} + \left(\frac{s^2}{\bar{x}} - 1 \right)\left(1 + \frac{s^2}{q\bar{x}^2} \right). \tag{7.2}$$

The sampling variance of $\overset{*}{x}$ is

$$\text{var}\left(\overset{*}{x}\right) = \frac{1}{q} \cdot \frac{\overset{*}{x}s^2}{\bar{x}^2}\left(\overset{*}{x} + \frac{2s^2}{\bar{x}} \right). \tag{7.3}$$

Now consider a numerical example. The data are from Lloyd (1967) and relate to the patterns of the litter-inhabiting cryptozoa of an English beechwood. We shall discuss here only the distribution of the number of individuals in each of 16 quadrats of the pseudoscorpion *Chthonius ischnocheles*. The 16 variate values were

$$0, 1, 2, 7, 2, 21, 5, 2, 8, 1, 4, 17, 24, 1, 7, 6.$$

The sum is

$$\Sigma x = 108.$$

The sum of squares is

$$\Sigma x^2 = 1560.$$

The sample size is

$$q = 16.$$

Therefore

$$\bar{x} = 108/16 = 6.75$$

and

$$s^2 = \frac{1560}{15} - \frac{108^2}{15 \times 16} = 55.4.$$

Substituting these values in equation (7.2) gives

$$\overset{*}{x} = 14.51.$$

---

* Only one of the possible estimators proposed by Lloyd is given here. Another that he gives is more troublesome to compute and is applicable only when certain assumptions hold, though if they do hold it is more precise.

This implies that every pseudoscorpion shared its quadrat with an estimated 14.51 other pseudoscorpions on average. From equation (7.3), $\mathrm{var}\,(\overset{*}{x})$ = 34.08 whence the standard error of $\overset{*}{x}$ is 5.85; (this is a large standard error but we are considering data from only 16 quadrats).

Mean crowding is obviously dependent on density; an organism in a dense population experiences more crowding than one in a sparse population. Therefore let us consider the ratio of mean crowding to density; (observe that one cannot assume without proof that this ratio is independent of density; we return to this point below). Lloyds calls the ratio the *patchiness* of the population. Writing m for the population value of the density, $\overset{*}{m}/m$ is the population value of the patchiness. A sample estimate of the patchiness is given by $\overset{*}{x}/\bar{x}$ and its sampling variance is

$$\mathrm{var}\left(\frac{\overset{*}{x}}{\bar{x}}\right) = \left(\frac{S^2}{\bar{x}^2}\right)^2 \frac{2\overset{*}{x}}{q\bar{x}} \tag{7.4}$$

For the pseudoscorpion data,

$$\frac{\overset{*}{x}}{\bar{x}} = 2.15 \quad \text{and} \quad \mathrm{var}\left(\frac{\overset{*}{x}}{\bar{x}}\right) = 0.397$$

so that the standard error is 0.63.

It should be recalled that we wish to define a measure of pattern intensity in such a way that it is unaffected by the population's density. In order to test whether patchiness meets this requirement, we must specify precisely what we mean by saying that two patterns have the same intensity although their densities differ. A convenient way of making this notion concrete is the following. Visualize any population of individuals and imagine that a proportion r of them, picked completely at random, is removed. Then, if the density was initially m it is now reduced to rm. However, since the individuals removed were a random selection, it seems reasonable to say that the density is the only thing that has changed and the pattern intensity is the same as before. If we stipulate that a measure of intensity is to be such that it is unaffected (apart from chance variation) by such *random* removal of some of the population members, then it can be proved (Pielou, 1969) that patchiness fulfils this condition.

To illustrate the effect of random removal, we return to the pseudoscorpion data and imagine that one half of them, randomly chosen, are removed. Figure 7.4 shows this conceptual experiment schematically; the black circles represent the individuals allowed to remain in the population and the hollow circles those that were removed. The decision to remove an individual was made by tossing a penny. The numbers of survivors in the 16 quadrats were:

$$0, 1, 1, 4, 2, 9, 2, 0, 3, 1, 2, 9, 13, 0, 4, 3.$$

Calculating the mean crowding and the patchiness as before gives

$$\overset{*}{x} = 6.85 \quad \text{and} \quad \frac{\overset{*}{x}}{\bar{x}} = 2.03.$$

As was to be expected, the mean crowding has been reduced to roughly one-half its initial value, and the patchiness is virtually unaltered.

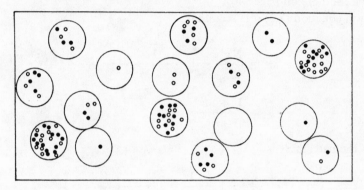

FIGURE 7.4   The effect of removing half the members, chosen at random, from an aggregated population. The fate of each individual initially present was decided by the toss of a coin; the black circles show the survivors and the hollow circles those that were removed

The fact that patchiness is unaffected by the removal of individuals at random from a population may make it a useful measure in studies on the natural regulation of populations. Imagine a population of organisms in which no births are occurring and in which the density is steadily declining because of deaths. The organisms are assumed to occupy discrete natural units and to be sedentary so that there is no migration from unit to unit. This might be approximately true, for example, of a cohort of immature insects all born at about the same time and still too young to reproduce. The foregoing arguments show that, if the deaths take place at random, the population's patchiness will remain constant. But if the deaths are density-dependent, i.e. if crowded individuals are less likely to survive than isolated ones, patchiness will show a downward trend as time passes and the population dwindles.

It is worthwhile exploring the range of possible values of the patchiness for different kinds of pattern. In what follows, only the true population values (as opposed to sample estimates) of mean crowding and patchiness will be considered and there is no risk of misunderstanding if, for brevity, we write Σx in place of $\sum_{j=1}^{Q} x_j$. The variate is the number of individuals per sampling unit and we denote its mean by m, as before, and its variance by V. That is,

$$m = \frac{1}{Q}\Sigma x \quad \text{and} \quad V = \frac{1}{Q}\Sigma (x - m)^2 = \frac{1}{Q}\Sigma x^2 - m^2.$$

(Note: Q and not $Q - 1$ is used in the denominator of the formula for V; see the footnote on page 140).

From equation (7.1) it is seen that

$$\overset{*}{m} = \frac{\Sigma x^2}{\Sigma x} - 1 = \frac{\frac{1}{Q}\Sigma x^2}{m} - 1$$

$$= \frac{V + m^2}{m} - 1.$$

Thus the patchiness is

$$\frac{\overset{*}{m}}{m} = \frac{V}{m^2} + 1 - \frac{1}{m} = 1 + \frac{1}{m}\left(\frac{V}{m} - 1\right).$$

It follows that

$$\frac{\overset{*}{m}}{m} > 1 \quad \text{if} \quad \frac{V}{m} > 1$$

and

$$\frac{\overset{*}{m}}{m} < 1 \quad \text{if} \quad \frac{V}{m} < 1.$$

Now, we have already shown (page 142) that for an aggregated pattern the variance exceeds the mean, or $V/m > 1$; and for a regular pattern $V/m < 1$.

Thus for an aggregated pattern, $\overset{*}{m}/m > 1$ and it can become very large; there is no theoretical ceiling value. For a regular pattern, $\overset{*}{m}/m < 1$; but since a negative variance (i.e. $V < 0$) is impossible, we must always have

$$\frac{\overset{*}{m}}{m} \geqq 1 - \frac{1}{m}.$$

Thus even when a pattern is extremely regular, if m is large the patchiness will be very close to unity.

These results accord with intuitive notions as to the values the patchiness should have in different circumstances. Since patchiness measures the intensity of a pattern, it is high when the pattern shows pronounced density contrasts, and is close to or somewhat below unity when there are negligible contrasts; indeed, $\overset{*}{m}/m < 1$ implies that the differences among the units in the numbers of individuals they contain is *less* than would be expected if the pattern resulted solely from chance.

For a random pattern, $V/m = 1$, and also $\overset{*}{m}/m = 1$. Notice, however, that one should *NOT* conclude that a pattern is random because its patchiness is close to unity; or, what comes to the same thing, that a pattern is random because the variance of the number of individuals per unit is close to the mean (i.e. the variance: mean ratio, V/m, is close to 1). This is an exceedingly common error, but logically it is on a par with the following elementary blunder: all crows are birds and therefore all birds are crows. In the present context, though all random patterns yield a V/m ratio close to 1, not all patterns with a V/m ratio close to 1 are random. Thus it is not difficult to construct a pattern having $V/m = 1$ by aggregating the individuals in clumps and spacing them regularly within the clumps.

A random pattern is one that can come into existence only in rather special, and precisely defined, circumstances. It has a number of properties, one of which is that its patchiness is unity. As we have argued above, it is absurd to conclude that any pattern with unit patchiness is *ipso facto* random. The conclusion that a pattern is random should only be drawn if there are ecological reasons for hypothesizing that it may be, and at least two statistical tests give no reason for rejecting the hypothesis. The word "random" should not be used casually to describe patterns that are neither strongly regular nor strongly clumped but somewhere between.

4 PATTERN STUDIES AND THE SPACING OF INDIVIDUALS

It was pointed out earlier in this chapter that when a pattern in a continuous region is sampled with arbitrarily defined units such as quadrats, the results, and their interpretation, are likely to be greatly influenced by the size of the units. Plant ecologists, especially, are conscious of this fact; conclusions based on quadrat observations are always tempered by the thought that they might have been different had quadrats of another size been used.

This difficulty can be avoided by so-called "plotless" sampling. This is done by measuring a sample of distances. Each distance may be measured from a randomly chosen individual to its nearest neighboring individual, or from a randomly located point to the individual nearest to it. The method is well suited to studying the patterns of populations of plant species that do not reproduce vegetatively; it is particularly useful in studies of forest patterns. Often the underlying reason for studying a pattern is to judge whether the individuals in a population are evenly spaced and, as a result, making optimum use of available resources of light, water and nutrients. When this is the objective, distances are of obvious concern.

Consider first the kind of result one would expect. In a truly random pattern (but in no other), it makes no difference whether distance is measured from a randomly located point or from a randomly selected individual; the expected distance to the nearest neighboring individual is the same in either case. At first thought this is surprising; it might be supposed that a randomly located point would tend to be closer to a population member than would another population member that had been there since before the observations were begun. This is not so, however. Selecting a random point from which to make a measurement (which is usually done by taking a pair of coordinates for the point from a table of random numbers) is conceptually no different from choosing a site for an imaginary new "individual". If a single individual were added in this manner to an indefinitely large population it would make no appreciable change to the density and pattern of the pre-existing population. We therefore see that, provided the pattern under investigation is random, a point-to-nearest-individual distance and an individual-to-nearest-neighbor distance are conceptually identical and must have the same expectation.

This is obviously not true, however, of non-random patterns. If a pattern is regular, like F of Figure 7.1 (page 142), the distance from any individual to its nearest neighbor clearly has a greater expectation than the distance from a random point to its nearest individual; the individuals are so evenly spaced that a random point placed among them is likely to be much closer to an individual than the individuals are to each other.

The opposite is true of the aggregated pattern D in Figure 7.1. Most of the individual-to-neighbor distances are short within-clump distances; but a random point is most likely to fall in one of the large spaces separating the clumps, and therefore to be a long way from the nearest individual.

We now investigate in greater detail the distances in a random pattern. For concreteness we shall assume that the individuals are trees and that each of the distances measured is from a random point to the tree nearest it. The diameters of the trunks are ignored and all measurements are to the centers of the trunks. We require a symbol for the density of the population and, as will be seen below, it is convenient to use a circle as the unit of area. Therefore we denote by λ the mean number of trees per circle of unit radius (i.e. of area π square units). Also, let r be the distance from a random point to the nearest tree center. Then the probability that r is less than or equal to some specified value, say r_1, is given by

$$\Pr\{r \leqq r_1\} = \Pr\{\text{a circle of radius } r_1 \text{ centered on the sampling point}$$
$$\text{contains } at \ least \text{ one tree center}\}$$

$$= 1 - \Pr\{\text{this circle is empty}\}$$

$$= 1 - e^{-\lambda r_1{}^2}. \tag{7.5}$$

This last follows since the pattern of the trees is random and thus the probability that a specified area contains 0 trees is given by the appropriate term of a Poisson distribution with (in this case) a mean of λr_1^2 trees per circle of area πr_1^2.

In what follows, for simplicity, we shall put $r^2 = \omega$. The original measurements are of distances, or values of r, but the mathematical analysis is concerned with the squares of these distances, or values of ω. Then equation (7.5) becomes
$$\Pr(\omega \leqq \omega_1) = 1 - e^{-\lambda\omega_1}.$$
Similarly
$$\Pr\{\omega \leqq \omega_2\} = 1 - e^{-\lambda\omega_2};$$

so that, if $\omega_1 < \omega_2$, it immediately follows that

$$\Pr\{\omega_1 \leqq \omega \leqq \omega_2\} = \Pr\{\omega \leqq \omega_2\} - \Pr\{\omega \leqq \omega_1\}$$

$$= e^{-\lambda\omega_1} - e^{-\lambda\omega_2}. \tag{7.6}$$

In this way, if λ is known or can be estimated, we may calculate the expected probability distribution of ω and compare it with the observed distribution yielded by a sample.

As an example, consider again the population of lodgepole pines already described (page 137). It occupied a square area of 2500 sq. yards. One hundred sampling points were placed at random in the area, their locations being selected by picking a pair of coordinates for each point from a random numbers table. From every point the distance (in feet and tenths of a foot) to the center of the nearest tree was measured and this distance was squared to give a value of ω. The sample of ω values was then grouped into classes and the observed frequencies of the several classes are shown in Table 7.2. The density of the trees had already been estimated from the quadrat data described earlier (page 138). In terms of number of trees per circle of 1 foot radius, the estimate is

$$\hat{\lambda} = 0.1144.$$

Using this estimate of λ, the expected frequencies of ω values in the different classes were calculated and the results are shown in Table 7.2. For example, the estimated probability that a value of ω will be in the class 4.00 to 5.99 ft^2 is, from equation (7.6),

$$\Pr\{4 \leqq \omega \leqq 6\} = e^{-4\hat{\lambda}} - e^{-6\hat{\lambda}}.$$

Putting $\hat{\lambda} = 0.1144$ this becomes

$$\Pr\{4 \leqq \omega \leqq 6\} = 0.6328 - 0.5034 = 0.1294.$$

The expected frequency is thus 12.94 since the sample size was 100.

TABLE 7.2   The observed and expected distributions of the distance from a random point to the nearest tree in a lodgepole pine population ($\hat{\lambda} = 0.1144$ trees per circle of 1 foot radius; estimate obtained by quadrat sampling)

$r^2 = \omega$	$\hat{\lambda}\omega$	$e^{-\hat{\lambda}\omega}$	Classes of ω values	Frequencies Expected	Frequencies Observed
0	0	1			
			0 –0.99	10.81	14
1	0.1144	0.8919			
			1.00–1.99	9.64	10
2	0.2288	0.7955			
			2.00–3.99	16.27	14
4	0.4576	0.6328			
			4.00–5.99	12.94	14
6	0.6864	0.5034			
			6.00–9.99	18.49	21
10	1.1440	0.3185			
			10.00–14.99	13.87	15
15	1.7160	0.1798			
			15.00–19.99	7.83	3
20	2.2880	0.1015			
			20.00–24.99	4.42	2
25	2.8600	0.0573			
			$\geqq$25.00	5.73	7
				100.00	100

$$X^2 = \Sigma \frac{(\text{Observed–Expected})^2}{\text{Expected}} = 6.377$$

With 8 degrees of freedom*, $P(\chi^2) > 0.5$

* The number of comparisons between observation and expectation is 9. No parameters were derived from the observations (λ was estimated from independent observations) and therefore the number of degrees of freedom is 8.

Judging the goodness of fit of the theoretical distribution to the observed distribution by a χ^2-test (see Appendix 7.2), it is found that the test statistic is $X^2 = 6.377$; the probability that this value would be exceeded by chance, given the hypothesis of randomness, is greater than 0.5. The result therefore lends confirmation to the conclusion already arrived at, namely that the pattern of the lodgepole pines is random.

In a theoretical random pattern, in which the individuals are geometrical points, there is no lower limit to the possible distance separating any two individuals. In practice, of course, when the individuals are objects of finite size such as trees, the center-to-center distance between a pair of them cannot be less than their mean diameter. But this obvious fact will have no effect on a population's observed pattern if the density of the trees is low enough; for then the probability of observing an impossibly short tree-to-tree distance is negligibly small and the absence of such distances from the data is, in fact, in full accord with expectation.

If a population of trees were known to have a random pattern, we could obtain an independent estimate of λ from a sample of ω values only, without recourse to quadrat sampling. It can be shown (see Appendix 7.3) that the expected value of ω is

$$\mathscr{E}(\omega) = 1/\lambda.$$

Thus, putting $\bar{\omega}$ for the observed mean of the ω's, an estimate $\hat{\lambda}_\omega$ of λ is given by*

$$\hat{\lambda}_\omega = 1/\bar{\omega}.$$

In the lodgepole pine example the mean of the ω's (calculated from the original ungrouped data which are not given here) was

$$\bar{\omega} = 8.097 \quad \text{and hence} \quad \hat{\lambda}_\omega = 0.1235.$$

This is very close to the value (0.1144) obtained by quadrat sampling.

Unfortunately the fact that, *if the pattern is random*, density can be estimated from a sample of distance measurements only is of no practical use. The reason is this: one can never assume without a test that a naturally occurring pattern is random and it is impossible to carry out a test for randomness using only distance measurements. To perform a test, an *independent* estimate of density is required and this has to be obtained by quadrat sampling or by a complete census of all individuals in the population; but obviously, when either of these things has been done, there is no further need to estimate density from a sample of distances.

---

* This estimator is biased; the unbiased estimator is $(n - 1)/n\bar{\omega}$. This fact has no bearing on the argument in the text, however, and will be ignored.

The expected frequencies shown in Table 7.2 were calculated by substituting $\lambda = 0.1144$, the density estimate based on quadrat sampling, in equation (7.6). It might be argued that we could, instead, have used $\lambda_\omega = 0.1235$, the estimate based on distances, and thus have carried out a self-contained analysis using distances only. This procedure is apt to be misleading, however. It is often found that the probability distribution defined by equation (7.6) (which is known as the negative exponential distribution) fits empirical distributions of ω obtained from *non*-random patterns as well as from random ones if λ_ω is used as the estimator of λ. But for non-random patterns this parameter is not equivalent to density. It is easy to see that for a regular pattern, $1/\bar\omega$ overestimates the density and for an aggregated pattern $1/\bar\omega$ underestimates the density.

We therefore conclude that the close correspondence between observed and expected frequencies shown in Table 7.2 can be taken as strong evidence that the trees' pattern was random because λ, the *independently* obtained density estimate was used to calculate the expected frequencies. If we had used λ_ω, the estimate yielded by the distance data themselves, the fit would again have been very close but this fact would not constitute evidence for randomness.

Yet another test for randomness can be done by comparing λ and λ_ω. If the pattern is random the expected value of their ratio is unity. That is, if we put $\lambda/\lambda_\omega = \lambda\bar\omega = \alpha$, say, then

$$\mathscr{E}(\alpha) = 1.$$

Further, if the size of the quadrat sample is m, and of the distance sample is n, it can be shown (Mountford, 1961) that the sampling variance of α is estimated by

$$\text{var}\,(\alpha) = \frac{m\lambda + n + 1}{mn\lambda}.$$

In the lodgepole pine example.

$$n = m = 100, \quad \lambda = 0.1144 \quad \text{and} \quad \bar\omega = 8.097.$$

Then

$$\alpha = \lambda\bar\omega = 0.926$$

and var $(\alpha) = 0.098$ so that the standard error of α is $\sqrt{\text{var}\,(\alpha)} = 0.314$. Clearly the observed α is not significantly different from unity.

The result of this test does not, of course, supply additional evidence on the randomness of the pattern since it uses the same data as the goodness of fit test shown in Table 7.2. The two tests are simply alternative ways of analysing a single body of data and do not reinforce each other.

We have now discussed the distribution of distances in a random pattern in some detail. Next, consider regular patterns. Distance sampling (from individuals to their nearest neighbors) is especially useful if one wishes to judge whether a pattern departs from randomness in being regular, that is, in having its members more evenly spaced than they would be by chance. A regular pattern will occur when density is high and each member of the population requires a "territory" of appreciable size; or, what comes to the same thing, when the existence of a lower limit to the possible distance between two individuals is having an observable effect on the pattern.

Regular patterns are likely to be found among desert shrubs where there is strong competition for water. They are also to be expected in dense, even-aged forest; the trees, all of roughly equal size, must each have a certain minimum area of ground to supply the required water and nutrients and enough space in the canopy for its crown. A regular pattern would also be expected among plants that secrete toxic exudates which inhibit the growth of seedlings of their own species. Examples will be given below. First we discuss how regular spacing may be detected.

It has often been asserted that regular patterns are exceedingly rare in nature. Investigators have been surprised to find that the trees in a dense forest, for instance, or the xerophytic shrubs in an arid area, were aggregated in spite of the fact that the plants seemed, subjectively, to be evenly spaced and the competition among them was presumably intense. The probable explanation for this is that density is seldom consistently high throughout the whole of an area delimited for study. It seems likely that patterns such as that in Figure 7.5 are quite common. The pattern as a whole is aggregated and the customary sampling procedures would usually fail to show that in spite of the clumping, none of the interplant distances is excessively short.

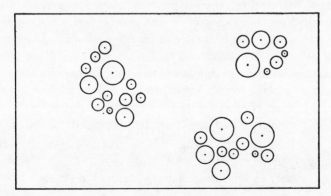

FIGURE 7.5   A clumped population of large individuals. The within-clump pattern is regular

The need of each plant for a certain minimum territory and the regular spacing that this entails are only apparent *within* dense clumps. Therefore, to determine whether competition among plants is in fact affecting their pattern, it is necessary to limit the sampling to regions where the density is high. At the same time, one must not prejudge the outcome by collecting only the evidence that supports one's hypothesis and ignoring any that does not.

A way out of the difficulty is to measure a sample of plant-to-nearest-neighbor distances that are shorter than some upper limit chosen beforehand. As before we shall be concerned with the square, ω, of each measured distance. Put c for the maximum value of ω to be included in the sample; i.e. $\sqrt{c}$ is the chosen upper limit of distance to be measured. We can compare the sample of observed values with the results that would be expected if there were no tendency for the plants to be evenly spaced because each needs a territory.

The test is performed as follows. (Only a recipe is given here; the derivation will be found in Pielou, 1962a). Imagine that the plants concerned are forest trees.

First a sample of distances, each less than the chosen upper limit, $\sqrt{c}$, must be obtained. Tree-to-nearest-neighbor distances (as opposed to point-to-nearest-tree distances) are what we now require and therefore a set of trees must be selected at random to serve as base trees from which to measure the distances. It should be noticed that it is *not* satisfactory to choose as a base tree the tree nearest a random point. This would give a sample of trees biased in favor of comparatively isolated ones. To see this, consider the clumped pattern D in Figure 7.1 (page 142). A point placed at random in the area is more likely to fall in an empty region than within a clump; consequently, a tree at the edge of a clump is more likely than one in the interior of a clump to be the nearest tree to a random point and the trees are therefore not all equally likely to be chosen. To make a truly random selection of base trees, we must ensure that every tree in the population has an exactly equal chance of being chosen. This could be done by marking all the trees in the area concerned with identifying labels and then choosing a sample of the desired size by picking from a hat containing a duplicate set of the labels. However, this is laborious and if the total population of trees is not too large it is probably simpler to measure *all* the tree-to-neighbor distances that are less than $\sqrt{c}$. Each measured distance is squared to give a value of ω.

Now suppose the sample of ω values is of size n, and that their mean is $\bar{\omega}$. First it is necessary to solve the following equation to obtain k.

$$\frac{\bar{\omega}}{c} = \frac{1}{k} - \frac{e^{-k}}{1 - e^{-k}}. \tag{7.7}$$

(The solution must be obtained by successive approximation; an example is given below.)

Next, calculate ω_l from

$$\omega_l = \frac{2.3026c}{k} \{1 - \log_{10} (9 + e^{-k})\}. \tag{7.8}$$

Then it can be shown that the expected number of ω values less than ω_l, if the pattern is random, is $n/10$. Let us put n_l for the observed number that are less than ω_l. If n_l is unexpectedly small, i.e. if the observed number of short tree-to-neighbor distances is surprisingly low, one would reject the hypothesis that the pattern is random and accept the alternative hypothesis that the trees are regularly spaced.

Now it can be shown that if we define a test statistic z, where

$$z = \frac{n - 10n_l}{3\sqrt{n}},$$

then

$$\Pr\{z > 1.645\} = 0.05. \tag{7.9}$$

Thus if it is found from the observations that $z > 1.645$ we can say that the probability of obtaining such a result, if the pattern were random, is only 0.05; that is, the discrepancy between observation and expectation is significant at the 5% level and we can reasonably conclude that the trees are interfering with one anothers' growth in a way that induces a regular pattern. For the test to be reliable it is necessary that n should not be less than 50.

As an example, consider the following observations (from Pielou, 1962a). They were made in semi-arid country in British Columbia in mixed woodland of Ponderosa pine and Douglas fir *(Pinus ponderosa* and *Pseudotsuga menziesii)*. The population studied consisted of all trees of these two species with trunk diameters of 10 cm or more (at 1.5 m above the ground) growing in an area of 1.6 hectares (about 4 acres). The population was noticeably clumpy but it was suspected that the pattern was like that in Figure 7.5 with fewer very short distances than would be expected in a random pattern of the same density. The population contained 220 trees and all of them (as opposed to a sample) were treated as potential base trees. It was decided to ignore tree-to-neighbor distances greater than 2 m, i.e. it was decided to put $c = 4.00$. Only 99 of the trees had nearest neighbors closer than 2 m and thus $n = 99$. The mean of the squares of the 99 admissible distances was

$$\bar{\omega} = 1.5591 \quad \text{and hence} \quad \bar{\omega}/c = 0.3898.$$

11*

We must now obtain k from equation (7.7). Appendix 7.4 gives a table of $\left\{\dfrac{1}{k} - \dfrac{e^{-k}}{1-e^{-k}}\right\}$ for a series of values of k and it will be found from the table that when $\bar{\omega}/c = 0.3898$, k lies between 1.3 and 1.4. Its value to two decimal places can now be obtained easily by trial and error. It will be found that

$$\frac{1}{k} - \frac{e^{-k}}{1 - e^{-k}} = \begin{cases} 0.3900 & \text{when} \quad k = 1.36 \\ 0.3893 & \text{when} \quad k = 1.37. \end{cases}$$

Therefore we take $k = 1.36$.

Next, ω_l is obtained from equation (7.8). It is

$$\omega_l = \frac{2.3026 \times 4}{1.36} \{1 - \log_{10} 9.2567\}$$

$$= 0.227.$$

Inspection of the original field observations showed that the number of ω values less than 0.227 was $n_l = 2$. This is considerably less than the expected number $n/10 = 9.9$.

We therefore calculate the test statistic z which is

$$z = \frac{99 - (10 \times 2)}{3\sqrt{99}} = 2.647.$$

Since $z > 1.645$ we conclude that where the density of the trees is high they are regularly spaced; the most likely explanation for this is that they are competing for soil moisture.

The test just described makes possible the detection of regular spacing among close-growing plants even when their density is variable over the region being studied and their pattern, as a whole, is therefore aggregated. The recognition of regular spacing when it occurs is of great ecological interest. It may result from one of two causes. Either the plants are competing for some limiting resource; probably this is the usual cause of regular spacing among forest trees. Or the cause may be *autotoxicity* (also called *homologous inhibition*). Plants are said to be autotoxic when they poison the soil in which they grow so that seedlings of their own species cannot become established close to them. The phenomenon is not common but more and more autotoxic species are being found and a review by Garb (1961) lists several.

Autotxicity is to be contrasted with *allelopathy* (or *heterologous inhibition*) which occurs when plants inhibit seedlings of other species than their own from growing in their vicinity. Both autotoxicity and allelopathy can have pronounced effects on vegetation patterns and some manifestations of allelopathy will be mentioned in Chapter 8. Here we are concerned with the regular patterns that autotoxicity can cause in stands of a single species. The apparently regular spacing often found among desert shrubs has in the past been ascribed to competition for water. Although this may be the explanation in some cases, there is increasing evidence that autotoxicity (and allelopathy as well) is common among the strongly aromatic xerophytic shrubs that are characteristic of arid country in the southwestern U.S.A. Muller (1966) describes the phenomenon in two abundant species of the California chaparral, *Salvia leucophylla* and *Artemisia californica*. It is frequently observed that in pure stands of either of these species young plants do not grow among mature ones so that full-grown and overmature shrubs are well spaced and separated by bare ground. Laboratory experiments have shown that established plants severely inhibit seedlings of their own species.

For adult members of a species to inhibit the growth of the succeeding generation might be thought extremely disadvantageous to the welfare of the species. In these shrubs seedling growth is inhibited by the heavy vapours of terpenes which are given off by the aromatic foliage of established plants and adsorbed onto colloidal soil particles. Muller believes that production of these toxins is beneficial to the plants notwithstanding that nearby seedlings of the same species are poisoned. He argues that plants must excrete their metabolic waste products and these desert shrubs do so by converting them to volatile byproducts which are diffused into the atmosphere; in this way toxic metabolites are eliminated instead of being permitted to accumulate in the plant's tissues. In his view excretion is the chief function of the terpenes produced by aromatic plants; if they also poison competing plants and repel browsing animals, these are additional, but subsidiary advantages.

It is interesting to notice that the majority of plants known to be autotoxic are xerophytes of arid country. If the production of easily diffusible volatile material is an efficient way for a plant to excrete its waste products, the question arises as to why the method appears to be uncommon among mesophytes. One possible explanation is this: in very dry ground each individual in a population of shrubs needs a large volume of soil if it is to obtain enough water. Therefore the shrubs must be widely spaced if they are to survive. But when they are widely spaced for this reason, there is no disadvantage to a plant's making its immediate surroundings uninhabitable by contaminating them with its own toxic wastes. Another possibility is that excretion of toxic wastes by mesophytes is not an uncommon phenomenon but that the wastes are soluble in water (unlike the insoluble terpenes produced by desert plants) and are leached out of the soil by rainwater. This is speculation, however. There is still much to learn about how plants have evolved to make the optimum use of available space, and of how this is reflected in their patterns.

CHAPTER 8

Spatial Patterns II
The Patterns of Mosaics and Maps

STUDIES ON THE spatial patterns of populations have shown that the great majority are aggregated. Regular and random patterns are comparatively uncommon. Thus, more often than not, when a population consists of separate individuals scattered throughout a continuous area, the density of the individuals varies from place to place. There may be a number of more or less well-defined clumps of individuals, as in pattern D in Figure 7.1 (page 142); or an irregular patchwork of subareas within each of which density is fairly constant but between which it varies markedly (e.g., all the patterns in Figure 7.3). Or there may be scattered clumps with densities high at their centers and low at their peripheries, so that over the whole area the density variations are gradual. In this chapter we shall regard the clumps or patches as the entities whose pattern is to be studied. A map of a pattern is to show, not separate dots on a blank background (as in Figures 7.1, B, D, and F and Figure 7.3) but a mosaic of differently colored patches in which each density is represented by a different color. In some cases such a map would resemble that of an archipelago of scattered islands; in other cases the map would be more like one showing the political subdivisions of a country. And if the density variations were gradual they would be shown as are the altitude zones on a contour layer-colored map. A map of any of these kinds can be called a *mosaic*. A mosaic map, by definition, is a pattern consisting of two or more different *phases*, each distinguished by a particular color. Any phase may occur as separate fragments, or as a background surrounding the fragments of other phases; indeed, it may constitute "islands" in some parts of an area and "background" in other parts. The salient characteristic of a mosaic is that, taken together, the pieces of the different phases cover the whole area; and they do not overlap. Overlapping of phases is avoided by defining them in such a way as to make overlap logically impossible. It then follows that every point in the area must belong to one and only one phase. It is convenient to use the word *patch* for each of the separate pieces of a mosaic, background as well as islands; all the patches of one kind collectively form a phase.

Mosaics occur in nature in enormous variety. We have embarked on a discussion of them by mentioning the mosaic pattern exhibited by a plant species which occurs in patches of different density; but this is only one of the many kinds of natural mosaics that concern ecologists. Most ecological spatial patterns are mosaics. It will be worth while to list some examples if only to demonstrate the diversity of contexts in which mosaics may be studied.

Consider vegetation first. We have already remarked that even when the plants of a particular species seem at first sight to occur as scattered individuals, the pattern may really be a mosaic in which the phases are distinguished by having different densities of individuals. An example has been described by Cooper (1961) who investigated the pattern of forests of *Pinus ponderosa* in northern Arizona. He found that these pine forests were mosaics of even-aged groups of trees, each group occupying a patch of ground averaging one-fifth acre in area. The younger trees were, of course, smaller individually and present at greater densities than the older ones. The probable explanation for the mosaic pattern is this: in groups of full-grown pines, highly flammable materials (dead pine needles, twigs and branches) gradually accumulate until a natural fire kills the trees. Fires of small extent, started by lightning, are apt to recur regularly under natural conditions and they remove groups of mature and over-mature trees leaving patches of bare mineral soil exposed to full daylight. Each bare patch forms a seedbed for an even-aged group of seedlings; since ponderosa pine seedlings are very intolerant of shade they can become established only in such openings and not beneath the canopies of older trees. Obviously this mechanism permits indefinite persistence of even-aged groups of trees, representing several different age classes. The pattern is self-perpetuating. At any one moment each patch represents a different successional stage; and with the passage of time each patch goes through the whole gamut of stages from youth to old age.

Plants that form extensive colonies, or vegetative clones, are very common. One such clone has more of the properties of a population than of an individual, as Harper (1968) has stressed. A single seedling, of a species that can reproduce vegetatively, has the capacity for exponential increase in the number of its shoots and there is no fixed limit to the size it may attain. Sometimes a vegetative clone may appear to be formed of scattered individuals, as happens if the organic links between shoots are long segments of root or rhizome concealed below the ground or if they are slender stolons that easily break and wither; and sometimes its clonal nature may be obvious, as is true, for example, of alpine cushion plants. Further, vegetative clones exhibit an enormous range in size, from the lichen patches on a rock to clones of poplar trees (*Populus* spp) with areas measured in acres. For a given species the geometry of its clonal mosaic pattern may be fairly predictable and will depend on the mode of growth of the plant concerned. Many detailed investigations of these "morphological patterns" have been made, especially by Greig-Smith and Kershaw and their followers (for examples see Greig-Smith, 1964, Kershaw, 1964, and their bibliographies).

To mention an example, Kershaw (1958) described the morphological pattern of the grass *Agrostis tenuis*, growing on mountain slopes in North Wales. Excavations of the underground parts showed that the rhizomes branched and rebranched several times; consequently the ultimate clumps of above-ground shoots growing from their tips were themselves clumped into superclumps, which in their turn formed super-superclumps, and so on. The number of hierarchical levels that such a pattern of nested clumps can exhibit depends on the relative speeds at which new clumps grow and old clumps either die or become unrecognizable because their territory is invaded by shoots from other clones.

Clones of some species keep their distinctness for long periods. McNaughton (1968) has shown this to be true of the cattail *Typha latifolia*. He noticed that seedlings are very rare in cattail marshes, where the ground is covered by layers of dead, incompletely decayed plant material, and this led him to test whether the plants are autotoxic (see page 164). Laboratory experiments showed that *Typha* seedlings failed to develop when watered with aqueous extracts of dried *Typha* leaves, and that the main toxic agents are phenolic compounds. Thus autotoxicity precludes successful invasion of an established clone by seeds from other clones, and each clone can maintain its individuality for as long as the area remains hospitable to the species, that is, for as long as the soil surface is close enough to the water table for a cattail marsh to flourish. Since the clones have recognizably distinct morphological characteristics, several contiguous clones present a clear mosaic pattern, all of one species.

Chemical inhibition by one species of others, or allelopathy (see page 164) can lead to the formation of mosaics of more than one species. Desert shrubs provide particularly good examples. Muller (1966, 1970) has described how xerophytic shrubs of the California chaparral, notably *Salvia leucophylla* and *Artemisia californica*, form pure stands in which seedlings (of their own and other species) are inhibited by the toxic exudates of the adult plants. We have already discussed (page 165) how this can lead to regular spacing among full-grown shrubs. It also leads to the establishment and persistence of pure stands, each of which is a patch in a vegetation mosaic. Around these stands the other mosaic patches form concentric zones of, first, bare ground where nothing grows, then a zone of stunted grasses, and finally a zone of thriving grasses. The bare ground phase is particularly characteristic of these mosaics. Bartholomew (1970) has argued that the bare zones that typically separate patches of shrubs from patches of grassland in the chaparral do not necessarily owe their existence to plant toxins released by the shrubs. An alternative explanation is that the bare zone is denuded by animals, chiefly rodents, rabbits and birds, which find cover under the shrubs and feed on, and destroy, all seeds and seedlings in a zone immediately adjacent to the shrubs. Bartholomew showed that when herbivorous animals are experimentally excluded from the bare zone, annual plants can successfully grow there; the relative importance of animals and chemicals in maintaining the zone is unknown.

McPherson and Muller (1969) have investigated another phase of the chaparral mosaic, that in which the vegetation consists of herbaceous dicots. These grow in patches where the ground has been artifically disturbed (e.g., in road cuttings and firebreaks) and where it has been laid bare by fire. The shrub-free patches, unless artifically maintained, are suitable for herbs only temporarily. When the shrubs (in this case *Adenostoma fasciculatum*) grow back, their toxic exudates poison the herbaceous species.

Although present evidence suggests allelopathy is commonest among plants of arid regions, other examples are known. Mergen (1959) found that the introduced shade tree *Ailanthus altissima* (Tree of Heaven) is allelopathic. He showed that an aqueous extract of its leaves inhibited growth in seedlings of many tree species, among them pines, spruce, fir, larch and birch. Pure stands of *Ailanthus* occur fairly frequently in New York and Connecticut, often beside highways and along the borders of fields, and these stands preserve their distinctness by virtue of their allelopathy. If it were not for the toxic exudate, the *Ailanthus*, whose seedlings are intolerant of shade, would presumably be succeeded by more shade-tolerant species.

If mosaic patterns can be formed where successional replacement is prevented (as in *Ailanthus* stands), they can also appear as a consequence of succession. One of the examples already discussed (the ponderosa pines) falls in this category. The pattern may be called a cyclical mosaic. At any one time all the stages of a successional cycle are represented in an area by several separate patches of vegetation; and if we could follow the course of events at a given point on the ground for a long enough time, the vegetation at the point would be found to go through the whole successional cycle. Cyclical mosaics may involve more than one species. The phenomenon was originally described by Watt (1947) who discussed several examples. A typical one, found in Scottish mountain moorlands, is a three-phase mosaic of *Calluna vulgaris* (heather), *Arctostaphylos uva-ursi* (bearberry) and the lichen *Cladonia silvatica*. One phase consists of young, vigorous shoots of *Calluna;* the second phase is made up of patches where old, dead *Calluna* stems, behind the young shoots which form an advancing "front", support a growth of *Cladonia*. In time, the *Cladonia* mats disintegrate, leaving bare soil where the third phase, *Arctostaphylos*, can establish itself. The *Arctostaphylos* patches cannot persist indefinitely however, since they in their turn are overtopped, shaded and crowded out by spreading *Calluna* branches.

In a more recent discussion of *Calluna*-dominated heath mosaics Barclay-Estrup (1971) has described the way in which individual *Calluna* plants of different ages and sizes affect the substrate on which they grow and the microclimate in their immediate neighborhood; through these factors they influence the further progress of the cycle.

A cyclical mosaic has been described by Anderson (1967) in which the four phases are not pure one-species patches but four different microcommunities, each with several species. He found these mosaics growing on gravelly raised beaches in northwestern Iceland. All four phases contained *Dryas octopetala*, but in different amounts relative to the other species, among which were the sedge *Kobriesia myosuroides*, the grass *Agrostis canina* and two species of birch, *Betula nana* and *B. pubescens*. The fact that the mosaic was cyclical was shown by the discovery of dead but undecayed remains of plant parts such as the leaves of dwarf birch *(B. nana)* in the soil below living plants of succeeding phases.

There has been speculation that unrecognized cyclical mosaics are common. They are likely to be positively identified only where drought, a cold climate, or acid soil ensures the preservation of dead plant parts; only in these circumstances is one likely to find remains of the successive phases forming a sequence of layers at one spot. Another possibility is that cyclical mosaics can develop only where climatic conditions are harsh enough to ensure the periodic recurrence of patches of bare soil. This will happen if one of the phases (or successional stages) is such that at their deaths the plants comprising it are destroyed and removed, for instance by frost heaving and the blowing away of dried remains by strong winds; once bare soil is exposed the cycle can start again with the pioneer phase.

The mosaics just described can develop in areas of completely uniform habitat and owe their existence to interactions among several different plant species, or among age classes within one species. Mosaics that are caused by variation in abiotic environmental factors are probably much commoner. Any plant ecologist can think of examples from areas he knows. A freshwater marsh is a familiar example; the patches of different species of emergent plants correspond with the contrasting habitat patches afforded by water of different depths. Extensive seaweed beds on sheltered shores often form mosaics; freshwater streams from the land, and hollows where rainwater can collect, create a pattern of pools of various salinities, and these support a mosaic of patches of different algal species. Mosaic patterns of greater or lesser clarity are discernible in almost any tract of vegetation that is sufficiently low-growing for a large expanse of it to be visible at one time. Obvious examples are bogs, moors, beaches, grasslands and mountain meadows. Forests and other tall vegetation are found to exhibit mosaic patterns when seen from the air.

Mosaic patterns are not confined to terrestrial vegetation. The density of the marine phytoplankton varies enormously from place to place producing discolored patches, and hence mosaic patterns, on the surface of the sea. Bainbridge (1957) has reviewed the subject. He quotes Captain James Cook who, in an account (dated 1773) of his first great journey, described yellow streaks on the ocean caused by plankton. A nineteenth century description mentions "banks" of diatoms that gave the sea a slate-green color. Bainbridge also gives maps of plankton mosaic patterns showing that the component patches are usually long ellipses, of the order of 40 miles long by 10 miles wide. Examination of water samples shows that the organisms producing them are predominantly flagellates and diatoms, typically at densities of about three per cubic millimeter. Probably the patches mark areas where there are upwellings of nutrient-rich waters from great depths into the sunlit layer of the sea.

In the foregoing paragraphs we have, for the most part, discussed mosaics in which the boundaries of the component patches are clearcut and unmistakable. Many patterns of ecological interest have phases that merge into one another gradually with no abrupt discontinuities. However, it is convenient to regard these patterns, too, as mosaics. We shall, in fact, treat as a mosaic any pattern that results from the possession of different properties by different regions, whether or not the regions have sharp boundaries. This is a broad definition; it includes all patterns except those described in Chapter 7 in which individuals were treated as dimensionless points scattered on a plain background or confined to discrete natural sampling units.

The methods to be described in the following pages can be used for analysing, not merely ecological patterns in the narrow sense, but all maps having the form of mosaics. Maps of various kinds are often, of course, the first products of ecological field work, so that to make a distinction between maps and patterns is artificial. Maps of ecological communities, of soils, of habitat types, and of sea bottom materials, to mention a few examples, are all of mosaic form, and it is easy to extend the list.

We must now consider what sort of ecological investigations require the analysis of mosaic patterns. Given a mosaic map, what are the questions to be asked and how are answers to be obtained? More problems suggest themselves than have yet been tackled. Let us look at some of them.

First, consider "morphological pattern". As already remarked (page 168) it seems likely that the mosaic patterns of the clones of many vegetatively reproducing plants are properties of the species concerned, in much the same way that a tree of a particular species has a characteristic shape. It would be interesting to know how consistent a given species is in always exhibiting the same pattern, and how the pattern varies phenotypically in response to environmental influences.

Patterns are not static and the way vegetation patterns change during the course of succession suggests many challenging lines of research. In earlier chapters we discussed at some length the fluctuations in size and age composition of animal populations. Plant populations are equally subject to fluctuation, and because plants are sessile we ought to observe not only changes in numbers but also changes in pattern. Thus, when a plant population is increasing, do established clones expand their areas vegetatively, or do seedlings become established at previously unoccupied sites and initiate new patches? Similarly, what happens to patterns when a population is declining?

We should also expect the pattern of a species to depend on whether it is competing, successfully or otherwise, with other species in its community. It would obviously be interesting to know whether a plant species is occupying all the ground that is suitable for it and, if not, whether it is being excluded by competitors, or is being continuously destroyed by herbivores which may be vertebrates or insects.

In earlier pages, self-perpetuating vegetation mosaics were described that are formed entirely as a result of interactions among the constituent plants, and which can therefore develop in wholly homogeneous environments. They may be called *endogenous* in contrast to *exogenous* mosaics in which the phases are formed in response to habitat differences within an area. Cyclical mosaics are endogenous and so are those resulting from allelopathy. The patterns of exogenous mosaics are, by definition, governed by abiotic factors, but endogenous mosaics would be expected to have their own "geometry" so that separate examples of the same collection of phases would have the same mosaic pattern. Whether this is so has yet to be explored.

Maps are as much the concern of biogeographers as ecologists. Indeed, separation of the two disciplines is artificial and unfortunate. Geographical range maps are mosaics. The degree to which their patches (representing presence of the species concerned) are fragmented has obvious evolutionary consequences.

Vegetation mosaics interest zoologists as well as botanists. A fine-grained mosaic, in which the patches are small and closely spaced, provides a proportionally greater amount of "boundary habitat" in a given area than does a coarse-grained mosaic. The description *boundary habitat* connotes a zone in which animals have two contrasting vegetation phases to move between and it is often found that such zones support a greater number of animal species than do uninterrupted tracts of a single type of vegetation. One example has already been described (page 169); it will be recalled that a variety of birds and rodents find cover in the shrub phase of arid-land chaparral and go on foraging sorties from this sanctuary into adjacent grassland where food is available but protective cover absent. In any geographical region the border of a forest, where it abuts on grassland, is likely to be richer in species of birds and mammals than pure forest or pure grassland. Likewise in lakes, the borders of weed patches are particularly likely to harbor fish. Thus the number of different animal species an area can support is closely dependent on the pattern of the vegetation mosaic.

Finally, there are often important practical reasons for studying mosaic patterns. For instance, if an area is to be artificially revegetated and one wishes to establish a continuous cover of a desirable species, it helps to know how quickly and how dependably the artificially planted individuals will develop into patches that expand and coalesce. There are many contexts in which such knowledge is valuable, for example in plantings or sowings for range improvement, for dune stabilization, for covering bare areas such as abandoned strip mines, and for revegetating regions denuded by fire, flood or herbicides. Observing the spread of pests and diseases in forests also entails the study of mosaic patterns. Many forest pests cause the foliage of affected trees to be conspicuously discolored so that in remote areas the patterns of infected patches are easily observed from the air or on air photos. It may be of great practical importance to be able to predict whether, if the pest spreads, it will do so by enlarging its existing patches or by starting new ones.

We next consider some of the ways in which mosaic patterns are analysed.

1 THE ANALYSIS OF DATA FROM ROWS OF QUADRATS

Suppose we wish to investigate the pattern of one of the species in a vegetation mosaic and the species is present at different densities in different phases. It is assumed that the density contrasts are not sufficiently pronounced for the patches to be recognizable on inspection. We can sample the mosaic by observing the amount of the species in each of a number of sampling quadrats and, since conclusions about patterns are sought rather than merely an estimate of the total amount of the plant in the study area, the quadrats are to be systematically arranged instead of being placed at random. One way to proceed is to range the quadrats contiguously in a single long row; another is to place them in a rectangular array or grid. In what follows, methods of analysing data from a row of quadrats, in effect a belt transect subdivided into short sections, will be discussed in detail.

To begin with, of course, the data have to be collected. There are several ways of measuring the quantity of a plant species in a quadrat, depending on the plant's habit of growth. Two frequently used ways are to count the shoots, or to determine the *cover*. Cover is the areal proportion of the ground within the quadrat that lies directly below some part of the plant and its value can be estimated by eye or by "point sampling". The latter is done by observing a sample of true geometric points on the ground and noting the proportion that are covered. As to practical details, one method is to place a frame holding a number of retractable pins over the quadrat and, as the pins are lowered one by one, to note whether or not they touch some part of the species; alternatively, a sighting device with a telescope and crossed hairs can be used (Morrison and Yarranton, 1970).

The data provided by these observations consist of a list of the amounts of the species in each of a long row of contiguous quadrats. We shall describe three ways of analysing data of this kind: (1) Analysis of variance in blocks; (2) Counts of turning points; (3) Correlogram analysis.

1   Analysis of variance in blocks

This method of pattern analysis was first introduced by Greig-Smith (1952). He originally applied it to data from a square or rectangular grid of contiguous quadrats and subsequently Kershaw (1957) pointed out that it could also be used with data from a single row of quadrats.

The method is now widely used and for that reason it is described in detail here. It should be remarked, however, that it is debatable how the results of the analysis should be interpreted. At best the conclusions are uncertain and it is a waste of hard-won observations not to analyse them in other ways as well.

The argument underlying the method is this. Suppose the density of the plants, measured in whatever way is most appropriate to the species, has different values in different mosaic patches. Then, if we measure the quantity of the plant in each of a pair of contiguous quadrats and the quadrats are small relative to the average patch size, the variance of the two quantities will tend to be small. Probably most such pairs will lie wholly within one mosaic patch so that the quantities of the plant in the two quadrats of the pair are unlikely to differ very much.

If contiguous pairs of larger quadrats are used, the likelihood is greater that one member of the pair will lie in a mosaic patch with one density value, while the other lies in a patch with a different density value. As a result the variance of the two quantities relative to their mean is likely to be greater. In fact, the variance: mean ratio, V/m, would be expected to increase steadily with increasing quadrat size, until the area of the quadrats is equal to the average area of the mosaic patches; (or, until the length of the quadrats' sides is equal to the average diameter of the patches). Thereafter, further size increases will be accompanied by a fall in the V/m ratio because every individual quadrat will tend to sample parts of two or more patches; when this happens the contrast between the quantities in two contiguous quadrats will decline.

The analysis therefore consists in obtaining V/m ratios for quadrat pairs of a sequence of sizes. The raw data, which are the plant quantities in each of a long row of quadrats, are to be made to yield these ratios in the most economical manner possible. For reasons that will become clear, the number of quadrats in the row should be a power of 2.

Now consider a numerical example. To keep the account brief, we take a row of only $8 = 2^3$ quadrats. Let x denote the quantity of the plant in an individual quadrat. These values are shown schematically below. The total quantity of the plant in the whole sample is $\Sigma x = 54$ and thus the mean per quadrat is $m = 54/8$. The diagram shows the quadrats separated into pairs and below each is given the variance *of the pair*. Thus the variance of the first pair of values, which are 8 and 12 with mean $\bar{x} = 10$, is

$$\tfrac{1}{2}\sum (x - \bar{x})^2 = \tfrac{1}{2}\{(8 - 10)^2 + (12 - 10)^2\} = 4.$$

Similarly, for the second pair, 11 and 3 with mean $\bar{x} = 7$, the variance is

$$\tfrac{1}{2}\sum (x - \bar{x})^2 = \tfrac{1}{2}\{(11 - 7)^2 + (3 - 7)^2\} = 16;$$

and so on. The contents of the eight quadrats and their pairwise variances are:

8	12		11	3		2	6		9	3

Variance of pair: 4   16   4   9

The average variance (averaged over all four pairs) is

$$\frac{4 + 16 + 4 + 9}{4} = \frac{33}{4}.$$

Thus for these quadrats (or "units" as they are usually called), the variance: mean ratio is

$$\frac{V}{m} = \frac{33}{4} \bigg/ \frac{54}{8} = \frac{33}{27}.$$

To obtain a V/m ratio based on larger quadrats we now pool the contents of each pair of units and treat the data as coming from four "2-unit blocks". (The blocks are not square, of course.) Taking pairs of blocks, the same computational procedure as before is carried out, as follows:

20	14		8	12

Variance
of pair: 9   4

Average variance,

$$V = \frac{1}{2}(9 + 4) = \frac{13}{2}.$$

The mean, m, is now $\dfrac{\Sigma x}{4} = \dfrac{54}{4}$; i.e. it is twice as large as before since the "quadrats" (now 2-unit blocks) are twice as large. For quadrats of this size the variance: mean ratio is therefore

$$\frac{V}{m} = \frac{13}{2} \bigg/ \frac{54}{4} = \frac{13}{27}.$$

The third and last step is analogous; (when there are 2^n units in the row, there are n steps to the computations) Combining the 2-unit blocks in pairs to give 4-unit blocks, we now have a single pair of blocks, as follows:

34	20

Only one variance value is obtainable and it is $V = 49$. With quadrats of this size the mean is $m = 54/2$. Thus the variance: mean ratio becomes

$$\frac{V}{m} = 49 \bigg/ \frac{54}{2} = \frac{49}{27}.$$

Summarizing, the results are as follows:

Block size	Average variance between pairs	Mean	V/m	Relative V/m
1 unit	33/4	54/8	33/27	33
2 units	13/2	54/4	13/27	13
4 units	49	54/2	49/27	49

The last column gives the relative values of the successive V/m ratios in whole numbers. We have now attained our original objective, the *relative* sizes of the V/m ratio for a sequence of quadrat sizes, the sequence consisting of units, 2-unit blocks and 4-unit blocks. If we had used a row of 2^n units originally, the largest blocks would have been of 2^{n-1} units.

The method of computation described above is laborious and has been used here to show the rationale of the analysis. In practice the same results can be achieved far more simply by an analysis of variance. The steps are shown in Table 8.1.

TABLE 8.1   An analysis of variance in blocks (The data are those used for the analysis described in the text)

Units per block n	$\dfrac{\Sigma x^2}{n}$	Sum of Squares	Degrees of freedom	Mean square
1	468	66	4	16.5
2	402	13	2	6.5
4	389	24.5	1	24.5
8	364.5			

The first column gives the block size, n, that is, the number of units per block. The second column gives the sum of squares of the quantity in each block divided by the block size. The third column, headed "Sum of Squares" shows the differences between successive entries in the second column. The number of degrees of freedom is shown in the fourth column; it is the number of pairs of blocks of the size concerned; (we treat single units as 1-unit blocks). The final column, headed "Mean Squares", is obtained by dividing each sum of squares by its degrees of feedom. It will be seen that these values are in the proportions $33 : 13 : 49$, as before. Thus the mean squares yielded by the analysis of variance are in the same proportions as the V/m ratios and these proportions are what we require.

The final results are usually displayed in the form of a graph of mean square versus block size. Two examples are shown in Figure 8.2a (page 182). These are the mean-square-versus-block-size graphs for the two sets of data (artificial) shown in Figure 8.1 (page 181). Each graph in Figure 8.1 represents the quantities of an imaginary species in every quadrat of a row of 64 contiguous quadrats; the quantity of the plant is shown on the ordinate and the position of the quadrat on the abscissa. These data have been devised to demonstrate the different results obtained with different methods of analysis.

It is clear from the two graphs in Figure 8.2*a* that the two populations yield very similar results when the data are subjected to an analysis of variance in blocks. Notice that the mean squares have been plotted on a logarithmic scale in order to accomodate the wide range of values observed (2.3 to 2300). Also, the successive block sizes, of 1, 2, 4, 8, 16 and 32 units, are equally spaced on the abscissa.

Greig-Smith (1952), Kershaw (1957) and other ecologists who favor this method of pattern analysis maintain that the block size at which a peak in the graph occurs is equal to the mean size of the clumps comprising the pattern. If there is a second peak, it is postulated that there are superclumps (clumps of clumps) whose mean size is given by the block size corresponding to this second peak; and analogously for additional peaks.

However, conclusions as definite as these seem unjustified. Any non-random pattern will inevitably yield different values of V/m at different block sizes, so that peaks and troughs in the graph are bound to appear. Also, naturally occurring clumps are likely to be so variable in size and shape and sharpness of boundary that a single number (the block size corresponding to a peak) cannot be expected to convey much information about them. Goodall (1963) has also pointed out that the greater the distance (center to center) between two quadrats, the more dissimilar they are likely to be and hence the greater the variance of the two quantities associated with them. Now, when adjacent blocks are combined in pairs to form new blocks double the size of the old ones, the distance between their centers is also increased. Thus an increase of size automatically entails an increase of distance and it is impossible to determine the relative importance of these two factors in bringing about changes in the V/m ratio.

The foregoing arguments show that a mean-square versus block-size graph is difficult to interpret. In particular, it seems unwarranted to conclude that a peak in the graph provides direct and unambiguous evidence for the existence of clumps whose mean diameter is that of the block size corresponding to the peak. However, there is undoubtedly some relationship, difficult though it may be to define, between the shape of the graph and the characteristics of the pattern. In very general terms, one would expect the first peak in the graph to occur at a smaller block size in a fine-grained pattern than in a coarse-grained one, so if two patterns are to be compared in respect of their grains this form of analysis is worth trying. But, as already remarked, given the data from a row of contiguous quadrats, it is wasteful not to analyze them in additional, equally informative ways besides doing a mean-square versus block-size analysis.

Before discussing these other methods an important point must be mentioned. Although the results of a pattern analysis of the kind we have just described can be tabulated as an analysis of variance (as in Table 8.1) it is futile to calculate variance ratios and to try to determine their "significance". This is because the individual measurements used in the analysis are not independent of one another as they are in an ordinary analysis of variance. The list of n numbers that results when one measures the amount of a plant in n contiguous quadrats should not be thought of as n observations but as *one* observation. The row of quadrats is a single entity, a sample of size one from an indefinitely large number of such rows that might have been observed. It is not a sample of size n. In view of this fact it is impossible to make a statistical comparison between one population and another when only one row of quadrats has been examined in each. To attempt this would be analogous to trying to judge whether, for example, the mean height of the trees in a ten-year old white pine plantation differed from the mean height of the trees in a ten-year old red pine plantation by measuring exactly one tree in each plantation. There would be no evidence available on the variability within a plantation. Similarly, observations on one row of quadrats reveal nothing about the variability within a pattern, nor on whether the row examined happens to be "representative".

There are two ways out of the difficulty. Either one may collect data from several rows of quadrats, in effect from a sample of such rows. Or one may treat the observations on a single row as one would any other single observation. A single observation is seldom of much interest by itself, of course. But if, for instance, one were investigating how the pattern of an alpine plant varied with altitude, say, a single observation at each of several different altitudes would give the data required. Indeed, it is probably true to say that in population ecology in general, one is more often interested in examining interrelationships among the properties of populations and communities than in testing whether two sets of observations could be regarded as coming from the same parent population. Such tests are not appropriate unless there are commonsense grounds for postulating the existence of a single "parent population" (either real or conceptual), and often there are none.

We now turn to other ways of analysing the results obtained from a row of contiguous quadrats. The remarks in the previous paragraph again apply. However long a row of quadrats may be, it yields only one vector (cf. page 367) of measured quantities, that is, one "observation".

2 Counts of turning points

This is a straightforward statistical method, quick and easy to apply, for judging whether or not the numbers in a sequence are independent. Obviously any list of numerical observations arranged in random order will have several peaks and troughs. An observation represents a peak if it exceeds the neighboring observations before and after it in the list, and a trough when it is smaller than either of its neighbors. Peaks and troughs together are called *turning points*. Now suppose the observations are not in random order but exhibit gradual trends, partly obscured, perhaps, by chance variation superimposed on the trends. Then the number of turning points will be less than would be expected if the ordering of the observations were completely random. This suggests that the difference between the observed and expected numbers of turning points might be useful as a measure of the strength and persistence of trends in the observations. To obtain such a measure we need to know the expected number of turning points in a randomly ordered sequence of n observations.

This expectation is easily found. It is clear that three consecutive observations are needed to define a turning point, so let us consider any such trio of observations. Their absolute magnitudes do not matter, only their relative magnitudes, and therefore we may put 1 for the smallest, 2 for the middle-valued and 3 for the largest. These three numbers can be arranged in six permutations and if there is no dependence among the observations, the permutations are all equally probable. They are

$$\left.\begin{array}{ccc} 1 & 2 & 3 \\ 3 & 2 & 1 \end{array}\right\} \text{ with no turning point}$$

and

$$\left.\begin{array}{ccc} 1 & 3 & 2 \\ 2 & 1 & 3 \\ 2 & 3 & 1 \\ 3 & 1 & 2 \end{array}\right\} \text{ each with one turning point.}$$

Thus the probability that three randomly arranged observations exhibit a turning point is $\frac{2}{3}$. Write τ for the number of turning points in a sequence of n observations*. It now follows that, since the first and last observations cannot be turning points, the expected number in a random sequence is

$$\mathscr{E}(\tau) = \tfrac{2}{3}(n - 2).$$

It can also be proved (Kendall and Stuart, 1966) that the variance of τ is

$$\text{Var}(\tau) = \frac{16n - 29}{90}.$$

---

* In practice, if two adjacent numbers in the sequence are equal, they are treated as contributing half a turning point each.

As a standardized measure of sequential dependence we may use

$$T = \frac{\mathscr{E}(\tau) - \tau}{\sigma_\tau} \quad \text{where} \quad \sigma_\tau^2 = \text{Var}(\tau).$$

Recall the sequence of eight numerical values listed on page 175. The list is reproduced here with the turning points labelled with asterisks:

$$8 \quad 12* \quad 11 \quad 3 \quad 2* \quad 6 \quad 9* \quad 3.$$

Thus

$$\tau = 3.$$

Also

$$\mathscr{E}(\tau) = \tfrac{2}{3}(8 - 2) = 4;$$

$$\text{Var}(\tau) = 1.1 \quad \text{and} \quad \sigma_\tau = 1.049.$$

Therefore

$$T = \frac{4 - 3}{1.049} = 0.953$$

and, as already remarked, T can be used as a measure of sequential dependence among the eight numbers.

Now consider ecological applications. Clearly, in a population in which density variations are smooth and gradual (as in A in Figure 8.1, for example)

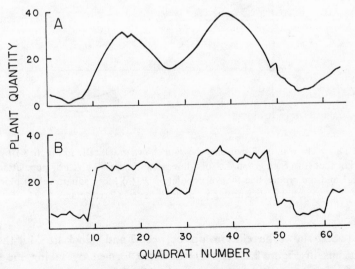

FIGURE 8.1 Each graph shows the data obtained by measuring the quantity of plant material in a row of 64 contiguous quadrats. The data (artificial) are used in the text to demonstrate different methods of pattern analysis

there is close dependence among the quadrats of a row when the quadrats are small and therefore close together; this dependence gets progressively less as larger, more widely spaced quadrats are used. In contrast to this is population B in Figure 8.1. In this case, though the density takes several different values the transitions from one value to another come as abrupt steps, and within any one patch density remains constant except for chance fluctuations. As a result sequential dependence is low even among small quadrats.

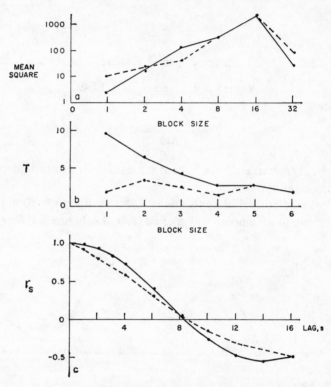

FIGURE 8.2   The results of analysing the data shown graphically in Figure 8.1 by three different methods. In each graph the solid line shows the result given by population A of Figure 8.1 and the broken line that of population B. (*a*) Mean square versus block size graphs; (*b*) Turning point analysis; (*c*) Correlograms

Figure 8.2*b* shows the relationships between T and "block size" for the two populations in Figure 8.1. The blocks were formed by taking the units singly to give 1-unit blocks; then combining them in pairs to give 2-unit blocks; then combining them in trios to give 3-unit blocks; and so on. The analysis does not require that block size be doubled at each step.

It is instructive to compare Figures 8.2a and 8.2b. It is seen that the turning point analysis is sensitive to a difference between the populations that is without effect on the variance analysis. Although the mean-square versus block-size graphs obtained from the variance analyses reveal that both populations have very variable densities, only the turning point analyses show how different the populations are in the gradualness of their density variations. This is not to argue that turning point analyses should be used in place of variance analyses since a turning point analysis gives no indication of the magnitude of the density variations in the way that a variance analysis does. But it is clearly worth while to do both analyses. The additional labor is negligible since both use the same data, and the two analyses complement each other.

Nor is there any reason to stop at two analyses. Any large body of ecological data can be looked at in a number of different ways, and the collecting of it is often motivated by the desire to answer a number of different, even if related, ecological questions. All the more reason, then, to analyse such data by several methods. There is no necessity, either, to confine oneself to methods that have been tried and described before. New problems are often best tackled by new methods devised especially for their solution.

We shall now discuss one more well-known method for treating the observations obtained from a row of quadrats.

3   Correlogram analysis

Serial correlation and the use of correlograms have already been discussed (page 89) in connexion with studies on wildlife cycles. It will be recalled that if we have a sequence of observations (whether from a sequence of times or a row of places obviously makes no difference) we can calculate a set of serial correlation coefficients. Thus the sth such coefficient, r_s, is an estimate of the average amount of correlation (and hence of interpendence) between any observation and the observation s units beyond it in the sequence. If r_s is calculated for several values of s (which is known as the *lag*), the results can be plotted as a correlogram. Figure 8.2c shows the correlograms for the two populations in Figure 8.1. They are not very different. This accords with what we should expect since the populations resemble each other in their major features.

It will be noticed that for lags from 8 to 16 (the largest lag for which the serial correlation coefficient was calculated) r_s is negative. This may seem, at first sight, to contradict the argument on page 89 where it was stated that negative serial correlations can occur only when there are regular "oscillations", or, in the present context, recurrent clumps of constant size at regular intervals. This is obviously not true of the populations in Figure 8.1.

Figure 8.3 explains the apparent discrepancy. In each of the figure's three sections is shown, diagrammatically, a transect through an imaginary plant population. The constant height of the hatched "clumps" signifies the presence of a constant amount of the plant within each unit of length (i.e. each quadrat of the row). The gaps between the clumps are empty. However, these simplifications do not affect the generality of the discussion that follows. Below each of the three diagramatic representations of a population's pattern is shown the corresponding correlogram.

It should be noticed, first, that population A has clumps of constant length separated by gaps that are also of constant length. It is assumed that the pattern continues in the same way indefinitely ("to infinity"). The popula-

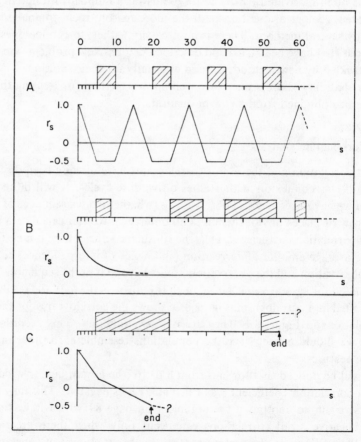

FIGURE 8.3   Three types of spatial patterns and the correlograms obtained from them

tion's perfectly regular pattern gives rise to the perfectly regular cyclical variation of its correlogram, shown immediately below it.

Next, consider populations B and C. Both populations have clumps and gaps of randomly varying lengths, but they differ from each other in one important respect. It is assumed that B, like A, continues as shown indefinitely; in other words, that though the length of any individual clump is unpredictable, each length is a variate picked at random from a precisely defined probability distribution. Thus one can specify exactly the probability that a patch will be of a given length (more precisely, of a length in a given interval); and likewise for the lengths of the gaps. In the case of pattern C, on the contrary, nothing at all is known or guessed about how the pattern continues beyond the last observed quadrat. Only one whole clump and part of a second are encountered in the row of quadrats actually examined; (the row ends at the 55th quadrat).

Now consider the correlograms.

Let x_i denote the ith observation in the sequence (i.e., the amount of the plant in the ith quadrat).

Let $\bar{x}$ be the mean of the x_i values.

And let $d_i = x_i - \bar{x}$ be the deviation from the mean of the ith value. Clearly some of the d_i are positive and some negative; their sum is $\Sigma d_i = 0$.

The sth serial correlation coefficient, r_s, (which is defined in Appendix 5.1) is proportional to a sum of products of the form $d_i d_{s+i}$. Thus, if the observed sequence consists of n observations, we have

$$r_s \propto \sum_{i=1}^{n-s} d_i d_{s+i}.$$

When s is small (i.e., when the quadrats are close together), it is likely that d_i and d_{s+i} will have the same sign; then, whether the sign is plus or minus, their product will be positive and hence r_s will be positive. As s increases, the interdependence between d_i and d_{s+i} obviously becomes progressively less and as a result r_s steadily dwindles as shown in the correlogram for population B.

It is worth examining in greater detail exactly what happens when s is moderately large in the case of population C. In this population the number of quadrats observed is $n = 55$. Quadrats 1 to 5 and 26 to 50 are in gaps and yield negative values of d_i; quadrats 6 to 25 and 51 to 55 are in clumps and yield positive values of d_i. Now let $s = 15$.

Then

$$r_s \propto \sum_{i=1}^{40} d_i d_{15+i}.$$

Let us look at the signs of the forty products in the sum. It is seen that long runs of positive products alternate with long runs of negative ones, thus:

$$r_s \propto (d_1 d_{16} + \cdots + d_5 d_{20}) + (d_6 d_{21} + \cdots + d_{10} d_{25})$$

negative   positive

$$+ (d_{11} d_{26} + \cdots + d_{25} d_{40}) + (d_{26} d_{41} + \cdots + d_{35} d_{50})$$

negative   positive

$$+ (d_{36} d_{51} + \cdots + d_{40} d_{55}).$$

negative

This sum turns out to be negative and consequently r_{15} has a negative value. However, it is clear that the sign of r_s is strongly influenced by the value of n, the length of the observed row of quadrats. This is because of the long runs of products of one sign that must occur alternately in the sum. Only when n is very large indeed can one assert that the contribution of positive terms to the sum will slightly exceed the contribution of negative terms and that the difference between the contributions will get less as s increases; that is, the relation between r_s and s will be as shown in correlogram B. When n is comparatively low, as in C, the portion of the pattern examined is too small to be "representative" and negative values of r_s occur in spite of the fact that the pattern is not cyclical like that of population A. To avoid misleading results in practice one should not use values of s, the lag, that exceed 10% of the total length of the data sequence.

Figure 8.4 shows two examples, taken from the ecological literature, of correlograms of the kind we have been discussing.

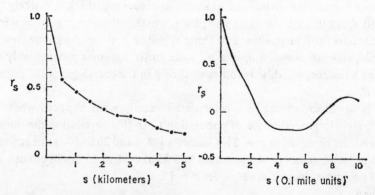

FIGURE 8.4 Two ecological correlograms. (*a*) From a mosaic formed by lakes and dry land (data from Matérn, 1960). (*b*) From a mosaic formed by patches of marine phytoplankton (redrawn from Platt, Dickie and Trites, 1970)

The correlogram in Figure 8.4*a* is based on the mosaic pattern formed by dry land and lakes in an area near Stockholm, Sweden. A full description will be found in Matérn (1960). He obtained the data from a map, on the scale 1 : 50,000, of a square area of 25 × 25 kilometers. The correlogam shown here is based on several transects across the map which were observed at points 10 mm apart (or 500 meters apart on the ground). Each point was assigned a "score" of 1 if it was on land and 0 if it was in water. This is obviously equivalent to sampling with a row of quadrats each 10 mm long on the map and recording a "measurement" of 1 or 0 according as the quadrat center is on land or in water.

The correlogram in Figure 8.4*b* is from Platt, Dickie and Trites (1970). The mosaic pattern it relates to was that of patches of marine phytoplankton in the waters of St. Margaret's Bay on the southwestern shore of Nova Scotia. The transect these authors examined was 8 miles long, and water samples were taken (from a depth of 10 meters) at sampling stations spaced at 0.1 mile intervals. Thus there were 80 stations. The quantity of phytoplankton in each sample of sea water was estimated by measuring the concentration of chlorophyll in the water; this is a method often used in oceanographic surveys for judging the "standing stock" of phytoplankton. From the various correlograms they plotted (Figure 8.4*b* is one example), Platt, Dickie and Trites inferred that, provided the lengths along the transect of the phytoplankton patches did not differ much from the lengths of the intervening gaps, then patches were present which were from 0.5 to 1.5 miles in diameter.

2 THE GRAIN OF MOSAIC PATTERNS

It has alredy been emphasized (page 179) that the vector of measurements obtained by observing one transect, or one row of quadrats, must be regarded as one observation. Therefore, to judge whether two populations differ in respect of some property that has been observed at points on a transect, at least two vectors of observations must be obtained from each population. If the number of sampling points (or quadrats) that are to be inspected has been decided beforehand, one is free to choose whether to observe a few long transects or many short ones. Indeed, there is much to be said for carrying the latter policy to its logical conclusion and taking a sample of a great many transects each consisting of a "row" of only two sampling points, or in other words, of a pair of points.

In what follows it is assumed that the mosaic pattern under study has two phases and that we merely observe which phase each sampling point of a pair belongs to; no measurements of quantities are entailed. Natural mosaics with only two phases are quite common. Thus, in Matérn's (1960) study of geographic patterns already mentioned (page 187), the two phases are land and water. Aerial photos of the transition zone between forest and prairie appear as a two-phase pattern with patches of trees and patches of grass. A salt marsh dominated by cord grass (*Spartina* species) often forms a very distinct two-phase mosaic with tracts of cord grass dissected by meandering tidal channels. Examples could be multiplied indefinitely. Any mosaic, in fact, regardless of the number of its phases, can be treated as two-phase by combining the phases into two contrasted groups, or by singling one phase out for special attention and lumping all the others together.

Now suppose that a two-phase mosaic is sampled at a great many pairs of points all having the same interpoint distance which will be treated as the unit of length. Thus each pair is equivalent to a line transect one unit long having sampling points only at its extremities. Let the two phases of the mosaic be labelled X and Y. Then each transect must yield one of three possible observations: XX, YY or XY. The paired observations can be classified even more simply as *self pairs* (XX and YY) and *mixed pairs* (XY). Obviously, for given areal proportions of the phases X and Y, the relative frequencies of self pairs and mixed pairs in a sample will depend on the length of interphase boundary in the area. If the phases are fragmented into numerous small patches, or if the patches (whatever their sizes) have complicated outlines with many lobes and indentations—in a word, if the pattern is fine-grained—mixed pairs will form a comparatively large proportion of a sample of paired observations.

Sampling a mosaic by this method therefore provides a way of measuring the grain of a pattern. The sampling procedure is the same as that envisaged in the famous Buffon's needle problem* of probability theory. The problem is as follows. Imagine a plane partitioned into parallel black and white stripes of equal width; let them be d units wide with $d > 1$. Now suppose a needle of unit length is dropped at random on the plane. What is the probability that it will fall with one end in a white stripe and the other in a black stripe? Put another way, what is the probability that the needle will lie across the boundary line between any two stripes? (Since the width of the stripes exceeds the length of the needle, it is impossible for the needle to straddle a whole stripe). The answer to the problem is $2/(\pi d)$; the derivation is given in most books on elementary probability theory.

---

* Named after the French naturalist and natural philosopher, the Comte de Buffon (1708–1788) who originated the problem in 1777. Buffon was the first biologist to postulate (in his monumental *Histoire Naturelle*) the occurrence of evolution in the animal kingdom.

This theoretical result permits us to represent any naturally occurring mosaic that we have sampled, however intricate its pattern, by a simple, standard "substitute pattern" of alternating black and white stripes whose width, d, is obtained from

$$P_{XY} = \frac{2}{\pi d} \quad \text{or} \quad d = \frac{2}{\pi P_{XY}}.$$

Here P_{XY} is the proportion of XY pairs (mixed pairs) in a sample of paired observations made with transects ("needles") of unit length. Some examples are shown in Figure 8.5. Beside the mosaic patterns are shown their substitute patterns of parallel stripes.

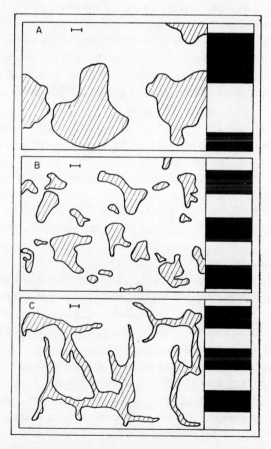

FIGURE 8.5 Three two-phase mosaics. To the right of each mosaic is shown its "substitute pattern". The scale indicator at the top of each mosaic shows the unit of length

To construct its substitute pattern, each mosaic was sampled with 100 randomly placed needles of unit length; (the length of this unit is shown at the top of the mosaic maps). The observed proportion of times the needle yielded a mixed pair of observations, XY, was taken as an estimate of P_{XY} for that mosaic and was used to determine the required stripe width, d, for the substitute pattern. In the figure the widths are: for mosaic A, $d = 4.90$ units; for mosaic B, $d = 2.36$ units; for mosaic C, $d = 2.05$ units.

The purpose of constructing substitute patterns of alternating black and white stripes is to obtain a graphic, easily visualized representation of a natural mosaic's grain. There are three points that should be noticed about these substitutes.

First: P_{XY} is the proportion of XY pairs and this is not necessarily the same as the proportion of times the sampling needle crossed the boundary of a patch. With a natural, irregular mosaic (as opposed to an artificial one made up of stripes of constant width) the needle may chance to lie across a lobe of one phase with both its ends in the other phase so that a self pair is observed. Obviously, however, the shorter the needle the smaller the probability of this occurrence. To measure the grain of an intricate, fine-grained pattern one must use a short sampling needle or, which comes to the same thing, one must work with small units.

Secondly: the substitute pattern portrays the grain of the mosaic, *not* the areal proportions of the X and Y phases. Black and white areas are inevitably present in equal proportions in the substitute pattern. In a mosaic, of course, the area of the X phase, say, may form any proportion of the whole. In principle the mosaic's grain is completely independent of the areal proportions of its phases. In reality we accept that any patch must have linear dimensions in excess of some preassigned minimum (i.e., we rule out such absurdities as, for instance, a patch of forest one millimeter wide). The effect of this restriction is to set a limit to the degree of fragmentation of the phases, or equivalently, to the fineness of the grain; but this constraint is unlikely in practice to falsify the assertion that phase areas and grain are independent, except when one phase forms an exceedingly small proportion of the whole. It should be noticed that the grain of a mosaic is unaffected by which phase is labelled "X" and which "Y".

Thirdly: the value of P_{XY}, and hence of d, used to construct a substitute pattern must in practice be arrived at by sampling. Hence it is only an estimate, subject to sampling error, of the true, unknown probability that a randomly placed needle will have its ends in different phases.

Consider next how the grains of two mosaics, say I and II, may be compared. Suppose each has been sampled with N needles and that the results are as shown in the following 2×2 table.

	Self pairs	Mixed pairs	Totals
Mosaic I	a	$N - a$	N
Mosaic II	b	$N - b$	N
Totals	$a + b$	$2N - a - b$	$2N$

Now hypothesize that the two mosaics have the same grain, in other words that the two batches of N observations could just as well have come from the same mosaic as from two separate mosaics. The tenability of the hypothesis can be tested by judging how well the observed frequency distributions of self and mixed pairs yielded by mosaics I and II respectively, are fitted by the expected distribution common to both. The expected distribution is estimated by pooling the observed results from both mosaics; this is permissible since, by hypothesis, the mosaics do not differ in grain. Thus in a sample of $2N$ pairs of observations the expected frequency of self pairs is estimated to be $(a + b)$, and of mixed pairs $(2N - a - b)$. To do a χ^2 goodness of fit test (see Appendix 7.2) we must therefore calculate the test criterion X^2 where

$$X^2 = \sum \frac{(\text{Observed frequency} - \text{Expected frequency})^2}{\text{Expected frequency}}.$$

In terms of the symbols in the 2×2 table this is found to be

$$X^2 = \frac{2N(a - b)^2}{(a + b)(2N - a - b)}. \tag{8.1}$$

It is known that if the hypothesis is correct, then X^2 has (approximately) the χ^2 distribution with one degree of freedom. The probability of obtaining as great a value of X^2 as that observed, given the hypothesis, can be found from a table of percentage points of the χ^2 distribution. A closer approximation to the required probability is obtained if, instead of taking X^2 as test criterion, we take an adjusted value, X_c^2, defined as

$$X_c^2 = \frac{2N(|a - b| - 1)^2}{(a + b)(2N - a - b)}. \tag{8.2}$$

Here $|a - b|$ denotes $(a - b)$ if $a > b$, or $(b - a)$ if $b > a$. That is, it is the difference between a and b taken with a plus sign.

Since, if the hypothesis is correct, X_c^2 has a probability appropriate to a χ^2 distribution with 1 degree of freedom, it follows from Table 7.4 (page 396) that a value of X_c^2 greater than 3.84 has a probability of occurrence of only 0.05 (or 5%). Therefore if our calculated X_c^2 exceeds this value (the critical value for a 5% significance test) we shall conclude that the hypothesis is *not* correct, in other words that the two mosaics do not have the same grain.

We now apply the test to observations made on the three mosaics in Figure 8.5.

First, compare mosaics A and B. The results of the sampling were as follows:

	Self pairs	Mixed pairs	Totals
Mosaic A	87	13	100
Mosaic B	73	27	100
Totals	160	40	200

Then

$$X_c^2 = \frac{200(|87 - 73| - 1)^2}{160 \times 40} = 5.281.$$

This greatly exceeds 3.84, the critical value for a 5% test. We therefore conclude that the mosaics do not have the same grain. This is, of course, obvious on inspection; the demonstration shows, however, the sort of results we can expect with mosaics such as these.

Now compare mosaics B and C. The relevant table is

	Self pairs	Mixed pairs	Totals
Mosaic B	73	27	100
Mosaic C	69	31	100
Totals	142	58	200

Then $X_c^2 = 0.219$ and we conclude that the grains of the two mosaics should be regarded as the same. This result, also, accords with intuition. At the same time, there is a definite difference, not related to grain as we have defined it, between the two patterns. To coin a new term, one might say that mosaic C has much greater "connectedness" than mosaic B.

The distinction between *grain* and *connectedness*, and their independence of each other, are intuitively clear from inspection of mosaics B and C. Let us attempt verbal definitions of the properties: the grain of a pattern is a function of the probability that a randomly located straight line of given length will cross the boundary of a patch. The connectedness of a pattern might be defined as a function of the distances that could be travelled from one point to another without crossing the boundary of a patch. The biological implications of a pattern's connectedness seem not to have been looked into. As an example, it is conceivable that the connectedness of the pattern of home ranges of a mammal species in a particular community might depend on the animals' behavior and the availability of food.

We shall not explore this property of a mosaic further. It has been mentioned only to show that maps have a great many different properties and that to measure one will not necessarily reveal anything about another. What might be called "the comparative anatomy of patterns" is, indeed, exceedingly complicated. We have already mentioned the properties of intensity (page 149), degree of aggregation (page 148), crowding (page 150), patchiness (page 150), relative areas of mosaic phases (page 190), grain (pages 149 and 172) and connectedness. Each of these things can be measured (some of them in more than one way) and it would presumably be possible to define, and devise ways of measuring, several other geometrical properties of a pattern. It would be pointless to do so, however. Geometry is not ecology and ecologists are concerned with ecological problems. The time to investigate geometrical matters is when they have direct relevance to an ecological problem; the relevant aspect of pattern and the way it should be measured is then dictated by ecological considerations and this is as it should be.

3 ECOLOGICAL PATTERNS AND ISOPLETH MAPS

In the foregoing pages we have discussed many different kinds of ecological patterns and several different ways of observing them. However, we have yet to come to grips with a constant source of difficulty: that of devising uniform methods of describing and measuring them so as to enable comparisons to be made among patterns of contrasting kinds. There are various separate and unrelated ways in which ecological patterns could be classified; a system that would be agreed upon by everyone and that would be suitable in all circumstances is probably unattainable. But it seems reasonable to say that there are three "ideal types" of patterns, and an unlimited number of confusing blends. The ideal types are:

1   An ideal dot pattern. A naturally occurring example is the pattern presented by a plant population of small, scattered individuals of a species that does not reproduce vegetatively. In a map of such a population, it would be reasonable to portray each separate plant by an infinitesimal dot.

2   An ideal mosaic pattern in which the phases are strongly contrasted and occur as patches with clearcut boundaries. A good example is the pattern of receding snowbanks in spring on low vegetation.

3   An ideal "smooth" pattern, in which some property of ecological interest varies smoothly and continuously. Patterns of this kind more often involve environmental features, such as depth of the water table, or soil temperature, or slope of the ground, than properties of the plant cover itself.

As any field ecologist knows, it is difficult to find natural examples of patterns conforming at all closely to these ideal types. The great majoriy of patterns are not easily classifiable. The pattern of many-species vegetation, for instance, is usually some sort of mosaic with indistinct boundaries separating the phases; with continuous variation in the underlying environmental factors revealed by variations in plant stature within the phases; with phases that exhibit qualitative differences in different parts of the mosaic; and with several dot patterns, formed of individual plants of several species (probably with different sizes and life forms), superimposed on the mosaic. Thus an ecologist who has drawn up a plan of campaign with pencil and paper often finds that the material he has to deal with in the field is a discouraging hodge-podge.

It is often helpful to visualize such a vegetation pattern as being made up of several separate, superimposed mosaics. Thus each plant species that grows in definite clumps has a pattern representable as a two-phase mosaic having as phases the clumps of the plant and the gaps separating them. Each continuously varying environmental factor (e.g. soil moisture, soil depth) can, at any rate in theory, be shown by an isopleth (contour) map. This can be transformed into a many-phase mosaic by defining each phase in terms of a range of values of the factor thus giving what is, in effect, a contour layer-colored map. It remains to consider those species of plants whose patterns would normally be shown as dot maps; we must find a way to convert a dot map to an isopleth map in which the isopleths connect points of equal plant density.

The construction of an isopleth map obviously entails two separate steps. The first consists in determining numerical values for the property to be mapped at a number of points on the ground; geographers call these points *control points* and they may be scattered haphazardly or else be arranged systematically at the mesh points of a regular grid. The second step is the drawing pf the isopleths. This latter process is the same, in principle, for any isopleth map, and nowadays can be done by computer. But when we are converting a dot map into an isopleth map, the first step is quite different from that entailed in making any other kind of isopleth map.

Thus, as an example of the usual kind of isopleth map, consider an ordinary topographic contour map in which the isopleths (or contours) are lines connecting points of equal altitude above a datum level. Obviously, every point on the ground (and hence every control point) *has* an altitude and, in theory, a "perfect" contour map could be drawn. Imperfections in an actual map are, of course, inevitable and result from two causes: errors in determining the exact altitudes at the control points; and errors of approximation occurring when contours are interpolated among the control points.

Now consider the contrast when a dot map is to be converted to an isopleth map. In this case we require the isopleths to join points at which dot densities are equal. But in fact every point in the area either does, or does not, coincide with one individual dot. Consequently, to assign to each control point a number intended to measure density involves an abstraction. What do we mean by "the density at a point" in a dot map?

The question is unanswerable as it stands. The best that can be done is to count the dots in a small area (a sampling quadrat) centered on the point. We then treat the number so arrived at which, strictly speaking, pertains to a finite area, as though it pertained to the infinitesimal central point. The unavoidable defect of the method is that the area of the sampling quadrat is arbitrary. Obviously it would not be satisfactory to use quadrats so small that each of them contained either one dot or none at all. At the other extreme, if the quadrats are too large, local density variations will be so blurred as to be undetectable. The choice of a quadrat size somewhere between these unsatisfactory extremes can only be made subjectively. There is no unique optimum size that can be defined objectively.

Thus the first step in constructing an isopleth map from a dot map, that of assigning numerical density values to a set of control points, requires two subjective choices. One must first choose the arrangement of the control points and then, independently, the size and shape of the sampling quadrats. There is no objection to allowing the quadrats to overlap. The closer the spacing of the control points, the more accurate the map; at the same time, as already explained, we must not make the quadrats too small. Therefore it will often happen that quadrats of satisfactory size centered on closely spaced control points will each overlap several neighboring quadrats. The density recorded for each quadrat is then analogous to a moving average.

As an example, consider Figures 8.6 and 8.7. Figure 8.6 shows a dot map (artificial) with, below, a schematic representation of two sampling procedures used to obtain data for isopleth maps. The resulting maps are shown in Figure 8.7. The same array of control points was used in constructing both maps. As shown in Figure 8.6b, they were at the mesh points of a square grid and in both cases density values were obtained by counting the dots in circular quadrats centered on the control points. The sampling procedures differed in using quadrats of different sizes. For the map in Figure 8.7a the diameter of each quadrat circle was equal to the distance between adjacent control points in a row (see Figure 8.6b, left); consequently adjacent quadrat circles were tangent to each other and dots in the interstices among them were not sampled at all. To make the map in Figure 8.7b the quadrat diameter was doubled to twice the distance between control points. As a result, adjacent quadrats overlapped in the manner shown in Figure 8.6b, right;

13*

no dots were unsampled but, on the contrary, all were tallied at least twice
and some three or four times. Also, of course, doubling each quadrat's diameter quadrupled its area. Therefore, so that the density measurements
should be relative to the same unit area, each dot count obtained with the
larger quadrats was divided by four.

Each batch of data (one for each map in Figure 8.7) consisted of an array
of numbers representing the densities at the array of control points. A map
showing all the control points labelled with their observed dot densities
would, indeed, be analogous to a topographic map marked with an array
of spot heights. The second step in map construction starts at this stage and
consists in interpolating the isopleths among the control points. As was
remarked earlier, the task is the same for all kinds of isopleth maps. There

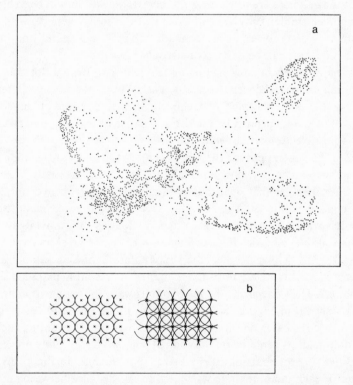

FIGURE 8.6 (*a*) The dot map from which the isopleth maps in Figure 8.7 were constructed. (*b*) Diagrams, on the same scale as the map, of the sampling methods used to
obtain data for the isopleth maps. The *X*'s show the control points. The circles show the
sizes of the quadrats. The small, non-overlapping circles (on the left) gave data for the
upper map in Figure 8.7; the large, overlapping circles (on the right) gave data for the
lower map

is an enormous number of techniques to choose from and one is again faced with the need to make decisions. Collectively, the techniques are known as trend-surface mapping and in recent years this has been a very active research topic among geographers, geologists and geophysicists. Chorley and Haggett (1968) have given a clear and comprehensive review of the rationale underlying the various methods.

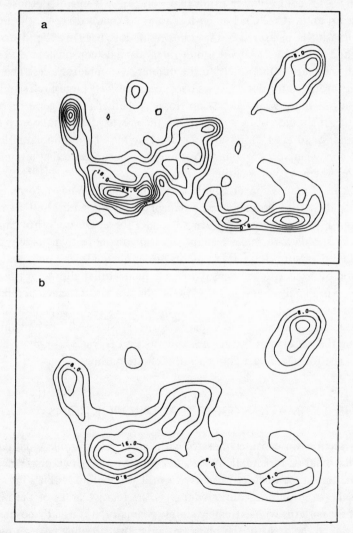

FIGURE 8.7   Two automatically plotted isopleth maps made from the dot map in Figure 8.6. (*a*) With the small quadrats; (*b*) with the large quadrats. The numerals show dot densities as numbers of dots per unit area; the unit of area is the same for both maps and is the area of the small quadrats

For ecologists to attempt to devise new methods without taking cognizance of the geographers' work in this field would be a wasteful duplication of effort. There are available, in any case, a number of computer programs that will determine the locations of the isopleths in conformity with one or another of the various techniques and yield a finished map, made by an automatic plotter, as print-out. The maps in Figure 8.7 are examples. They were made by a Calcomp plotter using that company's General Purpose Contouring Program (CGPC). For both maps, as explained above, the unit in which density is measured is the number of dots per small-sized quadrat circle; and the isopleths are at 4 unit intervals. The conspicuous difference between the maps is due solely to the difference in quadrat sizes. The large quadrats averaged the dot densities over comparatively large circles and thus flattened out sharp peaks and steep slopes; whether this is good or bad is debatable. It should not be assumed that the greater detail in the map made with small quadrats is necessarily an asset; some of the minor "hillocks" show places where dots are numerous merely by chance, so that attempts to "explain" them would be futile.

When a dot map is converted into an isopleth map, no *new* information is gained and there is probably some danger of seeing, and falsely ascribing significan e to, "patterns" that are not really there. On the other hand, in gathering the data to make a map, one can do more than merely count individual plants (or dots). If the plants are of unequal sizes, one can devise a method of measuring the density in each quadrat that allows for this; that is, instead of merely counting individuals, one can assess the actual quantity, or biomass, of the plant species in the quadrat. Isopleth maps plotted by a computer also provide what is perhaps the most satisfactory method of mapping the patterns of "stragglers", plants that reproduce vegetatively but that do not form compact clumps with definite outlines.

4   THE ESTIMATION OF AREAL PROPORTIONS

In this section we consider mosaics with easily recognizable, qualitatively dissimilar phases; for simplicity we shall concentrate on two-phase mosaics. The first and most obvious property to measure, before looking into subtleties such as grain and connectedness, is the relative areas of the phases. Except for patterns with extremely simple geometry, it is usually not feasible to determine the areas of the phases exactly. Instead, they have to be estimated by some sampling procedure and so there is inevitable sampling error. An intuitively obvious way to do the sampling is to inspect the mosaic at a large number of points and record which phase each point falls in; the relative proportions of points in each phase then give estimates of their relative areas.

So much is obvious. What is less obvious is how best to place the sampling points and how to estimate the magnitude of the sampling error. If the points are placed at random, which can be done by taking coordinates for them from a random numbers table, the position of each point has to be separately specified and located; this is very laborious and the method also suffers from the great disadvantage that the set of points may, by ill luck, constitute a very unrepresentative sample.

The alternative is to sample systematically; that is, the sampling points are arranged in some regular manner, say at the mesh points of a square grid. When this is done the sampled points will almost certainly give a better representation of the whole area than would an equal number of random points. As a result, the estimated proportions will be closer to their true, unknown values. But they will not, of course, equal these true values precisely and the observations will not permit calculation of the sampling error. This is because the datum obtained by sampling at a regular array of n, say, sampling points is one single observation with n component parts; it is *not* n observations. (See page 179 for analogous comments on the vector of values obtained from a row of contiguous quadrats.)

The best way of combining the merits of systematic sampling with those of random sampling is to sample systematically several times over, using a coarse grid of sampling points that is relocated at random each time. The practical details depend on circumstances. If one has a map (or air photo) of the mosaic, a grid ruled on clear plastic can be laid over it. When a mosaic is being studied on the ground the sampling points are usually located by measuring with a tape, or pacing. The method to use is dictated by the extent of the region being studied. For example, the mosaic pattern of low-growing plants on a small tract of smooth ground can be systematically sampled with a spiked wheel; as it is wheeled along each of several parallel transects, spikes on the wheel's rim touch the ground at regularly spaced points and these are the sampling points. Such a device has been used in semi-arid grassland in South Africa (see Greig-Smith, 1964).

Whatever method is used, each placing of the regular grid of points gives a different estimate of the proportional area, say p, of one of the phases. Since several such estimates are obtained, one on each occasion that the grid is relocated, the sampling variance of the estimate, and hence its standard error, can be found.

As an example, we shall estimate the proportion of the black phase in the mosaic in Figure 8.8. The pattern is reduced from part of a topographic map (National Topographic Series (Canada) 31C/7 West: Sydenham, Ontario) on the scale 1 : 50,000, in which wooded areas are shown green; they are the black patches in Figure 8.8. The area sampled was 5 × 10 kilometers on the ground, or 10 × 20 centimeters on the map; (the figure printed here is reduced; the check marks around the border are at two centimeter intervals in the original). The mosaic was sampled by laying over it a one centimeter square grid ruled on clear plastic, and counting those grid points that lay over the black patches. The process was repeated $N = 10$ times with the results shown in Table 8.2. The different numbers in the denominators of the estimates, p, arose from the following cause: when the sampling grid was exactly aligned with the map's edges, 200 of the grid points fell within the area; but slightly larger or smaller numbers were included when the sampling grid was randomly oriented.

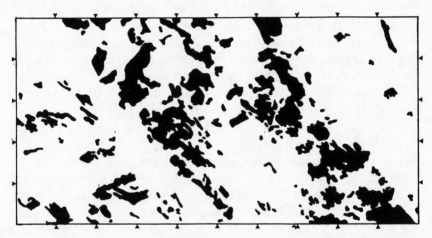

FIGURE 8.8   A two-phase mosaic map. The relative areas of the two phases were estimated as described in the text

Table 8.2 also shows $\bar{p}$, the mean of the 10 estimates of the proportion of wooded land in the 50 square kilometer region; the standard deviation of these estimates, s_p; and the standard error of $\bar{p}$, denoted by $s_{\bar{p}}$. We are now in a position to obtain a 95% confidence interval for the true value of the proportion (see page 381 and also Appendix 6.1, page 381). The required value of $t_{0.025}$ can be found in Table 6.11 (page 382). The number of degrees of freedom is $N - 1 = 9$ and therefore the appropriate value is $t_{0.025} = 2.262$. Thus the 95% confidence interval for p is

$$\bar{p} \pm t_{0.025}s_{\bar{p}} = 0.224 \pm (2.262 \times 0.007)$$

or 0.208 to 0.240.

TABLE 8.2 Estimating the areal proportion of the black phase in the mosaic in Figure 8.8

Sample No.	Estimate of p
1	$41/199 = 0.206$
2	$39/196 = 0.199$
3	$47/195 = 0.241$
4	$42/202 = 0.208$
5	$50/202 = 0.248$
6	$43/202 = 0.213$
7	$55/204 = 0.270$
8	$43/199 = 0.216$
9	$46/201 = 0.229$
10	$43/202 = 0.213$

$$\bar{p} = \frac{449}{2002} = 0.224$$

$$\text{var}(p) = \frac{1}{N-1}\left\{\Sigma p^2 - \frac{(\Sigma p)^2}{N}\right\}$$

$$= \frac{1}{9}\left\{0.507017 - \frac{(2.241768)^2}{10}\right\}$$

$$= 0.000496$$

Therefore

$$s_p = \sqrt{\text{var}(p)} = 0.022$$

and

$$s_{\bar{p}} = \frac{s_p}{\sqrt{10}} = 0.007$$

Writing $q = 1 - p$ for the proportion of the area that is not wooded (that is, the proportion of the mosaic shown white in the figure), it also follows that the 95% confidence interval for q is $0.776 \pm (2.262 \times 0.007)$ or 0.760 to 0.792.

CHAPTER 9

Competition I
Processes and Models

IN THIS CHAPTER we resume discussion of a topic first mentioned in Chapter 4, that of biological competition. In deriving the form of the logistic growth curve (pages 49 and 79 and Figure 4.1a) it was supposed that the individual members of a growing population inhibited one another so that, as the population's density steadily increased, the growth rate per individual steadily decreased. Ultimately growth slowed to a stop as population density reached an equilibrium, or saturation, level whose value was governed by the carrying capacity of the environment. The usual way in which a population's members inhibit one another is by depriving one another of the necessities of life, that is, by competing, and the earlier discussion did, in fact, introduce the study of competition. We shall now explore the matter in greater detail, as it is of central importance in ecology.

As a start, a definition of the word *competition* must be decided upon. That this should be necessary may be surprising since the meaning seems intuitively obvious. However, like many other terms used over the years by ecologists, its connotation has become progressively modified and augmented. Competition among living things has turned out to have so many manifestations and ramifications that a precise definition setting clear limits to the word's current meaning is called for. Periodic examination and modernization of the definitions of ecological terms are desirable provided they are not too frequent.

Milne (1961) carried out such an examination in the context of animal competition and his twenty-page paper is wholly concerned with definitions. The single, strict definition he finally advocates is as follows:

Competition is the endeavour of two (or more) animals to gain the same particular thing, or to gain the measure each wants from the supply of a thing when that supply is not sufficient for both (or all).

Notice that the two (or more) animals may belong to the same or different species; the definition thus includes both *intra*specific and *inter*specific competition. To extend the definition to cover plants as well as animals, one may be tempted to substitute the word "individual" for "animal". But such a definition is not satisfactory since with plant species that reproduce vegetatively it is meaningless to speak of individuals. Consider a large clone of such a species, whose parts were once interconnected by rhizomes or stolons; and suppose that many of these organic links have become decayed or broken. All the plant material of the clone is genetically identical and in that sense it collectively constitutes an individual. At the same time, if we disregard their genetic constitution, each of the unconnected parts can be treated as a separate individual. It therefore seems best to abandon the notion of "individual" entirely and use a definition of competition that avoids it.

McIntosh (1970) quotes a definition of plant competition first put forward by Clements, Weaver and Hanson in 1929. This states that "when the immediate supply of a single necessary factor falls below the combined demands of the plants, competition begins." Again the concept of "individual" is present; it is implicit in the word "plants". A slight modification enables this drawback to be overcome and I propose to use the following definition:

Competition takes place when the growth of a biological population, or any part of it, is slowed because at least one necessary factor is in short supply.

Here "part of a population" may connote one or more individuals in the ordinary sense, or part or all of a clone whether interconnected or not. Observe also that the definition says nothing about the results of competition, nor of how these may be recognized. As we shall see later (Chapter 10) the results are various and their detection often problematical. Before discussing how competition works in nature, however, we shall explore some simple mathematical models.

1 GAUSE'S COMPETITION EQUATIONS

We begin by recalling the model underlying the logistic growth curve. This is the simplest model for the growth of a one-species population in which intraspecific competition is occurring; it assumes that the population is growing in a limited space (so that crowding cannot be relieved by dispersal) and that at every instant the growth rate per individual is linearly related to population size. Expressing this assumption in the form of a differential equation, we have

$$\frac{1}{N}\frac{dN}{dt} = r - sN.$$

This is the same as equation (4.1) on page 47.

The model is still applicable when the population consists of a clone rather than of separate, countable individuals. Clearly, the argument on which the equation is based is unaffected if we let N denote the quantity of matter in the population (measured in suitable units such as grams of dry weight) instead of the number of individuals. For the sake of simplicity, however, it is assumed in what follows that populations consist of countable individuals.

The graph of the differential equation was shown in Figure 4.1a and is repeated here in Figure 9.1a. The point at which the line meets the N-axis, or equivalently the value of N at which $\dfrac{1}{N}\dfrac{dN}{dt} = 0$ and growth stops, will be denoted by K as before. That is, K is the population's maximum sustainable size in the limited space it is confined to. This size, the saturation level, can also be portrayed by a line K units long as shown in Figure 9.1b; the reason for using this representation will become clear shortly.

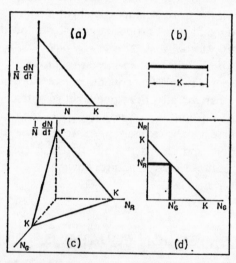

FIGURE 9.1   *Upper:* Logistic growth of a one-species population. (a) shows how the growth rate per individual, $(1/N)\,dN/dt$, varies with population size, N; (b) is a representation, in one dimension, of the equilibrium size of the population
Lower: Analogous figures for a combined population of two identical "species", the Reds and the Greens. See text for further details

Now let us perform a simple conceptual experiment. Starting with an unsaturated population of size N growing in the manner we have just described, imagine that N_R of its members are labelled with spots of red paint, and the remaining $N_G \,(= N - N_R)$ with spots of green paint. Imagine also that cross-breeding between the reds and the greens is prevented and that every newborn individual is given the same color label as its parents. Population growth is allowed to continue as before with the reds and greens mingling freely; the growth rate per individual, which is the same for both reds and greens is, as before, a linear function of $N = N_R + N_G$, the size of the combined population; and the population's equilibrium size, reds and greens together, is still K. Thus the growth of the population as a whole (i.e. disregarding the color labels) is still represented by the line in the two-dimensional graph of Figure 9.1a.

But if we wish to examine the growth of the two sub-populations separately, this graph must be replaced by the three-dimensional graph of Figure 9.1c. In the latter the vertical axis is $\dfrac{1}{N}\dfrac{dN}{dt}$ as before since it still represents the growth rate per individual of the combined population. The horizontal axes represent N_R and N_G, the numbers of reds and greens respectively.

The relationship shown in Figure 9.1a by the line

$$\frac{1}{N}\frac{dN}{dt} = r - sN$$

is shown in Figure 9.1c by the plane

$$\frac{1}{N}\frac{dN}{dt} = r - sN = r - sN_R - sN_G. \tag{9.1}$$

Thus the growth rate per individual (whatever its color) is a function of both N_R and N_G. Notice, also, that since the reds and greens are biologically identical, the growth rate per individual for each subpopulation is the same as that of the combined population. In symbols,

$$\frac{1}{N}\frac{dN}{dt} = \frac{1}{N_R}\frac{dN_R}{dt} = \frac{1}{N_G}\frac{dN_G}{dt}.$$

Thus the vertical axis of the graph can be taken to represent the growth rate per individual of the combined population, or of the reds, or of the greens. It makes no difference. The plane in Figure 9.1c can therefore be represented by equation (9.1) or equally well by

$$\text{either} \quad \frac{1}{N_R}\frac{dN_R}{dt} = r - sN_R - sN_G$$

$$\tag{9.2}$$

$$\text{or} \quad \frac{1}{N_G}\frac{dN_G}{dt} = r - sN_R - sN_G.$$

Now consider Figure 9.1d, a two-dimensional graph whose axes represent N_R and N_G. It is the "floor" of the graph in Figure 9.1c, at which $\dfrac{1}{N}\dfrac{dN}{dt} = 0$;

the line joining $N_G = K$ and $N_R = K$ is the line at which the sloping plane in Figure 9.1c cuts this floor. The equation of the line in Figure 9.1d is thus $N_R + N_G = K$. Every point on it represents a possible combination of reds and greens that, together, would constitute a combined population at the saturation level. An example is provided by the population of N_R' reds and N_G' greens shown in the figure. Thus we see that a point in two-dimensional space is needed to show graphically a possible equilibrium population of reds and greens; this is in contrast to the one-dimensional portrayal (as in Figure 9.1b) which sufficed for a single population.

The outcome of the conceptual experiment can now be summarized. Assume that two "species" (the reds and greens), whose biological properties are completely identical, form a combined population growing in a limited environment. Then, when the size of the combined population reaches the saturation level, K, and growth stops, the numbers in the two subpopulations, N_R and N_G, may have any values such that they sum to K.

We are now in a position to investigate what happens when populations of two different species compete. Unlike the reds and greens, which were hypothetical species with identical biological properties, two real species will certainly differ from each other to some extent. Thus if populations of each of them grew separately, under identical environmental conditions, the growth rates and saturation sizes of the two populations would not be the same. We wish to predict what would happen if they grew together.

Let the two species be labelled M and N and let these symbols also denote the sizes of the two populations. Since the growth rate per individual is different in the two species, equations (9.2) must be replaced by the following pair of equations:

$$\frac{1}{M}\frac{dM}{dt} = r_M - s_M M - u_M N$$

$$\frac{1}{N}\frac{dN}{dt} = r_N - s_N N - u_N M.$$

(9.3)

These are known as Gause's competition equations.

The meanings of the various constants are as follows:

r_M and r_N are the intrinsic rates of natural increase of species M and N respectively.

s_M and s_N measure the degree to which the growth of each species is reduced by *intra*specific competition. Thus the growth rate per individual of a population of M's alone, unmixed with N's, would be given by $\frac{1}{M}\frac{dM}{dt} = r_M - s_M M$, and would have a saturation level of $r_M/s_M = K_M$, say. Analogous relations hold for a population of N's growing alone which would have a saturation level of $r_N/s_N = K_N$.

u_M and u_N measure the degree to which the growth of each species is reduced by *inter*specific competition. Thus the growth rate per individual of a population of M's growing in competition with N individuals of species N is reduced by an amount $u_M N$. And correspondingly in the opposite case.

Now recall that both the formulas in (9.2) were equations of the plane shown graphically in Figure 9.1c. The two formulas in (9.3) differ from each other, however, and are therefore the equations of two different planes. The first represents the dependence of species M's growth on both M and N; and the second, that of species N. Both planes could be plotted on the same three-dimensional coordinate frame and if this were done the "floor" of the graph would show *two* lines where the planes cut it. One of the lines would show how M and N are related when species M is neither increasing nor decreasing, that is, when $\dfrac{1}{M}\dfrac{dM}{dt} = 0$; correspondingly, the other would show how M and N are related when $\dfrac{1}{N}\dfrac{dN}{dt} = 0$. We shall suppose that the two lines intersect.

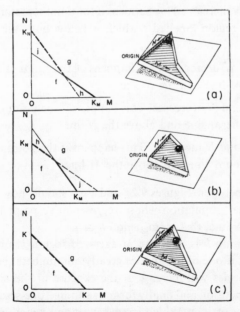

FIGURE 9.2   Each of the graphs on the left shows the relationship between M and N when M stops growing (solid line) and when N stops growing (broken line). The "ball in a box" models on the right show how a combined population of M's and N's behave. See text for further details

Figure 9.2 shows the possibilities. First consider the graphs in the left of each section of the figure. In all three the solid line (which we shall call the "M-line") is where $\dfrac{1}{M}\dfrac{dM}{dt} = 0$ and it reaches the M-axis (where $N = 0$) at K_M. The dashed line (the "N-lines") is where $\dfrac{1}{N}\dfrac{dN}{dt} = 0$ and it reaches the N-axis (where $M = 0$) at K_N. Notice particularly that there is a contrast between the graphs in Figures 9.2a and 9.2b although the two lines cross each other in both. The difference will become clear below.

Let us consider what would happen to any two-species population we cared to construct by putting together chosen numbers of M's and N's in the limited space. The make-up of any such combined population is representable by a point in one of the graphs. Now, if the point is below or to the left of the M-line (shown solid), species M will increase; if above or to the right of it, species M will decrease. Analogously, species N will increase or decrease according as the point is below or above the N-line (shown dashed).

We now limit attention to Figures 9.2a and 9.2b only. Clearly, an artificially constructed two-species population must behave in one of four different ways. These are:

1   Both M and N increase. This happens if the point representing the population falls in the region labelled f which is below both the M-line and the N-line.

2   Both M and N decrease. This happens if the point is in the region g, above both the M-line and the N-line.

3   M increases and N decreases. This happens if the point is in the region h which is below the M-line and above the N-line.

4   M decreases and N increases. This happens if the point is in the region j which is below the N-line and above the M-line.

The contrast between Figures 9.2a and 9.2b should now be clear. Thus, consider an initial population whose representative point lies between the M-line and the N-line, that is, in region h or j.

In case (a), region h is near the M-axis and region j is near the N-axis. The result is that if one of the species greatly outnumbers the other initially, the abundant species will increase at the expense of the sparse species. It will win the competition and finally form a saturated one-species population; the loser becomes extinct. Which species wins and which loses depends on the numbers of individuals of each species present initially. The set-up is shown pictorially on the right in Figure 9.2a. The fate of a combined population of M's and N's is modeled by the behavior of a ball rolling across a shallow box which can be tilted by raising the corner labelled "origin". Clearly the ball must come to rest in one or other of the two acute-angled corners which represent pure populations of one species or the other.

Now consider case (b) in which region h is near the N-axis and region j near the M-axis. If, initially, one species greatly outnumbers the other, it will dwindle while the sparse species grows. Whatever the composition of the initial population, the two species will in time come into equilibrium with each other and the numbers of M's and N's will thereafter remain constant. As the ball-in-a-box model (on the right in Figure 9.2b) shows, the rolling ball will come to rest in the obtuse-angled corner representing simultaneous presence of members of both species.

The composition of the two-species equilibrium population is easily found. When the combined population stops growing

$$\frac{1}{M}\frac{dM}{dt} = \frac{1}{N}\frac{dN}{dt} = 0.$$

Therefore, from equations (9.3), it follows that

$$r_M - s_M M_{eq} - u_M N_{eq} = 0$$

and

$$r_N - s_N N_{eq} - u_N M_{eq} = 0.$$

where M_{eq} and N_{eq} denote the numbers of individuals of the two species at equilibrium. Solving this pair of simultaneous equations for M_{eq} and N_{eq} gives

$$M_{eq} = \frac{r_M s_N - r_N u_M}{s_M s_N - u_M u_N} \quad \text{and} \quad N_{eq} = \frac{r_N s_M - r_M u_N}{s_M s_N - u_M u_N}. \tag{9.4}$$

Thus we have the equilibrium sizes of the two species-populations expressed in terms of the constants in equations (9.3). M_{eq} and N_{eq} are, of course, the coordinates of the point of intersection of the two lines in the graph of Figure 9.2b. Exactly the same formulas give the coordinates of the intersection point in Figure 9.2a. In the later case, however, the equilibrium is unstable. The ball-in-a-box models demonstrate plainly the contrast between *unstable equilibrium* [case (a)] and *stable equilibrium* [case (b)]. In case (a), the equilibrium "corner" is reflexive and the rolling ball cannot remain poised there for more than a moment; in case (b), this corner is obtuse and the ball can obviously remain there indefinitely.

Figure 9.2c has been included for the sake of completeness. It assumes that species M and N are biologically identical and consequently that the lines $\frac{1}{M}\frac{dM}{dt} = 0$ and $\frac{1}{N}\frac{dN}{dt} = 0$ are exactly coincident. Therefore there are no regions h and j but only f and g. The result is that the two species can coexist in *neutral equilibrium*. The box on the right of the figure is triangular and the ball can remain at any point on the hypotenuse where it chances to come to rest. It will be recognized that this is the same situation as that described earlier when the two hypothetical identical species were called reds and greens.

The foregoing discussion assumes that the two lines cross each other or, in the limit, coincide. There is another, less interesting possibility, of course, namely that the two lines do not cross. Then, if the M-line is everywhere farther from the origin than the N-line, species M will be capable of ousting species N in all circumstances. That is, whatever the initial composition of a mixed population, species N will ultimately disappear entirely leaving a one-species population of M's. And in the converse case species N will always win out over species M.

We now consider in detail how the two species-populations grow. Thus far only the final outcome has been discussed. But it is possible, as we shall see, to determine how the numbers of each species change with time. This is most easily done by obtaining from the differential equations in (9.3) a pair of difference equations (see page 79). These give M_{t+1} and N_{t+1}, the numbers of the two species present at time $t + 1$, in terms of M_t and N_t and constants which depend on the biological properties of the species. The difference equations were obtained by Leslie (1958) where details of their derivation may be found. Here we state them without proof. They are as follows:

$$M_{t+1} = \frac{\lambda_M M_t}{1 + \alpha_M M_t + \gamma_M N_t}$$

$$N_{t+1} = \frac{\lambda_N N_t}{1 + \alpha_N N_t + \gamma_N M_t}.$$

(9.5)

In terms of the constants in equations (9.3) (that is, the r's, s's and u's) the constants in these equations are:

$$\lambda_M = e^{r_M} \quad \text{and} \quad \lambda_N = e^{r_N}$$

$$\alpha_M = s_M(\lambda_M - 1)/r_M \quad \text{and} \quad \alpha_N = s_N(\lambda_N - 1)/r_N;$$

$$\gamma_M = u_M(\lambda_M - 1)/r_M \quad \text{and} \quad \gamma_N = u_N(\lambda_N - 1)/r_N.$$

Knowing these constants it is possible to determine what the outcome of competition between the two species will be; and, if it turns out that a two-species equilibrium population can persist in the limited space available, we can also find the numbers of individuals of each species that will comprise it. To make these predictions it is first necessary to evaluate the following four terms: $\alpha_M(\lambda_N - 1)$, $\alpha_N(\lambda_M - 1)$, $\gamma_M(\lambda_N - 1)$ and $\gamma_N(\lambda_M - 1)$. Then the four possible outcomes and the conditions for them are as follows:

Case 1: Species M will win if

$$\alpha_N(\lambda_M - 1) > \gamma_M(\lambda_N - 1) \quad \text{and} \quad \gamma_N(\lambda_M - 1) > \alpha_M(\lambda_N - 1).$$

(Or equivalently, if $r_M s_N > r_N u_M$ and $r_M u_N > r_N s_M$.)

As a rule of thumb, notice that in both the inequalities the term involving λ_M (or r_M) exceeds the term involving λ_N (or r_N). The λ's and r's are the growth rates of the species in unrestricted environments when they are not competing at all.

Case 2: Species N will always win in the opposite set of conditions: that is, if all the inequalities in Case 1 are reversed.

Case 3: There will be unstable equilibrium if

$$\gamma_M(\lambda_N - 1) > \alpha_N(\lambda_M - 1) \quad \text{and} \quad \gamma_N(\lambda_M - 1) > \alpha_M(\lambda_N - 1).$$

(Or equivalently, if $r_N u_M > r_M s_N$ and $r_M u_N > r_N s_M$.)

As a rule of thumb, notice that in both the inequalities the term relating to the effect of one species on the growth of the other (the γ's and u's) exceeds the term relating to the effect of a species on its own growth (the α's and s's). In this case either species may win, and to predict which the winner will be in particular circumstances it is necessary to know how many M's and N's are present intially and then to calculate the sizes of the two populations at a succession of times, using equations (9.5), until the outcome becomes apparent. Examples are given below.

Case 4: There will be stable equilibrium if all the inequalities in Case 3 are reversed. The sizes of the equilibrium populations, M_{eq} and N_{eq} are:

$$M_{eq} = \frac{\alpha_N(\lambda_M - 1) - \gamma_M(\lambda_N - 1)}{\alpha_M \alpha_N - \gamma_M \gamma_N}$$

and   (9.6)

$$N_{eq} = \frac{\alpha_M(\lambda_N - 1) - \gamma_N(\lambda_M - 1)}{\alpha_M \alpha_N - \gamma_M \gamma_N}.$$

These equations are arrived at by noticing that when equilibrium has been reached neither population undergoes any further changes in size, so that $M_{t+1}/M_t = N_{t+1}/N_t = 1$. Using this fact and equations (9.5) it follows that

$$\alpha_M M_{eq} + \gamma_M N_{eq} = \lambda_M - 1$$

and

$$\alpha_N N_{eq} + \gamma_N M_{eq} = \lambda_N - 1.$$

Solving this pair of simultaneous equations for M_{eq} and N_{eq} yields equations (9.6).

14*

There is one point that needs to be strongly emphasized before we describe numerical examples. It is this: all the foregoing discussions deal with *deterministic* predictions (see page 4). For brevity and to avoid repetitious qualifying clauses, we have spoken of "predicting" the outcome when two species compete, of "determining" how populations will grow, and have asserted that in specified circumstances this or that species will "always" win. In real life such flat statements are inadmissable, of course. There is always a chance that they will turn out wrong. The deterministic theory makes no allowance for chance (or stochastic) events, for unforeseen or unexpected births and deaths. When such allowance are made, the result is a *stochastic* theory of far greater mathematical difficulty. The use of high speed computers enables one to circumvent some of the difficulties and we shall take up the topic of stochastic predictions on page 221. Predictions based on deterministic theory should not be dismissed as valueless, however; they will turn out correct in the majority of cases. There is no reason to spurn them provided the innate uncertainty is always kept in mind. Quite often it is possible to measure, objectively, the amount of this uncertainty. In brief, interspecific competition can be treated as a sporting event of the same kind as a horse race or a cock fight.

As a numerical example consider equations (9.3) with numerical coefficients as follows:

$$\frac{1}{M}\frac{dM}{dt} = 0.100 - 0.0007M - 0.001N$$

$$\frac{1}{N}\frac{dN}{dt} = 0.075 - 0.0007N - 0.0007M. \tag{9.7}$$

These particular values of the constants have been chosen so that the equations represent the actual growth rates, in a specific restricted environment, of two species of flour beetle that were used in a classic series of experiments begun in the 1940's that did much to illuminate the process of ecological competition. The experiments, and the derivation of the constants, will be discussed in detail in Section 3 (page 221); for the moment we accept them without comment.

From the constants in the differential equations those required for the difference equations can be calculated.

Thus

$$\lambda_M = e^{0.100} = 1.105; \qquad \lambda_N = e^{0.075} = 1.078;$$

$$\alpha_M = 0.000735; \qquad \alpha_N = 0.000728;$$

$$\gamma_M = 0.001050; \qquad \gamma_N = 0.000728.$$

The difference equations (9.5) become

$$M_{t+1} = \frac{1.105M_t}{1 + 0.000735M_t + 0.001050N_t}$$

$$N_{t+1} = \frac{1.078N_t}{1 + 0.000728N_t + 0.000728M_t}.$$

(9.8)

We next enquire what the outcome of competition will be. This can be discovered using the constants in either the differential equations or the difference equations.

From the constants in the differential equations we have

$$r_M s_N = 7.0 \times 10^{-5}; \quad r_M u_N = 7.0 \times 10^{-5}; \quad r_N s_M = 5.25 \times 10^{-5};$$

$$r_N u_M = 7.5 \times 10^{-5}.$$

It is seen that

$$r_N u_M > r_M s_N \quad \text{and} \quad r_M u_N > r_N s_M.$$

Alternatively, using the constants in the differential equations,

$$\gamma_M (\lambda_N - 1) = 8.19 \times 10^{-5} > 7.64 \times 10^{-5} = \alpha_N (\lambda_M - 1)$$

and

$$\gamma_N (\lambda_M - 1) = 7.64 \times 10^{-5} > 5.73 \times 10^{-5} = \alpha_M (\lambda_N - 1).$$

It follows that a mixed population of the two species cannot persist (their equilibrium is unstable) and that one or other of the species will, after enough time has elapsed, be sole survivor. Which this species will be depends on the initial composition of the two-species population.

Now let us examine how a particular starting population will behave. Thus suppose 2 individuals of species M and 12 of species N begin growth together at time zero. It is desired to predict what the size of each population will be at a sequence of times $t = 1, 2, 3, \ldots$. The sizes are obtainable from equations (9.8):

Since

$$M_0 = 2 \quad \text{and} \quad N_0 = 12$$

we have

$$M_1 = \frac{1.105 \times 2}{1 + (0.000735 \times 2) + (0.001050 \times 12)} = 2.18$$

and

$$N_1 = \frac{1.078 \times 12}{1 + (0.000728 \times 12) + (0.000728 \times 2)} = 12.81.$$

In the same way

$$M_2 = 2.37, \quad N_2 = 13.66$$

$$M_3 = 2.58, \quad N_3 = 14.56$$

and so on.

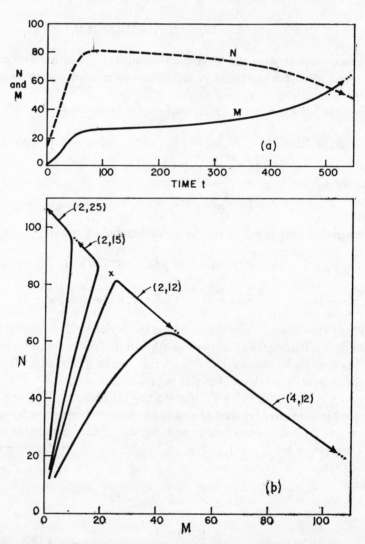

FIGURE 9.3 The behavior of a combined population of species M and N; the two species cannot coexist in stable equilibrium. (a) The change, with time, of the sizes of the two species-populations given that $M_0 = 2$ and $N_0 = 12$ initially. (b) The trajectories of four possible combined populations; the number pair in the label of each curve is (M_0, N_0), the coordinates of the trajectory's starting point. See text for further details

The results can be shown graphically in two ways. Figure 9.3a shows how the numbers of M's and N's change with time for the case in which $M_0 = 2$, $N_0 = 12$. Figure 9.3b shows the trajectories (cf. page 85) of several different populations, those for which (M_0, N_0) is, respectively, (2, 25), (2. 15), (2, 12) and (4, 12). Obviously in the first two cases species M becomes extinct and species N survives alone, and in the second two cases the opposite is true. Even though the differences among the chosen starting populations appear trivial, they lead to diametrically opposed outcomes.

The X in Figure 9.3b shows the unstable equilibrium point. From equations (9.6) its coordinates are:

$$M_{eq} = \frac{(0.000728 \times 0.105) - (0.001050 \times 0.078)}{(0.000735 \times 0.000728) - (0.001050 \times 0.000728)}$$

$$= \frac{-5.46 \times 10^{-6}}{-0.22932 \times 10^{-6}} = 23.8$$

$$N_{eq} = \frac{(0.000735 \times 0.078) - (0.000728 \times 0.105)}{-0.22932 \times 10^{-6}}$$

$$= 83.$$

The same values can be obtained by substituting the coefficients in equations (9.7) into equations (9.4).

It should be noticed that the graph of a trajectory shows only the relation between the number of M's and the number of N's. It does not give the relation between species composition and time. This could be shown on the same graph (though it would overcrowd it) by marking a dot on the trajectory to show the species composition at a succession of instants separated by equal intervals. Each of the sequence of points, (M_1, N_1), (M_2, N_2), ..., calculated as described above, would be shown by a dot; the more gradual the change in species composition, the more crowded the dots. It is found that the closer the trajectory lies to the equilibrium point the more gradual are the changes. As an example, compare the curves in Figure 9.3a with the corresponding trajectory [starting at (2, 12)] in Figure 9.3b; from the upper graph it is seen that more than 200 time units are required for M to increase from 25 to 30 while N varies between 77 and 81. These values are close to the equilibrium point.

Figure 9.4 resembles Figure 9.3 except that it describes changes in a two-species population whose equilibrium is stable. In this example the differential equations are:

$$\frac{1}{M} \frac{dM}{dt} = 0.10 - 0.0014M - 0.0012N$$

$$\frac{1}{N} \frac{dN}{dt} = 0.08 - 0.0010N - 0.0009M.$$

The difference equations are

$$M_{t+1} = \frac{1.105M_t}{1 + 0.00147M_t + 0.00126N_t}$$

$$N_{t+1} = \frac{1.083N_t}{1 + 0.00104N_t + 0.00093M_t}.$$

The equilibrium point is at

$$M_{eq} = 12.5, \quad N_{eq} = 68.8.$$

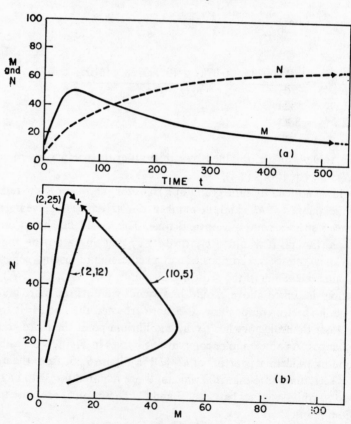

FIGURE 9.4 The behavior of a combined population of two species capable of co-existing in stable equilibrium

The foregoing discussion has been wholly theoretical and we have not paused to consider whether the theory is applicable to real life situations. In most cases one would not expect it to be applicable, at least without major modifications. The theoretical model rests on several unnaturally simple assumptions of which the principal ones are that the environmental conditions, and hence the coefficients in the equations, do not change with the passage of time; that the age distributions of the two species remain unchanged; and that the responses of the growth rates to overcrowding are instantaneous. The simple equations in (9.3) and (9.5) can be revised to allow for some, at least, of the many deviations of a natural population's growth from the idealized form predicted by the model; an ingenious theoretician can, indeed, devise a whole series of adjustments to the equations, which ought to allow for each of these deviations in turn.

However, if the equations are to be subjected to a succession of refinements designed to make them more realistic, it is obviously desirable that their applicability be checked at each stage. Ideally, each version of the model would be checked against a corresponding living population whose natural complexity matches, as closely as possible, that of the model currently under consideration. The first of these checks, the one entailing a comparison between a living population and the simple model of equations (9.3) and (9.5) must, to be fair to the model, employ an exceedingly simple living population. The earliest experimental tests were done by Gause (1934) using yeast cultures. A recent set of experiments has been described by Vandermeer (1969). He reared laboratory cultures of four species of ciliate protozoa (three of *Paramecium* and one of *Blepharisma*) which he combined in pairs to form two-species populations. These, in the form of 5 milliliter cultures kept in 10 ml test tubes, were maintained at controlled temperatures in the dark. They were sampled daily and it was found that Gause's equations described the observations quite well. Because each population studied was made up of simple short-lived organisms and was kept in a small, uniform environment under constant conditions, it is not unreasonable to infer that the mechanism symbolized by Gause's equations provided the true explanation for the observations. However, this conclusion cannot be accepted unreservedly unless it is also possible to show by direct experiment that the growth rate per individual of each species is a linear function of the numbers of the two species. It should be recalled (see page 52) that scarcely distinguishable growth curves for one-species populations can result from a variety of different models; the same is true for two-species populations.

This discussion shows how necessary it is to consider carefully whether one's simple, basic model (in this case the Gause equations) does, in fact, describe a truly simple biological system (in this case a small, uniform culture of two species of unicellular animals); only when one is satisfied that this is so (or, anyway, approximately so) is it reasonable to hope that refined versions of the basic model will describe more complicated biological systems.

2 THE EFFECT OF DELAYED RESPONSES

As a demonstration of how exceedingly complex a model can become when even the simplest "refinements" are made for the sake of realism, we shall consider the consequences of modifying Gause's equations to allow for delayed responses of the two species to crowding. We have already (page 79) discussed how delayed response (or lag) affects the growth of a one-species population; it may cause prolonged oscillations in population size. The difference equation describing the phenomenon appears as equation (5.4) on page 80. Equations (9.5) can be adjusted to allow for lag in a similar manner but there are now two lags to take account of, one for each species. Denote them by L_M and L_N. That is, the growth rate per individual of the M's is assumed to depend on the state of the combined population L_M time units previously; L_N is defined analogously.

The required equations are therefore

$$M_{t+1} = \frac{\lambda_M M_t}{1 + \alpha_M M_{t-L_M} + \gamma_M N_{t-L_M}}$$

$$N_{t+1} = \frac{\lambda_N N_t}{1 + \alpha_N N_{t-L_N} + \gamma_N M_{t-L_N}}.$$

(9.9)

Observe that the first, say, of this pair of equations has M_t, not M_{t-L_M}, in the numerator. The reason for this becomes clear if one considers what would happen if the growth of the M's were not inhibited by the presence of other individuals of either species. We should then have $\alpha_M = \gamma_M = 0$ and $M_{t+1} = \lambda_M M_t$; in other words, the population of M's would grow unhindered with a finite rate of increase λ_M. When competition is allowed for, this rate of increase is reduced by an amount that depends on the numbers of M's and N's that were simultaneously present L_M time units earlier. These values therefore appear in, and augment, the denominator.

As a numerical example, consider the growth of a combined population of M's and N's when the λ's, α's and γ's have the same values as in equations (9.8); we must also choose values for L_M and L_N. The computations are like those described in detail for the lag-free case (see page 213) except that it is now necessary to have several "initial values", not merely M_0 and N_0, to start the two sequences. The number of initial values required (for each species) is $L_M + 1$ or $L_N + 1$, whichever is the larger. The data for Figures 9.5 and 9.6 were obtained by using as initial values the first few terms of the sequences of M's and N's already computed to draw the curves in Figure 9.3a (equivalently, the trajectory starting at (2, 12) in Figure 9.3b).

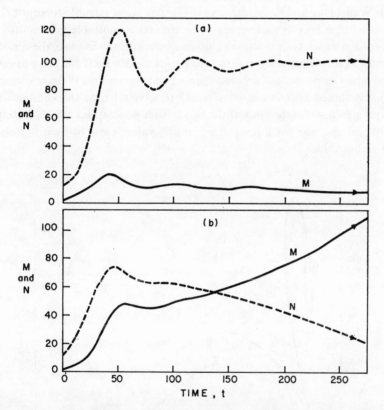

FIGURE 9.5 The behavior of combined populations of M's and N's when each species exhibits a lag in its response to crowding. In (a) $L_M = 5$ time units and $L_N = 15$; in (b) $L_M = 15$ and $L_N = 5$. The trajectories of these two combined populations are shown in Figure 9.6

When time lags are allowed for, a different trajectory is obtained for every pair of values assigned to L_M and L_N. The number of possibilities is thus enormous. Shown here are the results when $L_M = 5$, $L_N = 15$ (Figure 9.5a and the left branch of the trajectory in Figure 9.6); and when $L_M = 15$, $L_N = 5$ (Figure 9.5b and the right branch in Figure 9.6).

Some interesting conclusions can be drawn from the graphs shown. In the first place, it is seen that oscillations can be set up as was true in the one-species case described in Chapter 5. Secondly, the final outcome of competition depends on the lengths of the two time lags as well as on the sets of initial values. Thus in the present two examples, which are completely identical except for the lag times, the species with the slower response is the winner; (it should not be inferred that this is true in general). However, according to Wangersky and Cunningham (1957), it is never possible to adjust the lengths of time lags in such a way as to convert an unstable system into a stable one in which both species can survive permanently. Even so, the species destined to lose eventually may manage to maintain itself for a far longer period when responses are delayed than when they are not. If the trajectory of the combined population goes round in a spiral near the equilibrium point, it signifies that the sizes of the two species-populations are oscillating, slowly perhaps, and for a long time, around values close to their unstable equilibrium values.

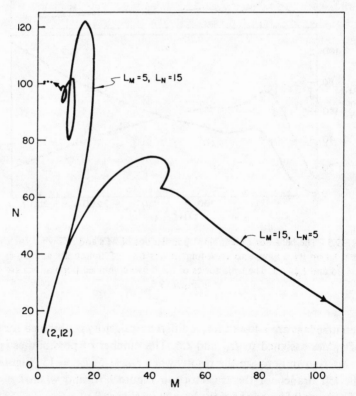

FIGURE 9.6   The trajectories of the two combined populations whose changes with time are shown in Figure 9.5. The different outcomes represented by the two trajectories are wholly determined by the different durations of the lags

Before leaving the subject of time lags still another complication needs to be mentioned. In the foregoing account it was assumed that the lag in response shown by each species (L_M time units for species M and L_N for species N) was due to a single cause and could be represented by a single number. In fact, this is an oversimplification. As Wangersky and Cunningham (1956) have argued, there are two separate sources of lag and ideally they would be kept separate in the mathematical analysis. The two lags are: (i) a reproductive-time lag equal to the length of the reproductive cycle; and (ii) a reaction-time lag equal to the time it takes for the animals to respond to crowding. As an example of reaction-time lag, imagine the decrease in a population's growth rate that must result from malnutrition caused by intense competition. It could well happen that the animals must be malnourished for an appreciable time before their reserves become so depleted that their death rate rises and their fertility wanes. Thus the reaction to malnutrition is not necessarily immediate. As Wangersky and Cunningham point out, in small populations just beginning a period of growth, reproductive lag will be more important than reaction lag. Later, when the population is approaching saturation, reaction lag is likely to predominate.

3 THE INFLUENCE OF CHANCE ON THE OUTCOME OF COMPETITION

So far we have treated competition as though it were a deterministic process; we have assumed that if Gause's equations described the process of competition between two species in some particular case and if, also, the parameters (constants) in the equations are known, then the outcome of competition is predictable with certainty. Let us now look at the equivalent *stochastic* process. That is, we shall recognize that changes in a population's size do not go on continuously but result from a sequence of separate elementary "events", namely births and deaths, which are not individually predictable. Consequently, future population changes cannot be foreseen precisely; the best that we can do is to state the probability that any given change will occur. In a few, very simple cases, these probabilities can be derived by mathematical reasoning. Thus if a one-species population is growing exponentially, it is possible to find the probability that it will be of a given size at a given future time; this is also true of populations growing logistically. Some of the results applicable in the case of exponential growth were given without proof in Chapter 1 (page 7 and Figure 1.3). Cases simple enough for mathematical solutions to be derived are few, however. In the majority of cases results must be obtained by what may be called *Monte Carlo simula-*

tion. We shall now describe how this is done to forecast a possible sequence of events in a population of two competing species. It should be ralized that simulation by various methods is applicable to the study of a multitude of biological systems, deterministic as well as stochastic; (the topic is discussed more fully on page 231). Monte Carlo simulation is the only feasible method for simulating any but the simplest stochastic process; it entails predicting a sequence of elementary events one after another in a manner that allows a chance mechanism to influence each event in turn. The object is to obtain a realistic simulation by allowing chance to affect the simulated process in the same way as it affects events in real life.

We shall now simulate the stochastic form of the process described by equations (9.7) which were

$$\frac{1}{M}\frac{dM}{dt} = 0.100 - 0.0007M - 0.001N$$

$$\frac{1}{N}\frac{dN}{dt} = 0.075 - 0.0007N - 0.0007M.$$

(9.7)

These equations give only the *net* rate of change of the two species-populations. For each species this net rate is the resultant of the birth rate and death rate, and to carry out a Monte Carlo simulation it is necessary to keep the two rates distinct so that we may generate each separate elementary event in turn.

Writing b_M and d_M for the birth rate and death rate of species M, and analogously for species N, let us now suppose that

$$b_M = 0.11 - 0.0007M - 0.001N,$$

$$d_M = 0.01,$$

$$b_N = 0.08 - 0.0007N - 0.0007M,$$

$$d_N = 0.005.$$

(These formulas* are believed to apply, at least approximately, to the two species of flour beetle mentioned on page 212. Further details will be given presently.) Thus the birth rate of each species is reduced by the presence of competitors of its own and the other species, but the death rates are assumed to be constant. Notice that for each species the difference between its birth and death rates gives its net rate of change per head as specified by equa-

---

* If M or N becomes so large that b_M or b_N is negative, the relevant birth rate is assumed to be zero instead of being given a negative value.

tion (9.7). That is,

$$b_M - d_M = \frac{1}{M} \frac{dM}{dt} \quad \text{and} \quad b_N - d_N = \frac{1}{N} \frac{dN}{dt}.$$

We are now in a position to start the simulation process. Because of the tremendous amount of repetitive computation involved this is, in practice, always done by high-speed computer but we shall here describe in full detail the simulation of the first event to illustrate how it is done.

To begin it is necessary to choose numerical values for the starting sizes, M_0 and N_0, of the two populations. Let us put $M_0 = 20$ and $N_0 = 40$. We now consider the four possible events that could happen and calculate their probabilities using the chosen values of M_0 and N_0. The events are:

Birth of an M with probability proportional to $Mb_M = 1.12$;

Death of an M with probability proportional to $Md_M = 0.20$;

Birth of an N with probability proportional to $Nb_N = 1.52$;

Death of an N with probability proportional to $Nd_N = 0.20$.

Since these four events exhaust the possibilities, the next event *must* be one of them. Therefore the separate probabilities must sum to unity and it follows that

$$\text{Pr (birth of an } M) = p_M = Mb_M/(Mb_M + Md_M + Nb_N + Nd_N)$$

$$= 1.12/3.04 = 0.368.$$

Similarly,

$$\text{Pr (death of an } M) = q_M = 0.066,$$

$$\text{Pr (birth of an } N) = p_N = 0.500,$$

$$\text{Pr (death of an } N) = q_N = 0.066.$$

Now, in imagination, divide a line of unit length into four non-overlapping intervals whose lengths are equal to these probabilities. We can then associate each event with an interval whose length corresponds with its probability of occurrence. Thus:

Birth of an M corresponds to the interval 0 to 0.368 of length p_M.

Death of an M corresponds to the interval 0.368 to 0.434 of length q_M.

Birth of an N corresponds to the interval 0.434 to 0.934 of length p_N.

Death of an N corresponds to the interval 0.934 to 1 of length q_N.

Now pick a random number between 0 and 1 from a table of random numbers and let the simulated event occur that corresponds to the number. For example, if the number is 0.713, the associated event is the birth of an N. As a result of this first event the number of N's in the combined population rises from 40 to 41 and the number of M's remains constant at 20. This completes the first event. To simulate the next one all the probabilities must be computed anew, taking account of the new population sizes following the first event. The huge amount of computing that has to be done to simulate even a short sequence of events is obvious.

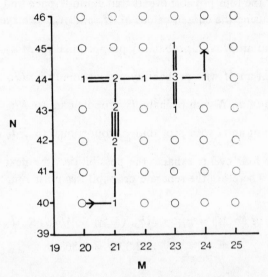

FIGURE 9.7 A small section of the stochastically simulated trajectory of a combined population of M's and N's. Each circle shows a possible species composition. The initial sizes of the populations were $M_0 = 20$, $N_0 = 40$. The simulation ends after 17 steps at $M = 24$, $N = 45$. The first six events were: birth of an M; birth of an N; birth of an N; birth of an N; death of an N; birth of an N

The trajectory of a combined population simulated in the way just described will resemble Figure 9.7. Observe the scale of the axes and compare them with those in Figure 9.3b which shows the deterministic trajectory of a population governed by the same model. It will be seen that Figure 9.7 gives an extremely small segment of the trajectory. Thus Figure 9.7 bears the same relationship to Figure 9.3b that Figure 1.3 does to Figure 1.2 (see pages 5 and 4). At the birth of an M the leading point of the trajectory advances one unit to the right, and at the death of an M one unit to the left. Likewise, at the birth or death of an N the leading point advances one unit up or down.

The numerals at each point along the trajectory show the number of separate occasions on which the combined population had the composition corresponding to that point. Clearly, if there chances to be a long run of alternating births and deaths in one species, the trajectory's leading point will go back and forth between the same two points many times over. As a result, appreciable changes in the population's composition may sometimes be very slow indeed. It would not be difficult to simulate a stochastic version of Figure 9.3a as well, although the result would be of less interest. To do so it would merely be necessary to simulate the stochastically varying time intervals between any event and the next.

We consider now what new information can be gained by allowing for the fact that in real life the competition process is stochastic. Recall that when we treated the process modeled by equations (9.7) as deterministic we were able, by means of the difference equations (9.8), to predict the fate of a mixed population exactly, provided we knew the starting conditions. That is, we needed to know the numbers of M's and N's simultaneously present at some one moment and this pair of numbers fixed the position of the beginning of a trajectory like those in Figure 9.3b. By predicting far enough into the future we could foretell which species would win and which lose; (since for this case the equilibrium is unstable, one species inevitably ousts the other). When we allow chance events to occur, as happens when the stochastic version of the model is simulated, it is no longer possible to make definite predictions about the outcome. Chance wanderings of a trajectory will sometimes bring about the opposite result to that expected in the deterministic case. Thus when chance is allowed for we can no longer assert that a trajectory starting from $M_0 = 2, N_0 = 12$, say, must end in victory for the M's and extinction for the N's. All that can be said is that there is some probability that the M's will win. The best way of finding the numerical value of the probability is by carrying out numerous computer simulations of the process, all starting with $M_0 = 2$ and $N_0 = 12$, and observing the proportion that end in victory for the M's. This proportion constitutes an estimate of species M's probability of winning.

An investigation along these lines has been done by Barnett (1962). He obtained computer-generated stochastic trajectories on a cathode ray tube and gives photographs of some examples. He also gives a table of estimated probabilities of victory for species M for an array of possible starting conditions. As examples we quote his results when the starting populations were the same as those used to begin the trajectories shown in Figure 9.3b.

It is interesting that for $M_0 = 2$, $N_0 = 12$, species M's probability of winning is estimated to be somewhat less than one-half notwithstanding the fact that according to the deterministic theory it is bound to win.

Initial number of M's	Initial number of N's	Estimated probability that the M's will win
2	25	0.078
2	15	0.099
2	12	0.456
4	12	0.823

The biological work that led Barnett to his theoretical investigations is a long series of experiments performed by Park and his associates on intra- and inter-specific competition in two species of flour beetle, *Tribolium castaneum* and *T. confusum*. The work was begun in the 1940's but recent accounts with references to the early work can be found in Park *et al.* (1964) and Leslie *et al.* (1968). The beetle populations were reared in vials of wheat flour to which a small amount of brewer's yeast had been added; the medium was changed monthly at which time a census of the beetles was made. In this way the histories of real populations of animals, living in restricted environments under controlled conditions, were accurately recorded. Population growth was allowed to proceed under constant physical conditions for long periods. One experiment of the series lasted for over $4\frac{1}{2}$ years which, in terms of numbers of generations, is equivalent to more than 1300 years of human history.

It was found that mixed populations of the two beetle species seemed to behave in accordance with Gause's equations with the values of the parameters depending, as would be expected, on the habitat conditions. Under some conditions of temperature and humidity *T. castaneum* won every time. Under other conditions *T. confusum* won every time. The most interesting results were obtained under intermediate conditions for which the outcome was indeterminate; *T. castaneum* won in a proportion of times and *T. confusum* in the remainder. It appeared that in these conditions—the temperature was 29°C and the relative humidity 70%—the combined population had an unstable equilibrium and it was inferred that the process could be described by equations (9.7).

Numerical values for the coefficients in the equations were obtained as follows (Bartlett, 1960). *Tribolium castaneum* is species M, and *T. confusum* species N.

The intrinsic rate of natural increase of *T. castaneum*, r_M, had been calculated from sets of experimentally observed m_x and l_x values (see page 12 et seq.) obtained by growing populations of this species by itself. It is $r_M \fallingdotseq 0.10$.

By allowing populations of each species to grow to saturation alone in the experimental environment, their one-species saturation levels, K_M and K_N, were estimated. They were $K_M \fallingdotseq 140$ and $K_N \fallingdotseq 110$.

The inhibitory effect of T. *castaneum* on itself, s_M, was given by $s_M = r_M/K_M \fallingdotseq 0.0007$. Intuitive evidence suggested that s_N was about the same whence $s_N \fallingdotseq 0.0007$ also. We can now find the intrinsic rate of increase of T. *confusum*, r_N. It is $r_N = K_N s_N \fallingdotseq 0.075$. (Some numerical rounding was done for convenience since it was known that subjective judgements had led to fairly crude approximations.)

As to the coefficients u_M and u_N, relating to the inhibitory effect of each species on the other, it was judged that u_N (the effect of species M, T. *castaneum*, on species N, T. *confusum*) was about the same as s_M and s_N, the effect of each species on itself. Hence $u_N \fallingdotseq 0.0007$. But u_M appeared somewhat greater and was set equal to 0.001.

Lastly, for stochastic simulation, the birth rates and death rates need to be known. It was guessed that for T. *castaneum* the death rate (assumed independent of the numbers of beetles of either species present) was $d_M = 0.01$. T. *confusum*'s death rate appeared to be about half as great, so $d_N = 0.005$. Then the birth rates, b_M and b_N (which are functions of M and N) were obtainable by subtraction since $b_M = \dfrac{1}{M} \dfrac{dM}{dt} - d_M$ and likewise for b_N.

Before leaving the subject of these experiments it should be mentioned that Lerner and Dempster (1962) do not agree that mixed populations of these beetles exhibit unstable equilibrium. Experimenting with genetically purer strains of the two species they found that the results of competition were not unpredictable, and they believe that the indeterminate results obtained by Park and his associates stemmed from variations, among the different experiments, in the mixtures of genotypes forming the initial populations. However, Leslie et al. (1968), after experimenting with inbred strains of the beetles, still report indeterminate outcomes.

4   SOME GENERAL REMARKS ON ECOLOGICAL MODELS

At this point enough "predictive ecological models" of various sorts have been described to make a general discussion of such models worth while.

Let us first define the term *predictive ecological model*. We use it here to connote an equation, or set of equations, intended to predict the future state of an ecological system. The predictions envisaged range from those of major practical importance to the often seemingly trivial ones on whose correctness a theory of population change stands or falls.*

There are three quite independent ways of classifying models dichotomously into two contrasted kinds. Thus one may distinguish deductive models from inductive ones; deterministic models from stochastic ones; and solube models from insoluble ones. We consider these distinctions in turn.

1. *Deductive* versus *inductive models*: (They may also be called, respectively, *theoretical* and *empirical* models.) A deductive model is one that purports to explain the processes going on in an ecological system and thus permit deduction of the system's future state. Examples are the models described in Chapter 4 (page 46 et seq.) which were designed to provide a reasonable explanation, as well as a means of prediction, of the growth of a one-species population.

In contrast to this, an inductive model aims merely to predict the future on the supposition that it will be like the past, without enquiring into underlying mechanisms. A simple analogy should make the distinction clear. If one proves Pythagoras's theorem (that for a right triangle on a plane surface the square on the hypotenuse is equal to the sum of the squares on the other two sides) by traditional Euclidean argument, that is deduction. If one reaches the same conclusion by measuring the sides of many right triangles and assuming that what is true of them is true of all, that is induction.

Deductive models are typically the province of the academic ecologist. Applied ecologists, whose work often requires that they produce a model without delay (instead of carefully reasoned arguments as to why this cannot be done), must often use inductive models. For those concerned with forestry, fisheries biology, game management, pest control, eutrophication problems, and numerous other branches of modern environmental science, a model of some sort is essential as a guide to action. It need not be elaborate. Even the crudest "educated guess", when formally propounded, constitutes a model. Some examples will be discussed at the end of this chapter. Others, together with a detailed discussion of the philosophy of model-building, may be found in Watt (1961, 1968).

Often the distinction between inductive and deductive models is less clear than we might wish. For example, if a symmetrical, sigmoid curve appears to fit an empirical growth curve, this may result more from good luck than

---

* Notice that this discussion is restricted to dynamic models, that is, those relating to growth and change. There are many other kinds of ecological models; ecologists often use the word *model* merely as a synonym for *theory* or *hypothesis*.

from any correspondence between actual biological processes and the simple logistic model (cf page 51). Similarly with the competing populations of flour beetles; the beetles have a complicated life history, they are cannibalistic, they contaminate the flour they occupy, and their age distribution does not remain constant for long. Nevertheless, there are good reasons, as we have explained earlier in this chapter, for supposing that the exceedingly simple Gause model describes the growth of two competing species of these beetles. Perhaps this is so because the effects of the complications we have disregarded happen to be negligible, in which case the model is indeed deductive. The same could be said if the many refinements the model appears to need to make it realistic tend to cancel out so that neglecting them again makes no difference. But it could be that the resemblance between the predictions of the model and the actual behavior of the living animals is only a coincidence—whether a happy one or not is debatable. We then have an inductive model masquerading as a deductive one. We have already argued (page 141) and it will do no harm to stress the point once more, that models should be subjected to a variety of different *kinds* of tests so that one may avoid being deceived by the apparent resemblance between an observed occurrence and *one* of the model's consequences. Unless the model is truly explanatory it is highly unlikely that all its consequences, or even two or three of them, will be duplicated by the real system.

2. *Deterministic* versus *Stochastic Models*: The distinction between these two kinds of models has been discussed earlier in this chapter and there is little more to say. Stochastic deductive models are unavoidably more complicated than deterministic ones but, because of their realism, are greatly to be preferred. Further, their use tends to discourage the unwarranted proliferation of special models to account for particular observations. Recall, for instance, the flour beetle experiments in which, even though conditions were identical, the victor was sometimes *Tribolium castaneum* and sometimes *T. confusum*. This phenomenon is inexplicable by any deterministic theory (if we accept that it was not due to the use of genotypically different populations in different experiments as Lerner and Dempster (1962) surmise). But as soon as we acknowledge that the outcome of an experiment is to some extent governed by chance, in other words as soon as we postulate a stochastic model, the difficulty disappears.

3. *Soluble* versus *Insoluble Models*: This distinction may be illustrated with models we have already described. The best known soluble model is that for exponential population growth (see page 11). The model itself consists of the assertion that the growth rate per individual remains constant as the population grows. The equivalent statement in symbols is the differential

equation
$$\frac{1}{N} \frac{dN}{dt} = r.$$

and the solution of the equation is

$$N_t = N_0 e^{rt} \quad \text{(cf page 11)}.$$

In saying that the differential equation is soluble, we mean that it is possible to obtain an exact mathematical formula expressing the dependent variable, N_t, in terms of the independent variable, t. Four other soluble models are exemplified by the sets of differential equations (4.2), (4.3), (4.4) and (4.5) of Chapter 4 (shown graphically in Figure 4.1, page 51), which constitute four different models that might explain the process of growth in a population whose members compete among themselves; their solutions are obtained in Appendix 4.1. Stochastic models, too, are sometimes soluble. Although they are usually of much greater mathematical difficulty, there is no intrinsic reason why they should be insoluble. For instance, the solution of the stochastic version of the exponential growth model is given by the formulas for $\mathscr{E}(N_t)$ and Var (N_t) on page 7. The model stipulates that the *probability* that an individual is born or dies remains constant as the population grows. Two formulas are required for the model's solution, one giving the expected (or predicted) future size of the population, and the other giving a measure of the precision of the prediction.

As an example of an insoluble model we may quote Gause's equations, (9.3). Even such an apparently simple pair of equations as this cannot be explicitly solved; that is, we cannot express M_t and N_t as functions of t. Indeed, of all the many models of population growth that could conceivably be envisaged, only a very small minority is soluble. This is because most models, when mathematically stated, consist of one or more differential equations; they amount to assertions about the *rates* at which population processes happen and how these rates vary. And only a minority of differential equations can be solved or integrated. However, the difficulty can always be circumvented by using so-called numerical methods. Even when the solution of an equation cannot be represented by a formula, it is nonetheless possible to find the numerical solution for any particular case, that is the numerical value of N_t for any given t. Large amounts of computation may be necessary and therefore, if a model is to be useful, its complexity must not be so excessive as to overtax the current capabilities of electronic computers; presumably these will continue to improve.

As an alternative to computing particular numerical solutions it is sometimes possible, as we have shown in several contexts (see pages 79, 83 and

210), to replace a differential equation by a difference equation that approximates it. The difference equation then serves as a formula for calculating the size of the population at a succession of instants separated by constant intervals.

The predictions of any model, whether it be soluble or insoluble, deterministic or stochastic, deductive or inductive, are most clearly portrayed in the form of a simulated growth curve. *Simulation* consists in forecasting the actual sequence of changes that a population would be expected to undergo if its behavior were governed by the model. We have already shown many examples, Figures 1.1, 1.2, 1.3, 2.1, 4.1, 4.2, 5.1, 5.2, and 9.3–9.7 (pages 3, 4, 5, 22, 51, 55, 77, 80, 214–224) all show simulations of various deductive models. Yet another example is given in Figure 9.8; it is redrawn from Garfinkel and Sack (1964) and shows the changes during one "year" of four of the six species in an imaginary community. The authors explored the feasibility of carrying out computer simulations of systems somewhat more complex than the one- and two-species populations usually modeled. The community they invented consisted of three species of plants (a tree, a shrub and a grass), a large and a small herbivore, and an omnivore; the three animals differed in their birth and death rates according as they were well fed or staved. The model is deterministic.

Goodall (1967) discusses computer simulation of a much more realistic imaginary ecosystem. His object was to test the value of simulation in predicting the effect of grazing on the vegetation of semi-arid sheep pastures in

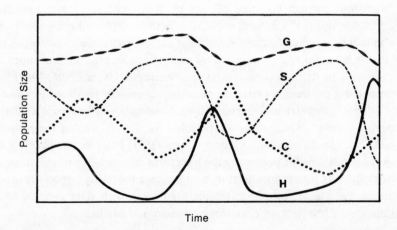

Time

FIGURE 9.8 Deterministic simulation of the behavior of an artificial many-species community. There were six species in the model community and the changes in population size over a period of time of four of them are shown here. They are: *G*, a grass; *S*, a shrub; *C*, an omnivore; *H*, a small herbivore. (Adapted from Garfinkel and Sack, 1964)

Australia. Several variables were allowed for, among them differences in the growth rate, palatability and other properties of five species of forage plant, and also differences in rainfall and in stocking rate.

Perhaps the most ambitious simulations are those carried out by fisheries biologists and foresters. Actual ecosystems (as opposed to imaginary examples) have been simulated and the results have been used as a guide to action. Fisheries biologists have to plan the wise exploitation of fluctuating populations of fish, and fish are capable of traveling long distances, are invisible most of the time, are of great economic importance, and are vulnerable to overfishing. Successful predictions about the effects of various management policies are difficult but necessary. Paulik (1967) has discussed the use of computer simulation for this purpose. He mentions, for example, that the simple logistic model of population growth has proved satisfactory in setting annual catch quotas for the yellowfin tuna fishery of the eastern Pacific. The same model has been used in the management of Antarctic whale stocks. These are examples of the use of the logistic equation as an inductive model; experience has shown that this simple model gives an acceptable approximation to the unknown, and presumably extremely complicated, true model.

The simulation of complicated models to predict changes in fish populations has been described by Larkin and Hourston (1964); they simulated the growth of stocks of Pacific salmon spawning in a large river system, and allowed for differences in environment, differences in age at spawning, and for the effects of intraspecific competition.

Projection matrices (cf. page 20) are proving useful in various branches of applied ecology. If a natural population is to be exploited one can, at any rate in theory, predict the consequences of various management stategies on the future behavior of a population by making appropriate modifications to the elements of its projection matrix. For example, Usher (1966, 1969) has constructed a projection matrix to describe the growth of a forest of Scots pine (*Pinus sylvestris*) in Inverness-shire, Scotland, and has estimated the maximum rate of exploitation compatible with the stability of the proportions of trees in different size classes. Jensen (1971) has discussed the use of projection matrices for predicting the long-term effect on fish populations of toxic pollutants that reduce fertility and increase mortality (especially in the youngest age class); he illustrates his arguments with data relating to a population of brook trout (*Salvelinus fontinalis*) in Michigan.

CHAPTER 10

Competition II
Field Observations and their Interpretation

ACCORDING TO THE mathematical arguments of Chapter 9 if populations of two species are growing together in the same limited space, one species is bound to survive at the expense of the other unless a stable equilibrium becomes established; and this can happen only if each species-population inhibits its own growth more than that of its competitor's. But when this is true, that is when *intra*specific competition is more intense than *inter*specific competition, it must follow that population growth in the two species is not governed by the same limiting factor. Thus when the further growth of each population is prevented because some necessary resource is in short supply, continued coexistence* of the two species is possible only if two different resources are involved. If a single resource, a particular food item for instance, is responsible for population regulation in both species, then ultimately one species must oust the other.

The foregoing is a statement of the famous "coexistence principle" or "competitive exclusion principle" of theoretical ecology. Other names for it are "Gause's hypothesis", "Grinnell's axiom" and "the Volterra-Lotka law" and it has been a constant topic for debate among ecologists throughout the twentieth century. Grinnell (1904) seems to have originated the principle when we wrote:

Every animal tends to increase at a geometrical ratio, and is checked only by limit of food supply. It is only by adaptations to different sorts of food, or modes of food-getting, that more than one species can occupy the same locality. Two species of approximately the same food habits are not likely to remain long evenly balanced in numbers in the same

---

* In discussing competition among plants, Harper and McNaughton (1962) use a different terminology. They say that two species *coexist* when their individuals grow sufficiently close to one another to hybridize (if they are interfertile). And that they *cohabit* when they are so close as to compete for the same resources. In this book the commoner terminology is used: here *coexistence* connotes what Harper and McNaughton would call *cohabitation*.

region. One will crowd out the other; the one longest exposed to local conditions, and hence best fitted, though ever so slightly, will survive, to the exclusion of any less-favoured would-be invader.

DeBach (1966) has collated recent opinions on the principle and considers that the following statement sums up the modern version: "Different species which coexist indefinitely in the same habitat ... must not be ecological homologues." This statement of the principle has aroused some fairly heated controversy. Cole (1960), for example, has called it a "trite maxim" and this seems a fair criticism. To call two species "ecological homologues" must mean, if it means anything, that they are ecologically identical: that is, that their habitat requirements and tolerances are identical. But it is obvious that two different species, in virtue of the fact that they are different species, must differ in some respect and the difference, however small, must affect their ecological properties in some way. If this is accepted it follows that there can be no such things as ecological homologues and hence that the problem of their coexistence does not arise. The whole argument thus degenerates into a "neat circular package" (Cole, 1960).

But if we go back to the principle in its earlier, sturdier form, as propounded by Grinnell, it is obviously sensible. It states no more than that competition between two (or more) species for the same resource will not continue unresolved indefinitely. In this form the principle is not a tautology but a testable hypothesis susceptible (if it is erroneous) to disproof. However, it must be admitted that (as we shall see later) disproof may be very difficult.

The recent literature is full of examples of competitive exclusion and we shall summarize a varied sampling of them. It will be noticed that the competing species dealt with are always closely related and often congeneric. There is, of course, nothing in the principle that restricts it to closely related species. It is simply that such species are more likely than unrelated ones to depend for their livelihood on the same resource. And their coexistence, when it is observed, is more likely to catch the eye of ecologists and invite investigation than is the coexistence of dissimilar species.

1 EXAMPLES OF COMPETITIVE EXCLUSION

We now look at eight examples chosen tor their variety.

1] Pocket gophers

Miller (1967) has described a clear case of competitive exclusion among pocket gophers (Geomyidae) in the foothills of Colorado. Two of the species he studied were the congeners *Thomomys bottae* and *T. talpoides*. Both species require space for their burrows and both prefer deep soil of fine texture. However, burrows of the two species are not found intermingled on a hillside. *T. bottae*, which is the superior competitor, takes possession of the lower slopes with better soil and excludes *T. talpoides* which is forced to move up-slope to sites where the soil is shallower and coarser. This phenomenon, the "banishment" of an inferior competitor to a less favorable habitat is quite often observed. The winning species excludes the loser from a habitat known to be favorable to both of them and thus demonstrates its superiority in overt competition. The loser, however, whose requirements are less specialized than the winner's, is able to exist in inferior environments where the winner cannot follow it. As a result both species survive, but in different locations.

2] Redwinged and Yellowheaded Blackbirds

(Miller, 1967, 1968; Nero, 1964; Orians, 1961; Robertson, 1972).
This is another example of the banishment of an inferior competitor to a less favorable habitat. Miller (1967) has described the course of events in a Saskatchewan marsh in the blackbirds' breeding season. Both species breed in deep freshwater marshes, building their nests in emergent vegetation such as cattails (*Typha*), bulrushes (*Scirpus*) and reeds (*Phragmites*); but the redwing (*Agelaius phoeniceus*) can also nest successfully in the shallower parts of a marsh and in the weeds and shrubs on the surrounding dry land, whereas the yellowhead (*Xanthocephalus xanthocephalus*) cannot. Although its needs are less specialized, the redwing prefers the deeper part of a marsh and Miller found that in early spring this species established territories throughout the whole of the marsh he studied. But a week or two later, when the yellowheads reached the area from their wintering grounds in the south, they drove the redwings out of the deep marsh and took possession of the most favored sites for themselves. However, the redwings were able to continue breeding in the less favorable peripheral sites.

3] Barnacles

Connell (1961*a*, 1961*b*) has described another example of the banishment of an inferior competitor to a less favorable zone. Presumably the reason why cases such as these are so often reported is that the less successful competitor is still present and observable in its environmentally inferior refuge. If one species ousted another entirely and there were no adjacent zone in which the loser could persist undisturbed, there would be no evidence that competitive exclusion had taken place. Connell's observations were of barnacles, which he studied in the Isle of Cumbrae, Scotland; the two species are *Balanus balanoides*, the winner, and *Chthamalus stellatus*, the loser. The competition is for space, for rock surfaces on which the motile larvae (cyprids) can attach themselves for their lives as adults. Connell found that the two barnacle species showed clear zonation, with *Chthamalus* in a higher zone, between the high water levels of spring and neap tides, and *Balanus* occupying the rest of the intertidal zone below. The *Chthamalus* zone was too dry for *Balanus* but the lower zone could have been occupied by *Chthamalus* if *Balanus* had not excluded it. Connell demonstrated this experimentally. In certain areas he protected *Chthamalus* individuals from competition by removing all the *Balanus* individuals surrounding and touching them, and it was found that survival was very much higher among the protected *Chthamalus* than elsewhere. Where both species settled and began their growth as adults together, *Balanus* smothered, undercut or crushed *Chthamalus* and managed to exclude it almost entirely.

4] Chipmunks

Two separate cases of competitive exclusion among pairs of chipmunk species (of the genus *Eutamias*) may be mentioned. In the first case Sheppard (1971), working in the Rocky Mountains in Alberta, found that *Eutamias amoenus* and *E. minimus* were competing. *E. amoenus*, which is the larger and more aggresive species, drives *minimus* from subalpine forest but the latter can persist in alpine habitats where its smaller size enables it to survive on smaller food supplies. A second chipmunk example is due to Brown (1971) who describes competition between the chipmunks *Eutamias umbrinus* and *E. dorsalis* in mountain ranges of the Great Basin region of Nevada. In this case each species was a "winner" in its own area. The more terrestrial species, *dorsalis*, which is very aggresive, is the superior competitor in areas where the trees are widely spaced and *umbrinus* must run on the ground to escape its attacks. But where tree growth is dense not only can the more arboreal *umbrinus* escape pursuit easily but the fact that it is less aggressive gives it a competitive advantage. This is because in the presence of numerous chipmunks of another species *dorsalis* spends so much of its time chasing them (unsuccessfully) that it has insufficient time to feed itself and to do so must retreat to its own area. Thus this example differs from the preceding ones in which the optimum habitat was the same for both competitiors and survival of the inferior competitor was possible only because its requirements were less specialized.

5] Salamanders

Competitive exclusion between two species of terrestrial salamanders has been observed by Jaeger (1970). One species, *Plethodon cinereus cinereus*, lives in woodland and is widely distributed in eastern North America. The other,

P. richmondi shenandoah, has a very restricted range and is known only from three talus slopes in the Blue Ridge Mountains of Virginia. Both species lack lungs and must live in close contact with moist materials (soil, leaves and moss) so that respiration through the epidermis is possible, but *shenandoah* is more resistant than *cinereus* to dehydration. Jaeger found that the common species, *cinereus*, has excluded *shenandoah* everywhere except from isolated pockets of moist soil occurring as islands in the talus; *shenandoah* has managed to persist in these few remaining refugia only because *cinereus* cannot traverse the dry rocks of the talus to invade them. Presumably *shenandoah* will not survive for long and its exclusion by the stronger competitor will end with its extinction.

6] Ants

Another example in which changes in the geographic ranges of two species, caused by competitive exclusion, can be seen in progress has been given by Crowell (1968). The Argentine ant (*Iridomyrmex humilis*) was introduced into Bermuda in 1953 and since then has steadily expanded its range replacing the native ant, *Pheidole megacephala*, in all its habitats. The way in which the two species compete has not been discovered.

7] Hydrida

Many experimental studies of competitive exclusion have been performed since Gause (1934) first investigated the consequences of trying to grow two species of yeasts or two species of paramecia together in closed cultures. Slobodkin (1964) has described interspecific competition between two hydrida, one brown (*Hydra littoralis*) and one green (*Chlorohydra viridissima*). He maintained populations of the two species in synthetic pond water in Petri dishes and fed them with brine shrimps. The competing populations were censused at four day intervals. For census purposes the size of each population was recorded as the number of mouths to be fed, since once a new bud has started feeding it is competing with all the other mouths for food, irrespective of whether it is still attached to its parent. When the experiments were carried out in the light *Chlorohydra* won invariably, presumably because its possession of symbiotic green algae in the endodermis gave it an advantage. But in the dark, when the two competitors were on more equal terms, competitive exclusion did not take place.

8] Slime molds

Horn (1971) has carried out experiments on competition among four different species of cellular slime molds (Acrasieae). In nature they exist together in forest soils and all of them feed on a wide assortment of soil bacteria. However, when the molds were cultured together on nutrient agar and fed with a single strain of bacteria for which they were forced to compete, it was found every time that one species of mold always outcompeted the other three. Ten bacterial strains were tested and it turned out that a particular mold species usually won on a particular strain of bacteria. Thus the result of any one experiment was highly, but not absolutely, predictable. In a small proportion of cases the "wrong" mold won unexpectedly. In this respect the experiments resembled those of Park and his associates with *Tribolium* beetles (see page 226); it will be recalled that in some conditions, though victory for the winning species was always complete, it was impossible to tell in advance which species the victor would be. Returning to the slime molds, it is clear that the coexistence of several species in a small volume of forest soil does not contravene the coexistence principle. The several strains of bacteria in the soil have patchy spatial patterns and each patch of a particular strain favors the localized success of a particular mold.

2 EXAMPLES OF COEXISTENCE

Good examples of competitive exclusion like the eight given above are not hard to find and the foregoing list could easily be increased tenfold. It proves convincingly that competitive exclusion can and does occur. Besides observed cases, in which populations of the excluding and excluded species are found living in contiguous regions or zones, there must also be an enormous number of inherently unobservable cases in which the losing species have been excluded so thoroughly as to be locally or globally extinct. To speculate on what species, if any, might be present at a place if they had not in the past been forced out by those now living there is futile. But the fact that many, perhaps most, cases of competitive exclusion must be presumed to be of this inherently unobservable type is worth keeping in mind.

The fact that competitive exclusion is common does not mean, of course, that it always occurs. If we avoid versions of the principle that rest on circular argument it states only that, of a group of competing species, one will *eventually* oust the others *if* all the species-populations are regulated by a shortage of some resource and *if* the limiting resource is the same for all. Further implicit assumptions are that the competitors, and also the environment, remain unchanged for a sufficiently long period for the competition to reach a conclusion; and that the community is closed to immigration so that the losing species cannot be reinforced by new recruits coming from an area with different conditions. It thus appears that for competitive exclusion to take place some fairly stringent conditions must be met; we should not be surprised to discover numerous cases in which closely similar species do, in fact, live together successfully. However, if the coexistence principle is true, it must follow that observed examples of coexistence can be accounted for only by supposing that one or more of the necessary conditions for competitive exclusion listed above are not fulfilled.

Observed cases of coexistence very often are explicable in this fashion and we now consider a few examples. It should not be assumed that any observed instance has only one cause; in many cases several causes act in concert and it may be difficult to assess their relative importance. We label each group of examples with its chief cause or causes.

Insufficient time for equilibrium to be reached; populations not closed to immigration. The large numbers of coexisting species of similar ecology, especially diatoms and desmids, in the phytoplankton of large lakes has been discussed by Hutchinson (1961). The environment is very homogeneous in a body of turbulent open water but no one species adapted to this environment seems able to exclude the others; this in spite of the nutrient deficiency which develops in lakes in summer and which no doubt leads to intense competition. Hutchinson ascribes this seeming failure of the coexistence principle to two causes. The first is that the organisms' active growing season is too short for an equilibrium state (establishment of the strongest species as sole survivor) to be reached. The winter die-back always intervenes before the struggle has been resolved and when growth starts again the following spring, the individuals that emerge from dormancy to found new populations belong to many species. The second probable cause of coexistence is that many plankton species sink to the bottom of the water when turbulence is too slight to support them. There they exist as benthic organisms and do not compete because the benthic environment is sufficiently heterogeneous for each species to survive and multiply in a microhabitat to which it is especially adapted. During their benthic periods the several species can therefore build up big populations which serve as sources of immigration to the plankton.

Similar cases in which a struggle among competing species never has time to reach a conclusion are probably common in temperate latitudes. Hutchinson (1961) has argued that continued coexistence of competing species is likely whenever the time required for the exclusion of unsuccessful competitors is of the same order of magnitude as the growing season. Probably this accounts for many cases of coexistence among insects. An example is the occurrence of eight species of aphid of a single genus (*Dactynotus*) on a single species of plant (the goldenrod *Solidago canadensis*) in Ontario (Pielou, 1972c). The season in which aphids can multiply is too short for the most abundant aphid, *D. nigrotuberculatus*, to oust the others. Moreover the aphid species which are subordinate to *D. nigrotuberculatus* on *S. canadensis* are not subordinate on other *Solidago* species. Populations that serve as reservoirs for reinvasion of *S. canadensis* are therefore common on neighboring plants of related, but different, species.

Populations not resource-limited. Competitive exclusion is only to be expected when continued growth of the competing species is halted by lack of a necessary resource. When populations are *not* so regulated, many can coexist without competing. Thus coexistence can occur if species-populations are kept small by the attacks of predators. This has often been experimentally demonstrated. In Connell's (1961*b*) studies of competition between the barnacles *Balanus* and *Chthamalus* (see page 235) it was found that predation by the Atlantic dogwinkle, *Thais lapillus*, greatly reduced the intensity of competition between the two barnacle species. In Slobodkin's (1964) experiments with hydrida (see page 237) it was found that the invariable victory of *Chlorohydra* over *Hydra* (when cultures were kept in the light) no longer occurred when the mixed populations were subject to indiscriminate "predation"—actually the removal of a proportion of newborn animals, selected without regard to species, by the experimenter. When this was done, both species persisted in the cultures.

Another experimental demonstration of the way in which predation can prevent competition has been described by Paine (1966). Along an eight meter stretch of rocky shore in Mukkaw Bay, Washington, he repeatedly removed all individuals of the starfish *Pisaster ochraceus*, a carnivore of fairly catholic tastes. The fates of the littoral communities in the experimental area, and in an adjacent undisturbed control area, were then observed. In the natural community predation by *Pisaster* prevented populations of animals such as barnacles (*Balanus*), mussels (*Mytilus*) and goose-necked barnacles (*Mitella*) from appropriating all the available space. But in the area where *Pisaster* was artificially excluded, it was found that populations of two species of chitons, two of limpets and three of benthic algae, that had maintained themselves successfully in the undisturbed community, were progressively crowded out by the barnacles and mussels. The overall result was the reduction of what had been a fifteen-species system to an eight-species system.

Competitive advantage resulting from sparsity. In a contest between two species it is obvious that for one to oust the other it must maintain its competitive advantage at all densities. It is easy to visualize cases that do not meet this requirement: that is, in which the advantage alternates between the two species so that the temporarily sparser one is always favored at the expense of the temporarily commoner one. This must happen whenever a predator tends to adapt its hunting tactics to the pursuit of whichever happens to be the commoner at the moment of two possible prey species. It must also happen when the species whose density is higher is more susceptible to disease than its less dense competitor. An example (due to Hanna) has been quoted by Harper *et al.* (1961) in a general discussion of the coexistence principle as it applies to plants. Different varieties of wheat are attacked by different strains of the stem rust *Puccinia graminis;* as a consequence, a dense stand of wheat of one variety is more vulnerable to damage than is a mixture of varieties in which epidemics have less chance to develop. A similar explanation may account (in part) for the fact that several poppy species (of the genus *Papaver*) can live together in equilibrium (Harper and McNaughton, 1962): some of the mortality within each species-population is due to seedling blight and probably each *Papaver* species is attacked by a specific blight.

These are two examples of a phenomenon well known to economic entomologists and plant pathologists: if large areas are devoted to a single economically valuable species, be it an agricultural crop or a stand of forest trees, the risk from pests and disease is much greater than if several species had been grown intermingled.

Differences in density-regulating factors. The coexistence among *Papaver* species mentioned above merits further discussion. Harper and McNaughton (1962) investigated the phenomenon experimentally. Five annual species of *Papaver* were used in the experiments which were carried out in a garden divided into a number of small plots, each 35 centimeters square. The plots were sown with one, or a mixture of two, of the species and it was almost always found that mixed sowings yielded fewer adult plants than pure sowings. Moreover, in mixed sowings the mortality due to crowding was greater in whichever species happened to be the more abundant. These two observations considered together imply either that *intra*specific competition ("self-thinning") in the densely sown species was more marked in the presence of a few alien competitors than in a pure stand; or that *inter*specific competition ("alien-thinning") in a mixture brought about a disproportionately great reduction of the majority species by the minority species. Whichever may have been the cause, the result was that two-species mixtures were self-stabilizing. In a typical experiment, using species A and B, say, four sowings were made each of 225 seeds in which the seeds were present in the following proportions: All A; $8A/1B$; $1A/8B$; all B. Figure 10.1 illustrates the results. In the graphs the abscissa shows the composition of the seed mixtures, and corresponding to each sowing are two bars whose heights represent the numbers of plants of the two species that were alive at harvest time, four months later. Figure 10.1*a* shows what would have happened, but did not, if the number of full-grown plants of each species had been proportional to the number of seeds sown; Figure 10.1b shows an actual result and is typical. The way in which natural thinning tended to equalize the proportions of two species whose seeds had been sown in very unequal proportions was most striking. Harper and McNaughton do not attempt a full explanation of the phenomenon. As mentioned above it is probably partly due to seedling blight caused by the attacks of species-specific pathogens.

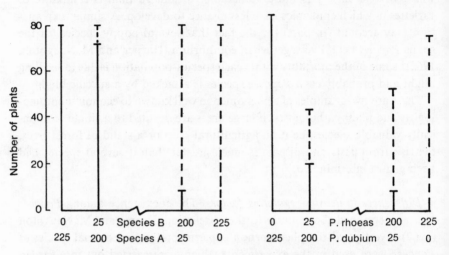

FIGURE 10.1   The number of mature plants obtained from mixed sowings of poppy seeds. (a) Hypothetical but never observed; the number of plants is proportional to the number of seeds. "Species A", solid bars; "species B", broken bars (b) A typical observed result. *Papaver dubium*, solid bars; *P. rhoeas*, broken bars. (Adapted from Harper and McNaughton, 1962)

But an additional important cause may be that seeds of the different species require slightly different microhabitats for successful germination. In Chapter 4 (page 63) we mentioned observations (by Sagar, using seeds of *Plantago* species) showing how exceedingly precise the germination requirements of a particular species may be. Seemingly trivial microhabitat differences, for example in the dampness and degree of compactness of the soil, can determine the success or failure of a seed; an area of a few square centimeters can contain a number of microhabitats each suited to the germination of one, and only one, plant species. These observations make it reasonable to suppose that the species of *Papaver*, also, differ among themselves in their precise site requirements for germination. When a mixture of seeds is sown, competition for the resources needed for germination is probably more intense within than between species and a limit to the number of seedlings of any one species that can become established is set by the availability of germination sites suitable for that species.

Genetic changes in the competitors. The coexistence principle presupposes that the competing species will not change genetically while the exclusion of an unsuccessful competitor is going on. Pimentel *et al.* (1965) have argued that natural selection automatically accompanies exclusion and that genetic change is therefore to be expected. Given two competing species, as the population of the losing species dwindles, those of its members that are competitively stronger are selectively favored. The proportion of strong competitors in this species therefore increases steadily until winners rather than losers predominate. At this point competitive advantage has switched from one species to the other and it may continue to alternate repeatedly. Thus at any instant natural selection favors the minority species, whose members are being selected for their ability to succeed in *inter*specific competition. Simultaneously, competition is going on within the majority species but its members are selected for their success in *intra*specific competition rather than for their ability to exclude the minority species. The advantage thus goes back and forth from one species to the other.

Pimentel and his associates demonstrated the process experimentally. The experimental animals were two species of fly, the housefly *Musca domestica* and the blowfly *Phaenicia serricata*. To begin with the houseflies were the more successful competitors, but later in the experiments success passed to the blowflies which finally excluded the houseflies. In separate experiments it was confirmed that the blowfly population used in the competition experiments has undergone genetic change; during the course of their competitive struggle they had evolved into more effective competitors against the houseflies.

Exclusion still in progress. The coexistence principle is a statement about populations that have reached equilibrium. Therefore observed cases of coexistence can always be accounted for merely by asserting that the competing populations have not reached equilibrium but that when they do only one species will remain. Equilibrium may require a very long time to become established. Thus, consider the simple two-species case. As was shown in Chapter 9, the occurrence of detectable changes in the relative sizes of two competing populations may be exceedingly slow and this because of three independent causes. First, if there are population sizes at which the combined two-species population could, theoretically, persist in an unstable equilibrium, actual changes will take place very slowly whenever the composition of the combined population happens to be not very different from the theoretical equilibrium composition. Second, if each species lags in its response to density changes both populations may slowly oscillate in size so that exclusion of the loser is long delayed. Third, stochastic (or chance) departures of the actual course of events from their theoretically predicted course may be such as to prolong the exclusion process.

The preceding paragraphs show that competitive exclusion cannot be counted upon to occur unless the following conditions hold:

i)   The sizes of all the competing populations must be regulated by the shortage of some necessary resource, the same resource for all competitors.

ii)   The competitors must remain genetically unaltered throughout the exclusion process.

iii)   The environmental conditions must be constant.

iv)   The community in which competition is occurring must be closed to immigration by incoming members of any of the competing species.

v)   Competition must have been going on long enough for equilibrium to have been reached.

Observe that competitive exclusion may sometimes happen even when one or more of the conditions listed above are not met. We are asserting only that the non-fulfillment of any one of them has on occasion been found to provide a perfectly satisfactory explanation for particular known cases of coexistence.

There is still another mechanism, originally suggested by Skellam (1951), that can account for the indefinitely prolonged coexistence of "complete" competitors. Indeed, given the mechanism, coexistence is a logically necessary consequence. We discuss it in the following section.

3   OUTNUMBERING OF THE STRONGER COMPETITOR

Imagine two species, such as annual plants, whose individuals are sessile consumers of resources (that is light, water, nutrients and space) only for one growing season. Between seasons the two species persist as seeds which are small, dormant inert bodies that demand no resources and are passively moved. Both species' seeds are dispersed in such a way that their pattern is random. Assume that the area we are concerned with contains a number of separate "safe sites", all of identical size and quality, which are the only places at which either species can grow. That is, all parts of the area except the safe sites are wholly inhospitable to both species. Further, any one safe site can support only one individual plant beyond the seedling stage so that if two or more seeds germinate in a site, all but one are bound to fail; and in a site where both species germinate, species A always succeeds at the expense of species B. Thus mature individuals of species A, the competitively stronger species, will be found in sites where seeds of its own species alone, or of both species, chanced to germinate. Whereas mature individuals of the weaker species, species B, will be found only in sites where they were able to grow unhindered because seeds of species A chanced not to fall there.

We now see that the two species will be able to coexist indefinitely provided species B is sufficiently fertile relative to species A. At the beginning of the growing season the safe sites can be allotted to four different classes depending on the seeds they contain. These classes are:

(i) Unseeded sites where no plants grow;

(ii) Sites with species B's seeds only; a species B plant will grow in each site.

(iii) Sites with species A's seeds only ⎫ a species A plant will grow in each

(iv) Sites with seeds of both species ⎭ site.

The set-up is shown in Figure 10.2. Figure 10.2a shows the patterns of the two species' seeds before germination. Figure 10.2b shows the locations of the mature plants. It is seen that species B is able to maintain itself, in spite of its weakness in direct competition, because its greater fertility ensures that some of its seeds will fall in sites from which seeds of species A are missing by chance. Thus species B compensates for its inferiority in direct competition by outnumbering its more powerful competitor.

It might be argued that this mechanism could ensure the continued coexistence of the two species regardless of their relative fertilities since at the time of seed dispersal there are always likely to be at least a few safe sites from which species A's seeds are absent. But species B's persistence year after year cannot be counted upon unless its fertility exceeds species A's as was shown by Skellam (1951). He also determined the minimum possible value for the ratio of species B's fertility to species A's if coexistence were to occur. This minimum, or "critical", ratio is not a constant but depends on the fertility of species A, the stronger competitor, for the fewer the sites left unseeded by species A, the greater must be species B's fertility if it is to have a good chance of colonizing the vacancies.

The derivation of the critical ratio is as follows. We shall need a unit for measuring area and it is convenient to use as unit the area of a safe site. Let there be a total of N safe sites. First it is necessary to notice that if species A were the only species present the size of its population would, when the population had been established long enough, be at a permanent equilibrium level such that a constant proportion (less than 1) of the safe sites was occupied by the plants in each succeeding season. Let this proportion be Q. We begin by determining Q.

Let F denote the density, in terms of number per unit area, of viable seeds yielded by one mature plant of species A. Then at the end of a growing

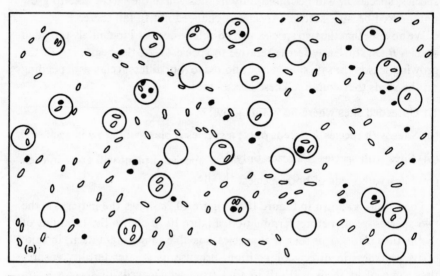

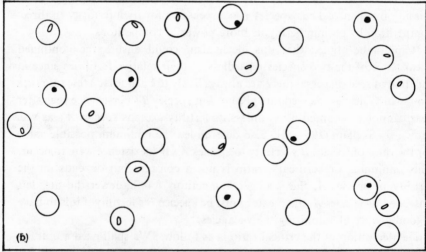

FIGURE 10.2 A mechanism whereby a competitively weak but highly fertile plant species can coexist permanently with a stronger one. Seeds can germinate only in "safe sites" (shown as circles) which can support only one plant of either species. (a) The pattern of newly fallen seeds; (b) the pattern of mature plants. The strong and weak competitors are shown by solid and hollow symbols respectively

season, when seeds of this species are dispersed at random over the whole area, the density of the seeds is NQF per unit area.

Therefore (see page 135) the probability that any sampling quadrat of unit area will contain r seeds is given by the Poisson probability

$$p_r = \frac{(NQF)^r}{r!} e^{-NQF}$$

Consequently, the probability that a unit sampling area or, which comes to the same thing, a safe site will contain no seeds is

$$p_0 = \frac{(NQF)^0}{0!} e^{-NQF} = e^{-NQF}$$

since $0! = 1$.

The complement of this probability, namely

$$1 - p_0 = 1 - e^{-NQF}$$

therefore gives the probability that a safe site will contain *at least* one viable seed of species A and this is the probability that the site will be the location of a mature plant.

But we have already specified that Q denotes the proportion of safe sites that are occupied by mature plants at equilibrium. Therefore we can now put

$$Q = 1 - e^{-NQF}. \tag{10.1}$$

Next it is necessary to consider species B, the weaker competitor. Let q be the proportion of safe sites that support plants of species B when the two competing species are in equilibrium. Also, let f be species B's fertility in terms of number of viable seeds per unit area per parent plant. (Thus q and f, which relate to the weaker species, correspond to Q and F which relate to the stronger species.)

The species B population will have an equilibrium size in the same way that species A has; but only a proportion $1 - Q$ of the total number of safe sites is available for species B since it is always crowded out by species A in sites where seeds of both species fall. Thus for species B the equation analogous to (10.1) is

$$q = (1 - Q)(1 - e^{-Nqf}). \tag{10.2}$$

It is seen that the right side of this equation gives the proportion of safe sites that, at the beginning of a growing season, contain at least one seed of species B and none of species A.

We now wish to express F and f in terms of Q and q.

From equation (10.1)

$$1 - Q = e^{-NQF};$$

taking natural logarithms,

$$\ln(1 - Q) = -NQF$$

and hence

$$F = \frac{-1}{NQ} \ln(1 - Q).$$

(Recall that the logarithm of a number between 0 and 1 is negative.)
From equation (10.2)

$$1 - \frac{q}{1 - Q} = e^{-Nqf}$$

or

$$\ln\left(1 - \frac{q}{1 - Q}\right) = -Nqf$$

whence

$$f = \frac{-1}{Nq} \ln\left(1 - \frac{q}{1 - Q}\right).$$

It now follows that if the two species are in equilibrium, with the weaker species in a proportion q of the safe sites and the stronger in a proportion Q (so that the ratio of the two kinds of plants is q/Q), the ratio of their fertilities must be

$$\frac{f}{F} = \frac{Q \ln\left(1 - \dfrac{q}{1 - Q}\right)}{q \ln(1 - Q)}. \tag{10.3}$$

The value of this expression when $q \to 0$ thus gives the minimum value of f/F required to ensure that the weaker species shall just escape local extinction. However, putting $q = 0$ in the right hand side of equation (10.3) yields the indeterminate fraction $0/0$. To obtain the desired limit it is necessary to use l'Hospital's Rule (see Appendix 10.1) from which it is found that

$$\lim_{q \to 0} \frac{Q \ln\left(1 - \dfrac{q}{1 - Q}\right)}{q \ln(1 - Q)} = \frac{-Q}{(1 - Q) \ln(1 - Q)} = R_C, \quad \text{say,}$$

where we have written R_C for the critical ratio of fertilities. In words: if the weaker competitor invariably loses in direct competition with the stronger, then its fertility must be at least R_C times that of the stronger if it is not to be excluded from the area.

Figure 10.3 shows the relationship between R_C and Q; (observe that R_C is plotted on a log scale). The shape of the curve shows that until a large proportion of the safe sites is preempted by the stronger competitor R_C remains low; for instance, even when $Q = 0.7$ the weaker competitor need only be twice as fertile as the stronger to coexist with it. But as Q approaches 1, R_C increases very rapidly indeed, as one would except intuitively. If Q ever reaches 1, the weaker species is automatically excluded.

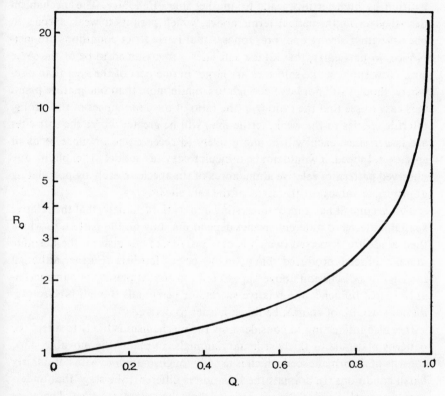

FIGURE 10.3 The relationship between R_C, the critical ratio of species fertilities needed to ensure permanent survival of a weak competitor, and Q, the fertility of the stronger competitor. See text for details

The mathematically rigorous argument set out above leads to an inescapable conclusion: namely, that the coexistence of competing species *is* possible even when all the conditions for competitive exclusion listed on page 244 hold. Therefore, to plug this loophole in the coexistence principle we must amend it by imposing yet another condition, the sixth, which states: the fertilities of the competitors must not be too dissimilar.

It should now be clear why disproof of the principle is likely to be so difficult. To prove it wrong (if it is wrong) would entail demonstrating that in an observed case of coexistence all the six conditions we have listed were fully complied with. This seems a quite impracticable task and one it would be a waste of time to attempt. Far more ecological insight is to be gained by investigating the way in which various mechanisms have evolved which (sometimes) enable similar species to avoid direct competition.

The fact that a "weak" species may make up for its failure in direct competition by high fertility leads to another line of enquiry. The mechanism described in mathematical terms above, which permits a weak species to coexist with a strong one, presupposes that fairly strict conditions are met. Notice, in particular, that all the safe sites were assumed to be of the same size. Now suppose the safe sites are larger in one part of the area than elsewhere, though still not large enough to contain more than one mature plant. It is easy to see that the ratio Q/q (the ratio of the abundances of the strong, infertile species to the weak, fertile one) will be greater where the safe sites are larger since each will be more likely to receive one or more seeds of species A. Indeed, it would not be difficult to devise a model to "explain" any observed pattern of relative abundances of the species merely by postulating appropriate values for the areas of the safe sites.

This argument has a most interesting implication, namely, that the relative spatial patterns of different species depend not only on the pattern in which their seeds are dispersed (which is obviously true) but also on the absolute numbers of seeds produced, that is, on the species' fertilities. Further, although the rigorous discussion above applies only to annual plants, the same mechanism must influence the relative success of perennials though precise predictions would, of course, be much harder to derive.

It is also interesting to consider how the mechanism is likely to affect the patterns of sessile or fairly sedentary animals. In plants the movable stage consists of a dormant seed which is usually tough enough to withstand fairly harsh conditions. In animals the situation is different; the stage that undergoes dispersal is usually more delicate than the sedentary stage. There are also contrasts among different classes of animals as to the stage of the life cycle during which they disperse. Thus in insects it is the immature forms which are sedentary: consider, for example, caterpillars and butterflies, maggots and bluebottles, grubs and bees. Whereas in marine benthic invertebrates a sedentary mature form often has a planktonic larva: barnacles, sea anemones and tube worms are examples. It seems likely that in all these animals, as in plants, high fertility may enable a species to survive in the presence of a stronger competitor. But among animals of the kind just mentioned, the sedentary, actively competing individuals are more robust than the passive, dispersible individuals. The way in which these contrasting life styles modify the processes of interspecific competition and their effects on spatial pattern constitute an enormous field of research which has scarcely been touched.

4 THE DETECTION OF COMPETITION

Hitherto we have tacitly assumed that whenever two individual organisms compete for a limiting resource the loser will be eliminated; either it will die of undernouishment at an early age, or it will be driven away by an aggressor or physically crowded out. But actual exclusion of the loser by the winner is only one of the possible outcomes of competition. The other possibility is that the loser will hold its ground in a stunted form. The ability to do so, to remain alive in extremely unfavorable conditions is common in plants and also occurs, though more rarely, in animals.

Thus competition can take two distinct forms and these have been named (by Nicholson, 1954) "contest" and "scramble".

Contest (sometimes called *interference*) is the form of competition we have so far discussed. It is the winner-take-all kind of confrontation and results in exclusion of the loser. It is typical of highly evolved animals that are capable of fighting. Even when overt fighting does not take place, the winner's superiority usually stems from its possession of superior weapons which enable it to stake out and defend a territory. A representative example was described earlier (page 235), that of yellowheaded blackbirds excluding redwinged blackbirds from the deepest part of a marsh.

Scramble (sometimes called *exploitation*) occurs when neither of two competing species can oust the other and a stalemate results. It is also common within one-species populations which have become so dense that the available resources are insufficient for all the individuals to thrive. Instead of sorting themselves into clearly-defined classes of victors and vanquished, the mass of the population persists in stunted form. Scramble is especially typical of plant populations. Plants are far more "plastic" than animals; that is, successful individuals of a given plant species may have a wide range of sizes and shapes whereas the individuals of a given animal species usually resemble one another much more closely and the adults are of comparable size. The ability of some (not all) plant species to survive crowding is well known. It is obvious to anyone who has sowed a row of carrot or lettuce seeds and neglected to thin the seedlings; very large numbers of small, deformed plants are obtained. The same phenomenon occurs with forest trees. Populations of lodgepole pine (*Pinus contorta*) and jack pine (*P. banksiana*) are especially likely to exhibit it. Baker (1950) has described a stand of 70-year old lodgepole pines so dense and uniform that the trees could be pulled up like weeds; on average their height was four feet and their diameter, just above the ground, one-third of an inch. Such a stand, in which competitors are so evenly matched that none can thrive, is described by foresters as "locked" or "stagnated".

Scramble is less common in animals but a few cases have been described in inactive or sedentary species. Examples are: the larvae of crowded laboratory populations of *Drosophila* (Miller, 1967) and of the blowfly *Lucilia* (Nicholson, 1954); barnacles (Knight-Jones and Moyse, 1961); and planarians (Reynoldson, 1964). Reynoldson, observing planarian populations in experimentally manipulated pools, obtained an exceptionally clear demonstration of the effects of scramble. Planarians can avoid starvation by shrinking to about one-third of their initial size during a foodless period, after which they grow again when food becomes available. The result is that when food shortage forces a density-dependent response on a saturated population, the size of the population in terms of numbers of individuals does not change much; instead, a reduction in total biomass is brought about by shrinkage of the individuals.

Scramble is easier than contest to observe since its effects on the competitors are visible and persistent. When the competitors are sessile one can even measure, rather crudely, the intensity of competition. This may be done by measuring the dependence of individuals' sizes on their distances from neighbouring individuals.

We consider an example (from Pielou, 1961) in which a two-species population of trees was investigated. The trees were ponderosa pine (*Pinus ponderosa*) and Douglas fir (*Pseudotsuga menziesii*) which form a very open forest on the lower slopes of mountains in the arid interior of British Columbia. The way in which the sizes of the trees depended on the proximity of their neighbors was revealed by making two observations on each tree in the area delimited for study. These were: the distance from each tree to its nearest neighboring tree; and the sum of the circumferences of the trunks of each tree and its neighbor.

The presence of two species of trees meant that the pairs formed by a tree and its neighbor were of three kinds: a pine with a pine; a Douglas fir with a Douglas fir; or one tree of each species. The observations were therefore sorted into corresponding classes which were treated separately. Here we consider only competition within the pines, and between pines and Douglas firs. Figures 10.4a and 10.4b, respectively, summarize the results. Both are scatter diagrams showing the relationship between the measured quantities for each tree, and both axes are scaled logarithmically. It should be noticed that in those cases when a tree was the nearest neighbor of its own nearest neighbor the pair of measurements relating to the tree and the neighbor were identical and contributed a pair of coincident points to the relevant scatter diagram; such points are shown by solid symbols. A hollow symbol represents a tree whose nearest neighbor had another tree as *its* nearest neighbor.

The scatter diagrams suggest that the variables are related, though not strongly, and we wish to judge whether the apparent relationship is likely to be merely the outcome of chance, and hence spurious; and if not, whether

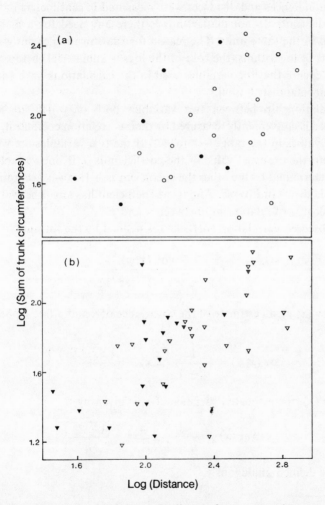

FIGURE 10.4   The relationship between the distance from a tree to its nearest neighbor (on the abscissa) and the sum of their circumferences (on the ordinate). All lengths were measured in centimeters and are plotted on logarithmic scales. Solid black symbols show pairs of coincident points. (*a*) Pine-pine pairs; (*b*) "mixed" pairs

there is any difference in the intensity of the relationship shown by the two scatter diagrams.

The procedure is as follows.

Put x for the log of the sum of trunk circumferences;

put y for the log of the distance between each tree and its neighbor;

and put n for the number of pairs of x, y values observed.

The circumferences and distances were measured in centimeters but in what follows the units do not matter, nor is there any need for x and y to be measured in the same units. The reason for transforming the original measurements to logarithms (the base of the logs is immaterial) before proceeding is to ensure that the variables used in the calculations have an approximately normal distribution.*

The relationship between two variables both of which are normally distributed is conveniently measured by their correlation coefficient, r, which can take values in the range -1 to $+1$. If the two variables are wholly independent, the expected value of the coefficient is 0. If one variable is completely determined by the other the coefficient is $+1$ or -1 according as the relation is direct or inverse. And if the coefficient has any other value, there is some degree of relationship between x and y.

The observed correlation coefficient is defined by the formula

$$r = \frac{\text{cov}(x, y)}{\sqrt{\text{var}(x) \cdot \text{var}(y)}}$$

where cov (x, y), an estimate of the covariance of x and y (see Appendix 3.1) is given by

$$\text{cov}(x, y) = \frac{1}{n-1} \left\{ \Sigma xy - \frac{(\Sigma x)(\Sigma y)}{n} \right\};$$

also, var (x), the estimated variance of x, is given by

$$\text{var}(x) = \frac{1}{n-1} \left\{ \Sigma x^2 - \frac{(\Sigma x)^2}{n} \right\};$$

var (y) is defined analogously.

---

* The fact that many statistical tests are applicable only to data having a normal distribution is often forgotten. Any good textbook of elementary statistics gives a description of the normal distribution and how to recognize it; and also warnings as to which tests are inapplicable unless the data are normal, and what to do if they are not.

Finally therefore

$$r = \frac{\Sigma xy - (\Sigma x)(\Sigma y)/n}{\sqrt{\{[\Sigma x^2 - (\Sigma x)^2/n][\Sigma y^2 - (\Sigma y)^2/n]\}}}.$$

Writing r_w and r_b for the values of r obtained from the within-species (pine-with-pine) and the between-species (Douglas-fir-with-pine) data respectively, the numerical values were

$$r_w = 0.416 \quad \text{based on } n_w = 30 \text{ pairs of observations}$$
$$\text{and having } n_w - 2 = 28 \text{ degrees of freedom};$$

and

$$r_b = 0.659 \quad \text{based on } n_b = 61 \text{ pairs of observations}$$
$$\text{and having 59 degrees of freedom.}$$

From tables giving critical values of the correlation coefficient (to be found in most books of statistical tables) it is seen that each of these calculated values is large enough for us to conclude with reasonable certainty that a true relationship exists between the variables. Thus the probability that a value of r_w as great as 0.416 would arise by chance if the measured variables were independent is less than 0.01; and the corresponding probability for r_b is 0.001.

The following test may be used to decide whether r_b significantly exceeds r_w or whether values as discrepant might equally well have been yielded by two batches of data from the same source. Here we give merely a recipe for the test and refer the reader to a statistics textbook for an account of the rationale. What are compared are not the r values themselves but values of a quantity z calculated from r by means of the formula*

$$z = \frac{1}{2} \ln \frac{1+r}{1-r}.$$

Thus in the example we may calculate z_b from r_b to obtain

$$z_b = \frac{1}{2} \ln \frac{1.659}{0.341} = 0.7910.$$

Similarly it is found that

$$z_w = \frac{1}{2} \ln \frac{1.416}{0.584} = 0.4428.$$

---

* Many books of statistical tables contain a table in which one may look up the z corresponding to any r directly, for example Table M in Rohlf and Sokal (1969).

We now wish to judge whether the difference between z_w and z_b is significant. The conclusion reached applies also to the difference between r_b and r_w.

Put $z_b - z_w = d$; thus $d = 0.3482$ in the example.

Not it is known that if the difference between z_b and z_w is due merely to sampling variability, then d has an approximately normal distribution with zero mean and variance given by

$$\text{var}(d) = \frac{1}{n_b - 3} + \frac{1}{n_w - 3} = 0.0543$$

in the example.

It therefore follows that if $|d/\sqrt{\text{var}(d)}| < 1.96$, the difference could have arisen by chance with a probability of 0.05 or more and to infer that there is any real difference between the two correlation coefficients is unjustified. Conversely, if $|d/\sqrt{\text{var}(d)}| > 1.96$, the probability is less than 0.05 that the observed difference was due to chance and it may thus be thought reasonable to conclude that the difference is real.

In the present case

$$d/\sqrt{\text{var}(d)} = \frac{0.3482}{0.233} = 1.494$$

and we shall accordingly treat z_b and z_w (and likewise r_b and r_w) as not differing significantly.

A statistical digression is necessary at this point. It should be noticed that in doing a test like the one above, there are three possible questions we could ask *before* making the observations. These are:

The "two-sided" question:

(i) Is r_b significantly *different from* (either greater or less than) r_w?

Or else one or other of the two possible "one-sided" questions:

(ii) Is r_b significantly *greater than* r_w?

(iii) Is r_b significantly *less than* r_w?

If we choose which question to ask before obtaining the data, any of the three questions (but only one of them) can reasonably be asked.

If (i) has been chosen we accept the difference as "significant at the 5% level" (since it has probability of less than 0.05, or 5%, of arising by chance) if

$$d/\sqrt{\text{var}(d)} < -1.96 \quad \text{or} \quad d/\sqrt{\text{var}(d)} > +1.96.$$

If (ii) has been chosen and d turns out to be positive, it is significant at the 5% level if $d/\sqrt{\text{var}(d)} > 1.645$.

If (iii) has been chosen and if, as before, d turns out to be positive, no test is called for since the answer is indubitably No. Obviously, $d > 0$ can happen only if $z_b > z_w$ (or, equivalently, $r_b > r_w$) so that it cannot possibly be inferred that r_b is significantly less than r_w.

Now suppose it is not until *after* the correlation coefficients have been calculated that we decide to compare them. Then one of the one-side questions is automatically nonsense, question (iii) in the case of the present example. When this is so, that is, when the data have already supplied an answer (albeit unasked for) to one of the one-sided questions, it is cheating to ask the other one-sided question as it stands since we already know part of the answer. The only question that can fairly be asked is: *Given* that $d > 0$, is it *significantly* greater than zero? For the difference to be treated as significant at the 5% level it is necessary that $d/\sqrt{\operatorname{var}(d)} > 1.96$. Thus it is seen that the critical value of $d/\sqrt{\operatorname{var}(d)}$ depends on whether we choose which question to ask before or after calculating d.

Returning to the example, we see that since $d/\sqrt{\operatorname{var}(d)} = 1.494 < 1.96$, the data give us no reason to conclude that $r_b > r_w$, and this may be interpreted as implying that the interspecific competition (between Douglas firs and pines) was no more intense than the intraspecific competition among pines alone.

It can be said, however, that the trees as a whole (disregarding their species) were competing. It could be argued that the reason why the combined sizes of two neighboring trees is positively correlated with the distance between them is, at least in part, that young, and hence small, trees tend to be more closely spaced than older ones. But this in itself may result from competition.

CHAPTER 11

The Spatial Patterns of Interacting Species

IN CHAPTERS 9 and 10 we have considered ecological competition in as broad a context as possible; the examples described range from sessile or very sedentary organisms (such as barnacles, poppies and slime molds) to active, fast-moving ones (chipmunks, blackbirds). A distinction was also made between the two contrasting modes of competition, contest and scramble. Contest is more typical of active, highly-evolved animals and scramble of plants and sedentary invertebrates, but there are many exceptions to this broad generalization. Thus though it is arguable whether we yet have any satisfactory "theory" of biological competition to underlie and unify the whole subject, it is unquestionable that field biologists studying competition are faced with a bewildering variety of phenomena which call for correspondingly varied methods of observation.

The strongest contrast is between methods appropriate to the study of fast-moving vertebrate animals and those appropriate to the study of plants. With vertebrate populations, the competing individuals do not remain in one place and are often inconspicuous. To observe the effects of competition it is usually necessary to estimate the sizes of the competing populations at a succession of times. With plant populations change is necessarily slow and the course of competition has to be inferred from the welfare of the competing individuals and from their spatial pattern; the latter is easy to observe since the competitors do not move.

Summarizing, the two kinds of competition are as different as a guerilla skirmish and a pitched battle. This chapter is concerned with the latter kind of competition, which is certainly not restricted to plants, and with the measurement and interpretation of the patterns of two or more interacting species-populations. Often the emphasis is more on studying two- or many-species patterns for their own sakes and only subsequently inquiring into their bearing on interactions among the species.

1 ASSOCIATION BETWEEN PAIRS OF SPECIES

When two species at the same trophic level occur in the same area it is natural to enquire whether they are "associated". They are said to be *positively associated* if the presence of one of the species in any small space or sampling plot makes it more likely that the other will also be found; conversely, they are *negatively associated* if the presence of one of the species makes that of the other less likely. Studies to determine whether and in what way two species are associated are so common in ecological work that we begin by describing how such a test is done. After that, however, it will be necessary to point out how difficult it is to interpret the result and how ambiguous the results of association tests often are.

The usual test is the χ^2 test for independence in a 2×2 table. Let us call the two species whose association is to be tested species J and species K. If the species are plants, the sampling units observed in the field are usually small plots of ground (quadrats); if they are species of invertebrates each individual of which dwells in or on a particular "habitable unit" (which might be a plant organ or, for a parasitic species, the body of a larger animal) then the habitable units constitute natural sampling units.

Now suppose N sampling units have been examined and each has been recorded as containing species J alone, species K alone, both species, or neither species. The results can be displayed in a 2×2 table as follows

		Species K		
		Present	Absent	Totals
Species J	Present	a	b	$a+b$
	Absent	c	d	$c+d$
	Totals	$a+c$	$b+d$	$N = a+b+c+d$

Here a, for example, denotes the number of units that contained both species, and the other three cell frequencies, b, c and d, are defined analogously.

Consider, next, what we should expect these frequencies to be if the two species did not interact in any way and if, also, habitat conditions were identical in all the sampling units. The expected cell frequencies are obtainable by considering the marginal totals (i.e., the row totals and column totals) of the table. For example, species J is present in $(a + b)$ units and absent from the remaining $(c + d)$ units so we can say it is present in a proportion $(a + b)/N$ of them. Therefore, if the two species were independent, a proportion $(a + b)/N$ of the $(a + c)$ units that were found to contain species K would contain species J as well. Thus the expected frequency of units with both species is $(a + c)(a + b)/N$. The same result could be obtained by arguing that of the $(a + b)$ units that contain species J, a proportion $(a + c)/N$ would be expected to contain species K also.

17*

Arguing in the same was we can arrive at expected frequencies for all four of the observed classes of unit, and can tabulate them in the following "expected" 2×2 table:

		Species K		
		Present	Absent	
Species J	Present	$(a + b)(a + c)/N$	$(a + b)(b + d)/N$	$a + b$
	Absent	$(c + d)(a + c)/N$	$(c + d)(b + d)/N$	$c + d$
		$a + c$	$b + d$	N

To judge whether it is reasonable to suppose that the observed frequencies are "virtually the same" as their expectations, in the sense that the discrepancies between them might well result merely from chance, we do a χ^2 goodness of fit test (see Appendix 7.2). The test criterion is

$$X^2 = \sum \frac{(\text{Observed frequency} - \text{Expected frequency})^2}{\text{Expected frequency}}.$$

Substituting the observed and expected frequencies shown in the two tables above it is easily found that

$$X^2 = \frac{(ad - bc)^2 N}{(a + b)(c + d)(a + c)(b + d)}. \tag{11.1}$$

This result should be compared with the one on page 191, in which a completely different problem (that of comparing two mosaic patterns) yielded a 2×2 table of identical form to those shown here. It will be found that the formal identity extends to the formulae also. Equation (8.1) on page 191 and equation (11.1) above are the same, as is easily checked by converting one set of symbols to the other.

As in the earlier example, we require to know the probability of obtaining as large a value of X^2 as that observed given the null hypothesis (in this case that the occurrences of species J and K are independent). A close approximation to the required probability is found by first calculating an adjusted value of the test criterion, namely

$$X_c^2 = \frac{(|ad - bc| - N/2)^2 N}{(a + b)(c + d)(a + c)(b + d)}. \tag{11.2}$$

(This is equivalent to equation (8.2) on page 191.)

Then, if the hypothesis of independence is correct, the probability that the calculated value of X_c^2 would be exceeded by chance can be looked up in a table of percentage points of the χ^2 distribution with 1 degree of freedom.

In particular, a value of X_c^2 in excess of 3.84 has a probability of occurance (under the null hypothesis) of 0.05 or less. In other words, 3.84 is the critical value of X_c^2 for a 5% significance test.

If the observed numbers of units with both species and with neither species exceed expectation (i.e., if $ad > bc$), the association is positive. Conversely, if the observed numbers of units containing either species without the other exceed expectation (i.e., if $ad < bc$), the association is negative.

As a numerical example, consider the following data which we shall look at again in another context later. Observations were made on the different species of frit flies (Chloropidae) breeding in fruiting bodies ("brackets") of a species of bracket (or shelf) fungus, *Polyporus betulinus*, which grows on the trunks of dead birch trees. Each bracket of a collection of 60 was separately caged so that newly emerged adult insects could be caught and identified, and the results for two of the frit fly species (*Conioscinella melancholia* and *Tricimba trisulcata*) were as follows:

		T. trisulcata		
		Present	Absent	
C. melancholia	Present	13	14	27
	Absent	2	31	33
		15	45	60

For this table

$$X_c^2 = \frac{(|13 \times 31 - 14 \times 2| - 30)^2 \times 60}{15 \times 45 \times 33 \times 27} = 11.87.$$

The probability of obtaining so large a value of X_c^2 if the occurrences of the two species were independent is less than 0.005, so it is presumably correct to conclude that they are not independent. If they were, only $15 \times 27/60 = 6.75$ of the fungus brackets would be expected to contain both species as against the observed 13; we conclude that the species are positively associated.

(The procedure for testing whether two separate 2×2 tables show the same degree of association is described in Appendix 11.1.)

Tests such as this are so easy to do that ecologists are sometimes tempted to overdo them. However, while there is nothing mathematically wrong with the test just described, 2×2 tables pose many traps for the unwary and it is worth attempting a fairly systematic account of them. There are six chief sources of misinterpretation or error.

1) Because two species turn out to be positively associated it does not necessarily mean that one is beneficial to the other. There may be no active interaction between them at all. Possibly their unexpectedly frequent co-occurrence signifies no more than that some of the sampling units provide

better conditions than others for both species. Thus there is a clear contrast between what may be called "active association" and "passive association", and testing a 2 × 2 table provides no way of distinguishing between them.

2) When the presences and absences of each of several species in each unit of a sample have been listed, interpretation of the data becomes difficult. If k species have been tallied, $k(k-1)/2$ different 2 × 2 tables can be compiled and it is *not* permissible to conclude that there is "true" association (either active of passive) between every pair of species whose 2 × 2 table yields a "significantly" high value of X_c^2. The test criterion, X_c^2, is a random variable: even if all the k species are wholly independent of one another and all the sampling units are completely identical, some of the 2 × 2 tables will yield high X_c^2 values by chance. In fact, 5 % of them would be *expected* to give X_c^2 values in excess of 3.84, the critical value for a 5 % test, which is the reason why 3.84 is the critical value. One would still be tempted, of course, to accept as significant a value of X_c^2 very greatly in excess of the critical value, and there is a way of testing whether this is permissible (Cooper, 1968).

Suppose n different 2 × 2 tables are being tested simultaneously. One wishes to find what may be called a "supercritical" value for X_c^2 such that (assuming a test at the 5 % significance level is required) the probability is only 0.05 that it will be exceeded by *any* of the batch of calculated X_c^2 values. Denote this supercritical value by $[X_c^2]_n$. Then $[X_c^2]_n$ is approximately equal to the critical value for a single test at the $(5/n)$ % significance level. For example, if $n = 5$, $[X_c^2]_n$ is the same as the ordinary critical value for a 1 % test which is 6.64; and if $n = 10$, $[X_c^2]_{10}$ is the same as the ordinary critical value for a 0.5 % test which is 7.88.

The fact that some tables will usually chance to deviate markedly from theoretical expectation also leads to another common error in data interpretation. Suppose a cursory glance at a page full of these tables shows that in one of them the frequencies are suspiciously unlike what we would, informally and vaguely, have supposed them to be. One cannot now do a full-dress statistical test on this particular table since the probabilities used in performing standard statistical tests apply only to tests done at random (i.e., without prejudgment). The act of taking a cursory glance at the data (or, more candidly, of peeking) and choosing subjectively which table to test, means that the calculated X_c^2 is certainly not a random variable. The probability that it would be exceeded by chance is anybody's guess.

3) The χ^2 test, which we have just described, is a large sample test; that is, the probability it yields is only approximate though the approximation is extremely close provided the observed marginal totals in the 2 × 2 table are large enough. The test is unreliable if these totals are small and, as a rule of

thumb, it is commonly recommended that the test should be used only when the smallest expected frequency is at least 5. It is also sometimes recommended that an exact test, as an alternative to the χ^2 test, be used whenever any expected frequency is below 5. But the exact test is inappropriate for most ecological 2 × 2 tables and we shall not describe it here. Its use presupposes that the marginal totals have given values, decided upon by the observer before any data are collected. This is sometimes the case in experimental work but is certainly not so in ecological sampling. When field observations are made to judge whether two species are associated, the numbers of sampling units in which the species are present and absent are not known beforehand and are themselves random variables.

4)   In judging the association between a pair of plant species, observations have to be made with quadrats of arbitrary size and shape; this results in unavoidable ambiguity as is shown schematically in Figure 11.1. The outcome of a test depends on the size of quadrat used. Clearly, given a pattern like the one shown, samples obtained with large quadrats would suggest positive association, and with small quadrats negative association.

Another unavoidable difficulty that arises whenever association is tested is the dependence of the outcome on the boundaries of the area delimited for sampling. Consider Figure 11.1 again and suppose we treat the whole mapped area, including that to the right of the dashed line, as the study area; then many randomly placed quadrats will be empty and the frequency labelled d in the 2 × 2 table will be large; (d is the customary symbol for the number of

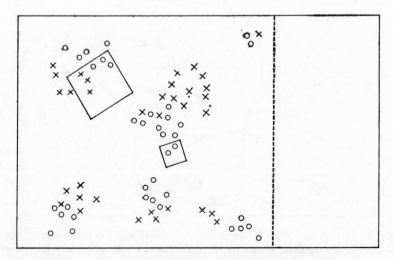

FIGURE 11.1   To demonstrate the effect on association tests of (i) different quadrat sizes; and (ii) differently defined study areas

units from which both species are absent). If the study area is redefined to exclude the empty area to the right of the dashed line, d will be reduced. It thus appears that the value of d, on which the outcome of an association test obviously depends, is not an objective observation at all; in effect, it is deliberately chosen when the boundaries of the study area are decided upon. Attempts have been made to overcome this snag by devising coefficients of "overlap" between two species that depend only on the frequencies a, b and c in the 2×2 table and are unaffected by d. But the difficulty cannot be made to go away simply by ignoring it in this manner and if the independence or otherwise of the species is to be tested, the value of d must be allowed for. Indeed, it must be accepted that the result of an association test applies to two species *and* a specified area *and* a specified sampling unit (for plants, a specified quadrat size).

5)   Figure 11.2 illustrates a source of error seldom recognized. Again it is most easily appreciated by visualizing plant populations but it is not confined to plants. Most plant species have naturally patchy patterns and if quadrats are too closely spaced several are likely to fall in the same patch. When this happens, the quadrats are not mutually independent and any test based on the observations is consequently invalid.

The study area mapped in Figure 11.2 is so small, relative to the patch size of the two species being tested, that it contains only one patch (with a solid

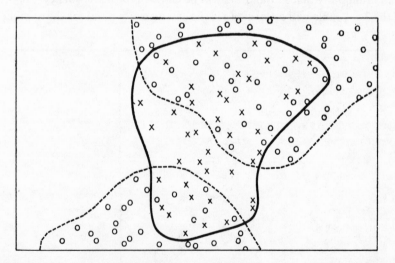

FIGURE 11.2   A map showing the pattern of individuals of two plant species. Because the plant patches (or clumps) are large relative to the total area, when the area is sampled with randomly placed quadrats several will fall in each patch. An association test cannot be done

outline) of one of the species and parts of two patches (with dashed outlines) of the other species. The arrangement of the patches, which is assumed to be governed solely by chance, is such that spurious evidence of association may be obtained when the area is sampled with quadrats. Of course, a pattern like that shown is not necessarily due to chance; the sizes and arrangement of the patches might be controlled by environmental factors. But whether this is so or not cannot be judged merely from a map of the patches. Thus the "association" (which may be positive or negative) suggested by quadrats spaced too closely to be mutually independent is as often as not an artefact of the sampling method. There is no way of telling.

6)   The only data needed for an association test are records of the presences and absences of the two species concerned in a number of sampling units. The observations are quickly and easily made. To learn more about how each species affects the other, it is often desirable to measure the quantities of each species in each unit and to judge how the quantities are correlated.

When this is to be done, one must first decide on a method for measuring quantity. For an animal species the obvious measure is the number of individuals, but it is rare for such a simple approach to give a satisfactory result with plants. If the plants reproduce vegetatively and form spreading clumps or if, though they are recognizably distinct as individuals, the individuals are very variable in size, counting is either impossible or meaningless as a means of determining quantity. Then the only solution is to measure (or estimate) the weight (fresh or dry) or the volume of material of each species in each quadrat. If the roots as well as the above-ground parts of plants are included as, ideally, they should be, the task of measuring plant quantities becomes very time-consuming.

If the realtionship between the two species is to be judged by calculating a correlation coefficient (see Chapter 10, page 254) it is nearly always necessary to transform the measured quantities in some way, by taking logarithms for example, to ensure that the distributions of the variables are at least approximately normal. It is also necessary to ensure that both species are present in all (or nearly all) of the quadrats or other sampling units; if, for either species, the list of quantities contains many zeros, there is no transformation that could convert the observed distribution to an approximately normal one. Ensuring that a sampling unit shall not lack some of each species may be comparatively easy when the organisms are plants, or soil or benthic animals, for then the usual sampling unit is a quadrat and the investigator is free to make it as large as required. But when the sampling units are naturally separate entities it may be impossible to obtain data of the sort needed for calculating a correlation coefficient.

Thus óne may have no choice between testing a 2 × 2 table for association and testing the quantities of two species for correlation. The method to use is often dictated by the nature of the data.

Correlation coefficients can be as hard to interpret as measures of association. Thus consider two species of plants. If their quantities are negatively correlated it may mean that both are drawing on the same limited reserves of water and mineral nutrients, and where one thrives the other cannot; in other words, they are competing. Alternatively, negative correlation could equally well imply that the plants differed somewhat in their needs and were growing in a heterogeneous habitat. Suppose the study area contained a habitat mosaic made up of patches of ground suitable for one or other of the species but not for both; the quantities of the two species in a quadrat would then depend on the relative areas of the two kinds of patch within the quadrat and hence, for certain quadrat sizes, they would be negatively correlated.

Positive correlation can also be given two diametrically opposed interpretations. It could indicate that, in a habitat that varies smoothly, conditions optimum for one species are adjacent to, but do not overlap, those optimum for the other. Alternatively, it could show that individuals of both the species, though they are competing fiercely for identical resources, can grow larger where the needed resources are relatively abundant than where they are sparse; in other words, the effects of competition are masked by the effects of environmental heterogeneity.

From the preceeding paragraphs it should be clear that association tests and correlation tests by themselves are seldom of great ecological interest. This does not mean, of course, that they are not useful. One of these tests may be an essential step in deciding the correctness of otherwise of some ecological hypothesis. But it is only a step. It is a mistake to believe that asking a statistical question of one's data will call forth an ecological answer. A statistical test by itself leads only to the acceptance or rejection of a statistical hypothesis and judgement as to whether the result justifies a more elaborate conclusion, with ecological implications, is a problem outside the scope of the test. Statistical tests are often necessary, but never sufficient, to solve ecological problem.

2 "MEAN CROWDING" BETWEEN SPECIES

As an alternative to measuring the association or correlation between two species Lloyd (1967) has suggested a way of measuring the degree to which each species crowds the other. His index is usable when the species are small animals for which one can count the numbers of individuals in a collection of sampling units. The units may be natural habitable units or arbitrary units such as soil cores. It will be recalled (see chapter 7, page 150) that Lloyd devised a measure of the mean crowding within a species which he defined as the mean number, per individual, of other individuals in the same sampling unit. The mean crowding between a pair of species is similarly defined and two coefficients are required: the first measures the degree to which species 1 is crowded by species 2 and the second the degree to which species 2 is crowded by species 1.

Suppose a sample of q soil cores (or other sampling units) is examined and the jth unit is found to contain x_{1j} individuals of species 1 and x_{2j} of species 2.

Then, as an estimate of the mean crowding *on* species 1 *by* species 2 we put

$$\overset{*}{x}_{12} = \frac{\Sigma x_{1j} x_{2j}}{\Sigma x_{1j}} \tag{11.3}$$

where the sums are from $j = 1$ to $j = q$.

Similarly, the mean crowding on species 2 by species 1 is estimated by

$$\overset{*}{x}_{21} = \frac{\Sigma x_{1j} x_{2j}}{\Sigma x_{2j}}. \tag{11.4}$$

Thus $\overset{*}{x}_{21}$, for example, is the average (over the sample) of the crowding experienced by each species 2 individual from the species 1 individuals that occupy its unit. Clearly $\overset{*}{x}_{12} \neq \overset{*}{x}_{21}$ unless $\sum_j x_{1j} = \sum_j x_{2j}$, that is unless the two species are equally abundant; obviously, the commoner species crowds the sparser more strongly than the sparser crowds the commoner.

Comparing the formulae above with equation (7.1) on page 150, it is seen that $\overset{*}{x}_{12}$ and $\overset{*}{x}_{21}$ are estimates, based on data from q sampling units, of population values $\overset{*}{m}_{12}$ and $\overset{*}{m}_{21}$ which could be exactly determined only by inspecting the whole population. For $\overset{*}{m}$, the mean crowding *within* a species, Lloyd was able to derive an unbiased estimator [$\overset{*}{x}$ in equation (7.2), page 151] and a measure of its sampling variance [var $(\overset{*}{x})$ in equation (7.3)]. He has not been able to do the same for $\overset{*}{m}_{12}$ and $\overset{*}{m}_{21}$ and therefore warns that the estimates $\overset{*}{x}_{12}$ and $\overset{*}{x}_{21}$ given above in equations (11.3) and (11.4) are liable to sampling errors of unknown magnitude. All the same, provided the warning is not forgotten, the measures should prove valuable in many ecological contexts. Presumably the intensity of competition between two species at any moment depends on the degree to which each crowds the other.

As a numerical example we obtain $\overset{*}{x}_{12}$ and $\overset{*}{x}_{21}$ for the data shown in Table 11.1. These were the data which have already been used on page 261 to demonstrate the application of the χ^2 test to a 2×2 table. Table 11.1 shows the numbers of individuals of the two frit fly species that emerged from 60 brackets of the fungus *Polyporus betulinus*. The fly species are *Conioscinella melancholia* ("species 1") and *Tricimba trisulcata* ("species 2").

It is found that

$$\Sigma x_{1j} = 185, \quad \Sigma x_{2j} = 36, \quad \text{and} \quad \Sigma x_{1j}x_{2j} = 408.$$

Therefore, as estimates of the mean crowding of each species on the other, we have

$$\overset{*}{x}_{12} = 2.205 \quad \text{(the crowding on } C. \text{ melancholia by } T. \text{ trisulcata)}$$

and

$$\overset{*}{x}_{21} = 11.333 \quad \text{(the crowding on } T. \text{ trisulcata by } C. \text{ melancholia)}.$$

TABLE 11.1   The number of individuals of two frit fly species in a collection of 60 fungus brackets

Bracket #	Species 1	Species 2	Bracket #	Species 1	Species 2	Bracket #	Species 1	Species 2
1	1	–	11	2	3	21	1	1
2	4	–	12	1	–	22	10	1
3	1	–	13	9	10	23	–	1
4	2	2	14	1	–	24	1	–
5	2	2	15	12	4	25	83	2
6	1	–	16	2	–	26	–	1
7	1	–	17	7	2	27	1	–
8	1	–	18	13	1	28	1	–
9	1	1	19	15	1	29	1	–
10	9	4	20	2	–	30 to 60	–	–

3   SPATIAL SEGREGATION

Measures of the association, correlation, and mean crowding between two species occurring in a continuous area are necessarily based on observations made on arbitrary sampling units, typically quadrats. This inevitably leads to ambiguity in the results as illustrated in Figure 11.1. The difficulty can be overcome by "plotless" sampling (see page 155) as we now show.

Quite independently of the spatial pattern of each species, that is, their patterns relative to the ground, they have patterns relative to each other

(Pielou, 1961, 1965). Thus if both species occur as separate and distinct individuals (in contrast to extensive ill-defined patches), it is reasonable to enquire whether they are throroughly intermingled (i.e., spatially unsegregated), as shown in Figure 11.3a; or else incompletely mingled (i.e., spatially segregated) as in Figure 11.3b. A test to decide the matter is straightforward.

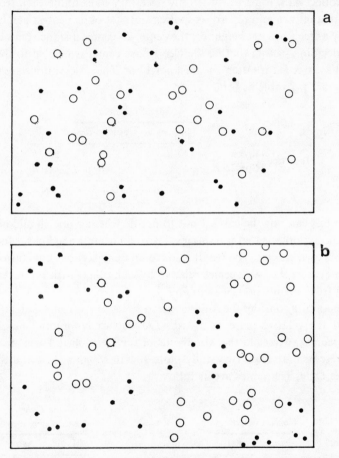

FIGURE 11.3   Patterns of two-species populations. In (a) the species are randomly mingled, or spatially unsegregated. In (b) they are not completely mingled, and are therefore spatially segregated

For concreteness, suppose the two species concerned are trees, and let them be labelled species J and species K. We also suppose that the study area is a forest of these two species only, with no other trees present; (this is not a necessary assumption but is made to clarify the discussion). Then if a K is as likely as a J to have a J as its nearest neighboring tree (and likewise if both species are equally likely to have a K as nearest neighbor) the two species are randomly mingled or unsegregated. Conversely if a J is more likely than a K to have another J as its nearest neighbor (and *mutatis mutandis*), then the J's are clumped relative to the K's (and also the K's relative to the J's) and the two species are not randomly mingled; on the contrary, they are segregated.

To collect data for a test it is only necessary to examine each tree in the study area in turn and record its species and that of its nearest neighbor. To identify a tree's nearest neighbor, the center-to-center distance between trees is the deciding factor; that is, the tree whose center is nearest to the center of the base tree is the latter's nearest neighbor. The observations are summarized in a 2 × 2 table as follows:

		Nearest neighbor		
		Species J	Species K	
Base tree	Species J	a	b	$a + b$
	Species K	c	d	$c + d$
		$a + c$	$b + d$	N

A χ^2 test can now be carried out to decide whether one should accept or reject the hypothesis that the species of a tree's nearest neighbor is independent of its own species. The result leads to an unambiguous conclusion about the patterns of the two species relative to each other, with no uncertainties arising from the method of sampling.

As examples consider the following two 2 × 2 tables from Okali (1966). He studied the relative patterns of ash (*Fraxinus excelsior*) and maple (*Acer pseudoplatanus*) trees in the woodlands of the Derbyshire Dales in England and presents data from two study plots, one in Monk's Dale and the other in Chee Dale. The tables are as follows:

		Monk's Dale Nearest neighbor					*Chee Dale* Nearest neighbor		
		Maple	Ash				Maple	Ash	
Base tree	Maple	26	39	65	Base tree	Maple	16	11	27
	Ash	30	149	179		Ash	9	73	82
		56	188	244			25	84	109

For the Chee Dale table the test criterion is

$$X_c^2 = \frac{(|16 \times 73 - 11 \times 9| - 109/2)^2 \times 109}{25 \times 84 \times 82 \times 27} = 24.13.$$

The probability, $P(\chi^2)$, of obtaining as large a value if the two tree species were randomly mingled is much less than 0.001; (the critical value for a 0.1% test is 10.83). Moreover, if the species were randomly mingled, $(27 \times 25)/109 \approx 6$ of the maples would have been expected to have other maples as nearest neighbors, as against the observed 16. It is almost certain, therefore, that the trees were relatively clumped rather than randomly mingled.

For the Monk's Dale table analogous calculations give $X_c^2 = 13.28$ and $P(\chi^2) < 0.001$. The much smaller value of X_c^2 suggests that the segregation of the species was less pronounced in the Monk's Dale than in the Chee Dale study area and this possibility can be tested. The method is described in full in Appendix 11.1 (page 395). The test consists in combining the two 2×2 tables into one 8-celled table (a $2 \times 2 \times 2$ table) and finding the expected frequencies in all eight cells on the assumption that the degree of segregation was the same in both study areas. This is the null hypothesis to be tested; (note that it is not hypothesized that the trees were randomly mingled in either area). The observed and expected frequencies are then compared by a χ^2 test for goodness of fit. In the present case it is found that the test criterion is

$$X_c^2 = 3.147 \text{ with one degree of freedom.}$$

Since X_c^2 falls short of the critical value (3.841) for a 5% test, there is no reason to reject the null hypothesis. It does not follow, of course, that the null hypothesis is true; all we are asserting is that if it is true the observed results would have a probability of occurrence in excess of 0.05 and hence are not particularly improbable. It seems more likely, however, that the null hypothesis is false. Okali hypothesizes that in both areas spatial segregation was due to the fact that ash trees tend to grow in "family clumps" where many seeds from one parent have all germinated in a small space. If this is so we should expect a difference in segregation in the two study areas. The trees were dense in the Chee Dale area (0.127 per square meter) where the observed segregation was pronounced; and less dense (0.078 per square meter) in the Monk's Dale area where the observed segregation was weaker. If family clumps of ash trees contained fewer members in the Monk's Dale area this would account both for the lower overall density and for the less pronounced segregation there.

Observing how segregation changes with time in a growing two-species population can yield interesting results. Thus, suppose a two-species forest grows up in an area laid bare by fire and where a plentiful supply of seed and a good seedbed have ensured that the second-growth forest shall be very dense. The density is too high to persist and natural thinning is bound to occur as the seedlings grow to saplings, to poles, and ultimately to full-grown trees. As the density of the forest (in terms of stems per unit area) decreases with the passage of time, the degree of segregation of the species will almost certainly change too; but whether it will increase or decrease depends on the way in which natural thinning operates. There are two contrasting possibilities:

(i) If the initial dense growth consisted largely of family clumps, the two species would be strongly segregated to start with. Also, one would expect individuals of each species to be most severely crowded by, and to compete most intensely with, sibling individuals in the same clump. The chief consequence of natural thinning would thus be a reduction of the density within one-species clumps and hence a reduction of segregation.

(ii) If, initially, the stand of seedings had grown from widely dispersed, well-mingled seeds, segregation would be slight or non-existent to start with and the intensity of crowding (and competition) would be about the same within and between species. But habitat variations, however slight, will affect the outcome of competition, and the survivor of every interspecific contest will tend to be the individual favored by local microhabitat conditions. As a result, each species will tend to survive in, and finally be confined to, the localities particularly suited to it. Unless the habitat mosaic is so fine-grained that its patches are no larger than individual trees, natural thinning will therefore bring about a sorting of the trees into one-species groups, each in its "own" habitat, and segregation of the species will increase with the passage of time.

Inspection of a single 2×2 table often reveals another interesting property of a two-species population. It may be found that the table is not symmetrical, that is, that the frequency, b, in the upper right cell is not equal to the frequency, c, in the lower left cell; (for instance, for the Monk's Dale table these frequencies are, respectively, 39 and 30). This implies that the number of times a J has a K as nearest neighbor is not equal to the number of times a K has a J as nearest neighbor. The effect may be due merely to sampling variation; alternatively, it is possible for there to be a real difference between these frequencies.

Thus, consider the results that would be obtained if one inspected each individual of each species in turn and recorded the number of individuals of both species for which it served as nearest neighbor. Since the distances between trees are measured between the centers of their trunks, geometrical considerations show that a single tree can be the nearest neighbor of as many as five others (but not more); or, if it is isolated, it may not be the nearest neighbor of any other tree (Clark and Evans, 1955; Pielou, 1961). Figure 11.4 shows an extreme example: trees of species J (shown by large hollow circles) do not, because of their isolation, serve as nearest neighbors either to members of their own species or to members of species K (the small black circles). But, though they *are* not nearest neighbors, of necessity they *have* nearest neighbors, which in every case are individuals of species K. Consequently (in terms of the symbols) the number of times black symbols serve as nearest neighbors exceeds the number of black symbols in the map; and the number of times hollow symbols serve as nearest neighbors is less than their number in the map.

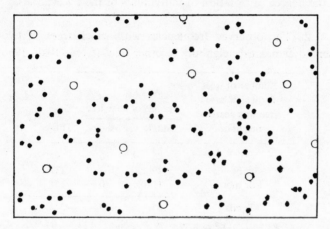

FIGURE 11.4   A two-species population for which the segregation table is unsymmetrical. Every individual, of both species, has a K (black symbol) as nearest neighbor and none has a J (hollow symbol)

Returning to a less extreme case, consider the results that would be obtained if one inspected, in turn, every member of species J, and then every member of species K, and recorded the number of other trees (without regard to species) for which each served as nearest neighbor. The result would consist of a pair of observed frequency distributions. Table 11.2, which uses Okali's (1966) data on ashes and maples in his Monk's Dale study area, is an example. Since only a small proportion of the trees served as nearest neighbors to two or more others, these frequencies have been pooled.

From the table it appears, for instance, that $25/65 = 0.38$ of the maples were not the nearest neighbors of other trees whereas the corresponding proportions for the ashes is $36/179 = 0.20$. This suggests that the maples were more isolated than the ashes, perhaps because their seedlings were less likely to occur as populous family clumps. To find the probability of obtaining two observed frequency distributions as dissimilar as those shown if the discrepancies were due only to sampling variation, we may test the distributions for homogeneity. This entails carrying out a χ^2 goodness of fit test to compare the observed frequencies with those expected on the null hypothesis that the distributions were homogeneous. Details are given in Appendix 11.2 (page 398). In the present example, the test criterion is

$$X^2 = 8.79 \text{ with 2 degrees of freedom,}$$

and its probability of being equalled or exceeded (given the null hypothesis) is less than 2%. Thus it is reasonable to conclude that there were real differences in the degree of isolation of individuals of the two species.

TABLE 11.2   The observed frequencies with which trees of two species served as nearest neighbors to other trees (from Okali, 1966)

Number of trees for which base tree was nearest neighbor	Species of base tree		
	Maple	Ash	Totals
0	25	36	61
1	29	103	132
2 or more	11	40	51
Totals	65	179	244

$$X^2 = 8.79; \quad P(\chi^2) < 0.02.$$

Suppose one wishes to determine whether two species are segregated for species whose individuals are not tall and conspicuous like trees; the task of finding each individual in a large study area and then finding its nearest neighbor (which must be done if a 2 × 2 segregation table is to be compiled) may be very laborious. An alternative way (Pielou, 1962b) of obtaining data for a segregation test is to lay out a narrow belt transect across the study area and list (by species) the sequence of individuals, of the two species concerned, in order of their occurrence along the transect. The transect should be made narrow enough to ensure that one never (or hardly ever) finds individuals of the two species occurring side by side within it; there is then no doubt as to the individuals' order of occurrence.

Let the two species be called species P and species Q. The observations on a transect will be a list such as:

PPPP QQ P QQQQQ PPP QQ PPP QQQQQQ P Q P QQ PPP Q PPP ...

and this is all the data required for a segregation test. No measurements are made of the distances separating adjacent individuals in the transect. Distances are deliberately disregarded since, in testing for segregation, the object is to focus attention on the pattern of each species relative to the other and to ignore the pattern of either of them relative to the ground. Thus if the individuals were represented by a row of knots in an elastic string, their segregtion would be unchanged however the string were stretched.

An uninterrupted sequence of individuals of one species is called a "run" of that species, and the "length" of the run is the number of individuals comprising it. To carry out the test, we need to know the mean length of the runs of each species, and also the number of runs of each species, within the transect. Let m_P and n_P be, respectively, the observed mean run-length and the observed number of runs of species P; and let m_Q and n_Q be the corresponding values for species Q.

Also, put

$$\frac{1}{m_P} + \frac{1}{m_Q} = Z.$$

Then, if the species are unsegregated, the expected value of Z tends to unity, provided n_P and n_Q are not too small; (in practice the smaller of them should not be less than 25).

The 95% confidence limits for Z, if the species are unsegregated, are given by

$$1 \pm 2\sqrt{\left\{\frac{m_P - 1}{n_P m_p^3} + \frac{m_Q - 1}{n_Q m_Q^3}\right\}}.$$

Therefore, when values of n_P, n_Q, m_P and m_Q have been observed, we may calculate the observed value of Z. Next, we may calculate the limits between which Z would be expected (with 95% probability) to lie on the null hypothesis that the species are unsegregated. Then, if the observed Z is outside these limits, we reject the null hypothesis and conclude that the species are significantly segregated (at the 5% level of significance).

In obtaining values for m_P and m_Q from transect lists, it is desirable to discard the first and last "runs" in a transect as they may be no more than segments of longer runs extending beyond the ends of the observed transect. For example suppose the list was

... PP QQ P QQQQQ PPP QQQ PPPPPP Q PPP QQQ ...

There are $n_P = 4$ whole runs of P's containing a total of 14 individuals whence $m_P = 14/4 = 3.50$. Similarly, $n_Q = 4$ and $m_Q = 11/4 = 2.75$. If it is found that a single transect across a study area yields too few runs for the test to be carried out, there is no objection to combining the data from several transects. All that one needs to know to do the test are the mean

18*

number of individuals per run (the mean run-lenght) for at least 25 whole runs of each species.

The following data (from Pielou, 1962b) illustrate the use of the test. The species concerned are two small herbaceous perennials that are common members of open grassland communities in the arid southern region of interior British Columbia. They are *Agoseris cuspidata* (Compositae) which does not reproduce vegetatively; and *Zygadenus venenosus* (Liliaceae) which, though it forms bulbs, nearly always occurs as isolated individuals. The order of occurrence of individuals of these species was observed in fifteen transects, each 25 centimeters wide and 75 meters long. It was found that: for *A. cuspidata* there were 303 individuals in $n_A = 111$ runs

whence $m_A = 303/111 = 2.730$;

for *Z. venenosus* there were 273 individuals in $n_Z = 111$ runs

whence $m_Z = 273/111 = 2.459$.

Therefore

$$Z = \frac{111}{303} + \frac{111}{273} = 0.7729.$$

If the two species are unsegregated we should expect, with 95% probability, to find Z within the interval

$$1 \pm 2 \sqrt{\left\{ \frac{m_A - 1}{n_A m_A^3} + \frac{m_Z - 1}{n_Z m_Z^3} \right\}} = 1 \pm 0.0812.$$

Since the observed Z is outside these limits it is concluded that the species are segregated.

Up to this point it has been tacitly assumed that if the members of a two-species population are not randomly mingled, then the species must occur as relative clumps. This is not the only possibility, however. It is also possible for two species to be mingled *more* thoroughly than they would be if their relative positions were solely a matter of chance in which case they can be described as *negatively segregated*.

Consider the two tests for segregation described above. The 2×2 table test, based on observations of the species of each individual and of its nearest neighbor, leads to acceptance of the null hypothesis of no segregation if $ad = bc$ approximately. If $ad > bc$ significantly, the conclusion follows that, for both species, the probability that an individual will have another member of its own species as nearest neighbor exceeds the proportion which the species forms of the combined population. Thus the species are relatively clumped, or positively segregated. Conversely, if $ad < bc$ significantly, the opposite conclusion follows: the species are so thoroughly mingled that each individual is *less* likely to have a member of its own species as nearest neighbor than their relative proportions would lead one to expect, and the species are therefore negatively segregated.

The contrasting results possible with the runs test are also worth noting. As explained above, if m_P and m_Q are the mean run-lengths of species P and species Q then $Z = (1/m_P) + (1/m_Q)$ has expectation unity (provided the numbers of runs are not small) when the species are unsegregated. Positive segregation causes the runs of each species to be unexpectedly long so that $Z < 1$; and negative segregation causes the runs to be unexpectedly short so that $Z > 1$.

Negative segregation is, as one might expect, much more rarely encountered than positive segregation. Circumstances which may lead to it, and possible examples, are described in section 5.

4 ASSOCIATION AMONG SEVERAL SPECIES

This chapter has so far been concerned with the spatial relations of pairs of species. Methods for investigating several species jointly are harder to devise and to interpret. Suppose one is interested in interactions among k, say, species. One can investigate the association and segregation between every possible pair of species in the group; there are $k(k - 1)/2$ such pairs. Or one can attempt to test for, and perhaps measure, the association and segregation exhibited by the group as a whole. First it is necessary to decide whether the concepts of association and segregation, as applied to groups of more than two species, have clear meanings.

In the case of segregation the answer is obviously yes. It is perfectly reasonable to ask whether the individuals of several (as opposed to two) species are thoroughly intermingled. Segregation among k species will be discussed in section 5 (page 283).

In the case of association (which will be discussed in detail in this section) the answer is not so clear. In a group of several species there are likely to be a number of intra-pair relationships, some positive and some negative and of varying degrees of intensity. These various pairwise relationships cannot be independent of one another. For example, consider a three-species group containing species A, B and C; if A and B, and also A and C are strongly negatively associated, then where A is present B and C will perforce be positively associated though elsewhere they could be independent.

As another example imagine a group of ten species. One could (if one had reason to) ask whether species A and B were positively, and C and D negatively associated in the presence of E, F and G and in the absence of H, I and J; or not. It is obvious that as soon as k, the number of species, becomes at all large a staggeringly large number of relationships can be envisaged within pairs and larger subgroups of the species. Thus if there are no preconceived hypotheses in need of testing, it is obviously useful to have some single measure of "interassociation" that applies to the whole group. The data used to calculate the measure are to be of the same kind as those used when the association between only two species is tested; that is, a list of the species present in each of a number of sampling units. Then the measure of *interassociation*, as it may be called, should measure the extent to which the field observations deviate from what would be expected if the species were all independent of one another.

Such a measure can be devised (Pielou, 1972a, b), as will now be shown. For clarity, its application to the measurement of association between only two species will be demonstrated first, and after that its extension to the measurement of many-species interassociation.

Consider the following 2×2 table which shows how the two species P and Q are associated in $N = 24$ sampling units. For conciseness, the sampling units will be called quadrats in what follows. The species are positively associated as is shown by the fact that 14 of the quadrats contain both species as against an expected number of $15 \times 16/24 = 10$ such co-occurrences if they were independent.

| | | Species Q | | |
		Present	Absent	Totals
Species P	Present	14	1	15
	Absent	2	7	9
	Totals	16	8	24

Imagine now what would happen if one could shift a species-occurrence from one quadrat to another. For instance, one of the 14 quadrats in which both species occur together could be chosen and all the species Q individuals in it transferred to one of the empty quadrats; as a result of the transfer the number of quadrats containing both species, and also the number of empty quadrats, would each be reduced by one. At the same time the quadrat from which species Q had been removed would now belong to the "species P only" category; and the quadrat into which Q had been placed would now belong to the "species Q only" category. Thus, as a result of transferring a single "occurrence" (i.e., all the individuals of one species in one quadrat), the 2 × 2 table's cell frequencies have been altered from

$$\begin{array}{c|c} 14 & 1 \\ \hline 2 & 7 \end{array} \quad \text{to} \quad \begin{array}{c|c} 13 & 2 \\ \hline 3 & 6 \end{array}.$$

The row and column totals are unchanged and have not been shown. If three more transfers are made, in every case from a two-species quadrat to an empty quadrat, the table obtained as a result of all the transfers is found to be the "expected" table on the hypothesis of independence of the two species. Thus four transfers, represented below by arrows, lead from the given "observed" table to the final "expected" table through the following stages:

$$\begin{array}{c|c} 14 & 1 \\ \hline 2 & 7 \end{array} \rightarrow \begin{array}{c|c} 13 & 2 \\ \hline 3 & 6 \end{array} \rightarrow \begin{array}{c|c} 12 & 3 \\ \hline 4 & 5 \end{array} \rightarrow \begin{array}{c|c} 11 & 4 \\ \hline 5 & 4 \end{array} \rightarrow \begin{array}{c|c} 10 & 5 \\ \hline 6 & 3 \end{array}.$$

It can therefore be said that 4, the number of species transfers needed to change the observed to the expected table, is a measure of the difference between these tables, and hence of the degree of interassociation (in this case positive) between the species.

Negative interassociation is possible, of course. Thus if the observed table in the example above had been

$$\begin{array}{c|c} 9 & 6 \\ \hline 7 & 2 \end{array},$$

its interassociation would be given by -1. This is because the number of co-occurrences in this table is less than expectation, implying negative association; and the transfer of a single occurrence will convert it to the expected table. The required transfer consists in choosing any quadrat with species Q only and shifting its species Q individuals into a quadrat with species P only (or vice versa). Obviously this transfer would decrease by one the number of quadrats containing only P, and the number containing only Q, and would increase by one both the number of two-species quadrats and the number of empties.

We now require a formula for calculating the number of transfers needed to convert an observed to an expected table. The following symbols will be used:

Let v be the number of transfers.

Recall that the total number of quadrats is N.

Let s_i be the number of species in the ith quadrat for $i = 1, 2, ..., N$.

Let n_j be the number of quadrats containing the jth species. In the case being considered j takes only the two values 1 and 2 but in general $j = 1, 2, ..., k$.

Let A be the total number of occurrence so that

$$A = \sum_{i=1}^{N} s_i = \sum_{j=1}^{k} n_j.$$

Then it may be proved (Pielou, 1972b) that

$$v = \frac{1}{2N} \left\{ N \sum_i s_i^2 + \sum_j n_j^2 - A^2 - AN \right\}. \tag{11.5}$$

As an example, recall the first of the "observed" tables discussed above (see page 278).

Species P was present in 15 and species Q in 16 of the quadrats, and therefore

$$n_1 = 15, \quad n_2 = 16, \quad \Sigma n_j = A = 31,$$

and

$$\Sigma n_j^2 = 481.$$

From the four cell frequencies it is seen that, of the $N = 24$ quadrats, 14 contained both species and 3 contained only one. Thus

$$s_1 = s_2 = \cdots = s_{14} = 2; \quad s_{15} = s_{16} = s_{17} = 1$$

and

$$s_{18} = \cdots = s_{24} = 0.$$

Then

$$\sum_i s_i = A = 31 \quad \text{(as already obtained from } \sum_j n_j\text{)};$$

and

$$\sum_i s_i^2 = 59.$$

Substituting these values in equation (11.5) gives $v = 4$ and this, as was found directly, is the required number of transfers.

Now consider the interassociation among $k > 2$ species. Instead of using a flat two-dimensional 2×2 table to tabulate the results, a k-dimensional $2 \times 2 \times \cdots \times 2 = 2^k$ table is needed. Thus any one of 2^k different species-combinations (including, as one of the possibilities, absence of all the species) could conceivably occur in a quadrat; though, of course, when $N < 2^k$, which is usually the case when k becomes large, only N at most of the theoretically possible species-combinations can be encountered.

As with a 2×2 table, any 2^k table can be converted to an "expected" table (whose cell frequencies are those to be expected if all the species-occurrences are independent) by a sequence of transfers of occurrences; the transfers do not affect the total number of occurrences of any of the species. However, some of these transfers may be positive and some negative. A transfer is positive if it entails transferring an occurrence from a quadrat containing both members of some particular pair of species to a quadrat containing neither of them; and negative if it entails the putting together of two occurrences that had previously been separate. Consequently, in a many-species table the index v, defined as in equation (11.5), gives the *net* number of positive transfers needed to convert an observed to the corresponding expected table. Depending on whether the number of positive transfers is greater or less than the number of negative ones, v is positive or negative. Since v depends on N, the number of quadrats observed, it is more convenient to use v/N as what may be called the 2^k table's *index of interassociation*.

The fact that v is the difference between the numbers of positive and negative transfers can, of course, lead to ambiguity. If observations on a group of species yielded a low value of v, it could mean either that the species-occurrences were almost or quite independent; or else that roughly equal numbers of positive and negative transfers were required to convert the observed to the expected table. This difficulty is the unavoidable price of trying to summarize in a single index the properties of an exceedingly complicated structure, in this case a 2^k table.

The difficulty crops up repeatedly when natural ecosystems, with all their natural complexity, are studied. When a coefficient or index is used to measure some single property of such a system, it is easy to overlook, or forget, the fact that one index can measure only one thing, and it is naive to except too much from a single number. We shall encounter this problem again in succeeding chapters (see, especially, page 290). Here it suffices to say that v/N, like a great many other ecological indices, will be most useful in intercomparing communities that are not too dissimilar. Differences in the index are then most likely to be truly indicative of differences in the property they are designed to measure (in the case of v/N, the degree of overall interdependence among the species of a group) and less likely to be the uninterpretable result of unsuspected causes.

To illustrate calculation of the index of interassociation in the many-species case, consider the body of data shown in Table 11.3 which is adapted from Culver (1970). He was investigating the animal communities in $N = 28$ limestone caves in West Virginia and each cave is treated as a single sampling unit. Seven species of animals were found and the table shows, for each animal in each cave, whether it was present (1) or absent (0). The names of the species appear as column headings. *Gammarus* and *Stygonectes* are amphipods, *Asellus* is an isopod, the *Cambarus* species are crayfish and *Gyrinophilus* is a salamander (its larvae were found).

The numbers of species in the several caves, which are the values of s_i for $i = 1, 2, ..., 28$, are the row totals of the table.

The numbers of occurrences of the species in the several caves, which are the values of n_j for $j = 1, 2, ..., 7$, are column totals.

TABLE 11.3   Occurrences of seven species of animal in 28 limestone caves in West Virginia. (Adapted from Culver, 1970)

Cave	Gammarus minus	Stygonectes spinatus	Stygonectes emarginatus	Asellus holsingeri	Cambarus nertriuse	Cambarus bartoni	Gyrinophilus porphyriticus	Values of s_i
#1	1	0	0	0	0	0	0	1
#2	1	1	0	0	1	0	1	4
#3	1	0	1	0	0	1	1	4
#4	1	0	0	0	1	0	1	3
#5	1	1	0	0	0	1	0	3
#6	0	0	0	0	0	0	0	0
#7	0	0	0	0	0	1	1	2
#8	1	1	1	1	0	0	1	5
#9	1	0	0	0	1	0	0	2
#10	1	1	0	1	1	0	1	5
#11	1	1	1	1	0	0	0	4
#12	0	0	0	0	0	0	0	0
#13	0	1	1	1	0	0	0	3
#14	0	0	1	1	0	0	0	2
#15	0	1	1	1	0	0	0	3
#16	1	0	0	1	0	0	0	2
#17	0	0	0	0	0	0	1	1
#18	1	1	1	0	0	0	0	3
#19	1	0	0	0	0	0	0	1
#20	1	0	0	0	0	1	0	2
#21	1	0	0	0	1	0	0	2
#22	1	1	0	1	1	0	0	4
#23	0	0	0	0	0	1	0	1
#24	1	0	0	0	0	0	0	1
#25	0	1	0	1	0	0	0	2
#26	0	1	0	1	0	0	0	2
#27	0	0	0	0	0	0	0	0
#28	0	0	1	0	0	0	0	1
Values of n_j	16	11	8	10	6	5	7	63

It is found that

$$A = \sum_{i=1}^{28} s_i = \sum_{j=1}^{7} n_j = 63,$$

$$\Sigma s_i^2 = 197; \quad \Sigma n_j^2 = 651.$$

Therefore

$$v = 7.75 \quad \text{and} \quad v/N = 0.277.$$

The expected values of v and v/N, given complete independence of the animal species, is zero. That is,

$$\mathscr{E}(v) = \mathscr{E}(v/N) = 0.$$

The fact that the observed values are low suggests that the cave animals may in fact be mutually independent, and this possibility can be tested. The procedure is described in Appendix 11.3; it leads to the conclusion that there is no reason to reject the null hypothesis of independence of the species.

5 SEGREGATION IN MANY-SPECIES POPULATIONS

There are many ways of using the evidence afforded by spatial patterns as an aid to understanding interspecies relationships. In this section we return to the topic of segregation and discuss its manifestations in plant communities of several species. There are two extreme cases. In one, each individual plant is distinct and clearly separate from all other individuals; a typical example is a many-species forest. In complete contrast are cases in which each plant species reproduces vegetatively to form one-species patches in which no individuals can be distinguished; then the patches themselves are the entities forming the pattern, which is a mosaic. Intermediate types of vegetation patterns, in which some species occur as patches or mats and others as separate individuals, are the commonest of all and no systematic method of studying them has yet been devised.

Testing whether the trees in a many-species forest are segregated is straight-forward and requires only an obvious extension of the first method described for the two-species case in section 3. Thus if, in a k-species forest, one observes the species of every tree and of its nearest neighbor, the observations can be tabulated in a $k \times k$ table. The null hypothesis, that the k species are randomly mingled, and thus that the species of a tree's nearest neighbor is independent of its own species, can be tested in the way described in Appendix 11.2 (page 398). As in the two-species case, segregation can be positive or negative. In a many-species community, any given species may be positively segregated from the rest, in which case it will tend to occur in groups from which other species are excluded. Or a species may be negatively segregated from the rest, in which case its members are so thoroughly mingled with those of other species that it rarely, if ever, has a member of its own species as nearest neighbor.

It seems likely that tree species in lowland tropical forests tend to exhibit negative segregation. Typically there are a great many species, each present at low density and so thoroughly intermingled that the mature trees of any one species are widely separated from one another. A probable explanation for this phenomenon has been given by Janzen (1970).

He suggests that the densities of young seedlings in these forests are reduced, in large part, by the attacks of host-specific insects which prey upon them. The attacks of these insect predators are likely to be most effective among seedlings clustered around their parent trees. Two independent causes can lead to this. In the first place, to use Janzen's terms, the predatory insect may be "density responsive", that is, its populations may thrive best where the seedling trees they feed upon are abundant (and hence the risk of starvation minimal) and often, except among autotoxic plants, seedling density is greatest around a parent tree. The second possible cause is that the predatory insect is "distance responsive". This occurs if it feeds on mature as well as on seedling individuals of its particular host plant; then the density of the insects will be greatest in and around an adult tree and seedlings close to the adult will be exposed to the greatest risk of attack. In either case, whether the insect is density- or distance-responsive, the seedlings most likely to survive are those at some distance from others of their own species. They are then likely to be surrounded by trees of other species and negative segregation results. Figure 11.5 illustrates schematically the mechanism just described. The solid line (the "population recruitment curve") shows the number of seedlings that will succeed as a function of distance from the parent tree. This number is proportional to the product of two factors: the probability that a seed will land at a given distance from its parent (shown by the dashed line) and its probability of survival at this distance (shown by the dotted line).

The concept of spatial segregation is not directly applicable to plants that do not occur as distinct individuals. When the ground is covered by a continuous layer of vegetation, made up of a mosaic of different species-patches all formed by vegetative reproduction, one cannot speak of one patch being the nearest neighbor of another; the patches have no definable centers and each is in actual contact with the patches surrounding it. And if a narrow belt transect is laid across mosaic vegetation, one cannot list the sequence of individuals encountered (since there are no individuals) but only the observed sequence of patches. In such a list it is impossible for a species to occur twice in succession; (it is assumed that each one-species patch is treated

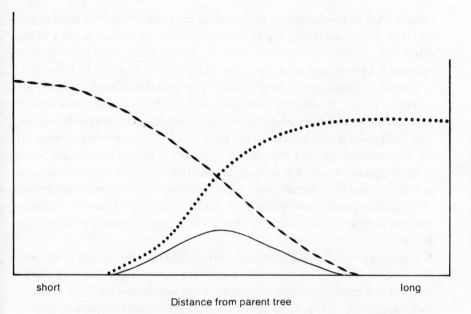

short   long
Distance from parent tree

FIGURE 11.5   If the density of seeds decreases (dashed line), and a seedling's probability
of survival increases (dotted line) with increasing distance from a parent tree, then the
"population recruitment curve" (solidline) has a maximum at some distance from the
parent tree. (Redrawn from Janzen, 1970)

as a homogeneous entity). Thus if there were five species, say, represented by
the letters A, B, C, D and E, a list of the sequence of patches cut through by a
transect would be of the form

$$A \quad B \quad C \quad A \quad D \quad B \quad E \quad D \quad E \quad C \quad E \quad D \quad A \quad B \quad C \quad E \quad D \quad A \quad C \quad A \ldots,$$

for example, in which each letter is necessarily different from those on either
side of it.

Given such a sequence, it is reaonable to enquire whether the different
species-patches are randomly mingled. If they are, it follows that the probability that a particular patch will be of species A, say, is unaffected by the
species of the patches to left and right of it. If they are not randomly mingled,
then some pairs of species must be juxtaposed unexpectedly often (and therefore others unexpectedly seldom). A method for testing whether mosaic
patches are randomly mingled has been given by Pielou (1967a).

The ecological processes that give rise to, and maintain, a vegetation
mosaic are of great interest. We have already considered two-phase mosaics
(Chapter 8, page 167); the problems pertaining to many-species mosaics,
except for their greater complexity, are of the same kind. In particular, one

often wishes to know whether the plants in adjacent patches are competing; and if so, which one is winning and enlarging its patches at the expense of the other.

From a dynamic point of view, three kinds of mosaics can be envisaged.

First, if a mosaic were "static", that is, if it maintained itself unaltered for a lengthy period, it would be reasonable to infer that it owed its existence to permanent heterogeneity of the habitat that was not modified by the vegetation itself; and also, that the boundaries of each species' patches were set by the tolerance limits of the species. Although it is possible to conceive of a static mosaic in which the patch boundaries represent "fronts" between actively competing species that have reached a prolonged equilibrium or stalemate, this seem unlikely in nature. The only plausible explanation for a truly static mosaic is that its geometry is controlled by abiotic factors.

A second kind of mosaic is the cyclical type and we have already considered several examples (cf. page 169 et seq.). Another example, which is especially relevant to a discussion of competition among similar species, has been given by Harper et al. (1961). They describe a raised *Sphagnum* bog in which semi-aquatic species of *Sphagnum* occupy, and gradually fill, the wettest hollows; other *Sphagnum* species form the beginning stages of hummocks; and still others, needing better-drained sites, establish themselves on, and add to, existing hummocks before they die. The result is a fine-grained mosaic of several congeneric species which, because they are closely related, presumably compete where they come in contact. But since, at any one spot, a cyclical succession occurs in which, for example, species A is displaced by species B, B by C, and then C by A, the bog as a whole exhibits long term uniformity. It is not static however, for its component patches are in continual migration.

The third kind of mosaic, also dynamic, is that exhibited by many kinds of vegetation at various stages in a natural succession. Nearly all many-species herbaceous communities that are past the pioneer stage are dominated by vegetatively reproducing, and hence patch-forming, perennials. Such communities are therefore dynamic mosaics which undergo changes in pattern, as well as in component species, as succession proceeds. A familiar example, which is no less a mosaic because its patches are often of great geometrical regularity, is provided by the concentric annular zones of vegetation closing in on a drying pond as it gradually fills with plant remains. In such dynamic mosaics the patches are likely to show sustained migration in one direction in contrast to the wandering migrations that occur in cyclical mosaics. It should not be inferred, however, that the replacement of a species by a successor is necessarily the outcome of competition.

As McIntosh (1970) has emphasized, some pioneer annual species are incapable of persisting for long at a site even when the invading perennials that customarily replace them are excluded. It is seldom easy to judge with certainty whether or not competition is taking place.

Summarizing, it seems clear that much can be learned from studying the patterns of vegetation mosaics, especially if attention is given to the way they change in time.

CHAPTER 12

Ecological Diversity

WHEN SEVERAL or many species-populations occur together and interact with one another in a small region of space, they jointly constitute an ecological "community", and the two final chapters of this book are concerned with the properties of communities considered as indivisible entities. A community's "behavior" can be thought of as the resultant of the behaviors of all its component, mutually dependent, species-populations. But it is not feasible, even if it were desirable, to investigate the population dynamics of each species in turn and the interactions among designated subsets of them. Such a pedestrian approach would lead too slowly and too indirectly to the topic of central interest, that of the behavior of the community treated as an entity. The ultimate objective in studying the ecology of a community is to determine the nature and the relative importance of the factors controlling its composition; also whether, to what extent, and why, the community is changing with time. To pursue this objective it is necessary to define some measurable properties of the community as a whole. If this can be done, making it possible to write down a short list of measurements that constitute a summary description of a particular community at a particular time, it then becomes possible to make quantitative comparisons among several communities. This is a necessary first step toward an understanding of how communities function. The definition and measurement of what has come to be known as the "diversity" of a community is the subject of this chapter and in the final chapter we shall consider some of the conclusions that diversity studies have led to.

It should be noticed that to define all the species-populations that occur together in one place as the "entity" to be studied is not to assert that they constitute a "natural" entity. The "community" that an ecologist delimits for research purposes may prove to be clearly separated from surrounding communities by abrupt boundaries; or it may merge into them gradually and imperceptibly; or it may be clearly distinct from some but not others of its neighbors. Usually the true state of affairs is unknown and is one of the matters to be examined. One cannot, therefore, insist that the boundaries of a study area should coincide with natural boundaries since these may be non-existent or unrecognizable. Thus in drawing the boundaries of the community to be studied a certain degree of arbitrariness is sometimes unavoidable.

Another decision to make before research can begin concerns the plants and animals that are to belong to the "community" under study. Although a purist might insist that the community in a specified region consists, by definition, of all the living organisms without exception inside it, it is clearly impossible to use so broad a definition in practice. Therefore, some clearly defined *taxocene* must be designated as the group of organisms under study. This term (Hutchinson, 1967) connotes all the members of any taxonomic group of higher level than a species. Thus a taxocene may be a group no larger than a family: for example, all the Phocidae (true seals) in a subarctic fjord. It may be an order, such as the Odonata (dragonflies and damselflies) in and around a body of water. Or a class, such as the birds (Aves) breeding together on a rocky oceanic island. A still more inclusive taxocene might consist of all the vascular plants and bryophytes within a study area. In any case, the student of a community must be no less precise than the student of a one-species population in specifying the group of organisms under study; there must be no vagueness as to its spatial and taxonomic limits. Temporal limits must be set also if the taxocene contains animals (migratory birds, for instance) that may be in the study area only part of the time.

1 DEFINITIONS OF "DIVERSITY"

The first and most immediately interesting property of a community to consider is, of course, the number of species it contains. It is in this respect that comparable communities show the most striking contrasts. Thus the number of species of vascular plants in an area of given size is far greater if the study area is part of an alpine meadow than if it is part of a salt marsh. Another example: the birds in a particular tract of woodland in temperate latitudes form a species-rich community in the period when numerous spring migrants are arriving or passing through; the same taxocene (all birds) in the same area forms a different community, with far fewer species, in winter.

In both these pairs of contrasting communities the one with the greater number of species could be said to be more diverse than the other, or to have greater *diversity*. This has led to the suggestion that the number of species, customarily* denoted by s, be used as the "index of diversity" of a community and many ecologists do use the word *diversity* in this sense.

---

* Notice that in this and the following chapter s denotes the total number of species in the sample, or community, or collection, being studied. In chapter 11 this quantity was denoted by k, and s was used for the number of species in a single, usually small, sampling unit. The changed meaning of s should lead to no misunderstanding; it seems desirable to stick to the customary usage in each context.

However, two communities having the same value of s may differ greatly in what one would, using the word colloquially, describe as their diversities. For example, imagine two forests, each with ten species of trees; suppose that in one forest the species are present in fairly equal proportions; and that in the other, one species is overwhelmingly dominant and the rest are represented by only a few scattered, hard to find, individuals. The former forest is certainly the more diverse in the ordinary meaning of the word and this has led to the devising of measures of diversity that take account of differences in the relative abundances of the species forming a community. The two best-known and most useful of these are described below.

The concept of *eveness* has also been introduced (Margalef, 1958) and a way of measuring it will be discussed. The evenness of a community is high or low according as its several species are more or less evenly represented.

The measures of diversity to be described depend both on s and on evenness and they attempt to summarize, in a single index, these two quite different properties of a community. It is futile to debate whether the loss in information that results from combining (or "confounding") these two properties is offset by the gain in simplicity resulting from use of a single index. Obviously there is no one answer to this question; it must always depend on the reason for which diversity indices and measures of evenness were calculated in the first place, that is, on the nature of the underlying ecological problem whatever it may be.

The two diversity indices we shall now consider in detail have become known as the *Shannon-Wiener Index* and the *Simpson Index*. Both are functions of the relative abundances (i.e., the proportions) of the species in the community and for the moment we shall assume that these proportions are known exactly so that no allowance need be made for sampling errors. It is also assumed that the community is "indefinitely large" (in other words, that no appreciable numerical errors are introduced by treating the number of individuals it contains as infinite). What to do when these assumptions do not hold will be discussed subsequently.

The *Shannon-Wiener Index*, H', is defined as

$$H' = -\sum_{i=1}^{s} p_i \log p_i \quad \text{units}$$

where p_i is the proportion of the community that belongs to the ith species.

The base of the logarithms is immaterial. Some authors use e, some 2 and some 10. The only effect of changing the base of the logs is to alter the units in which diversity is measured. When the base is e the unit is a "nat" (McIntosh, 1967); when the base is 2 the unit is a "bit" and when it is 10, a "decit" (Pielou, 1966b). No one unit has yet become standard and therefore, when numerical results are given, it is essential to state explicitly the units being used.

The index H' defined above has three properties that make it especially appropriate for measuring ecological diversity. Indeed, as can be proved (Khinchin, 1957), it is the *only* mathematical function of the p_i values that has these three properties. They are as follows.

1) For a given number of species H' takes its maximum value when $p_i = 1/s$ for all i; in other words, when all s species are present in equal proportions (i.e., are completely "even").

2) Given two completely even communities, the one with the larger number of species has the greater value of H'.

3) H' can be split into components that are additive. This is most easily explained in terms of a concrete example. Imagine that the community under investigation consists of all the adult mosquitoes in a given area. Suppose there are four species and that they are present in proportions p_1, p_2, p_3 and p_4. Imagine also that the mosquitoes tend to feed at different times of day and that data are collected for three observation periods of equal length, Periods 1, 2 and 3, which run from 6 p.m. to 2 a.m., from 2 a.m. to 10 a.m., and from 10 a.m. to 6 p.m. It is assumed that an individual mosquito will feed in only one period but that the members of one species do not necessarily all feed in the same period. Let the proportions of the total mosquito population active in the respective periods be q_1, q_2 and q_3. Finally, let u_{ij} be the proportion of species i mosquitoes that feed in period j; and let v_{ij} be the proportion of period j mosquitoes that belong to species i. These facts can be summarized in a 4×3 table as follows; (they are treated as "facts" rather than as "observations" since no allowance is to be made for the uncertainties that arise in practice from sampling errors).

		Feeding Period			
		1	2	3	Totals
	1	n_{11}	n_{12}	n_{13}	$n_{1.}$
Mosquito	2	n_{21}	n_{22}	n_{23}	$n_{2.}$
species	3	n_{31}	n_{32}	n_{33}	$n_{3.}$
	4	n_{41}	n_{42}	n_{43}	$n_{4.}$
	Totals	$n_{.1}$	$n_{.2}$	$n_{.3}$	N

Here the cell frequency n_{ij} denotes the number of mosquitoes of species i feeding in period j; and N is the size of the whole mosquito community. In terms of the symbols in the table the proportions defined above are

$$p_i = \frac{n_{i.}}{N}, \quad q_j = \frac{n_{.j}}{N}, \quad u_{ij} = \frac{n_{ij}}{n_{i.}} \quad \text{and} \quad v_{ij} = \frac{n_{ij}}{n_{.j}}.$$

19*

Notice, also, that all the following sums are unity:

$$\sum_i p_i = \sum_j q_j = \sum_j u_{ij} = \sum_i v_{ij} = 1.$$

Now consider the "diversity" in the mosquito community. Every mosquito has been doubly classified and therefore the community can be thought of as "diverse" in two different ways: first, in respect of the different species present, and second, in respect of the different feeding periods of the mosquitoes.

The species-diversity, which will be written $H'(S)$, is given by

$$H'(S) = -\sum_{i=1}^{4} p_i \log p_i.$$

The community also has "period-diversity", $H'(P)$ say, given by

$$H'(P) = -\sum_{j=1}^{3} q_j \log q_j.$$

We can also measure the species-diversity within each separate feeding period. Thus the species-diversity in the jth period may be written

$$H'_j(S) = -\sum_{i=1}^{4} v_{ij} \log v_{ij}, \quad \text{for } j = 1, 2, 3.$$

The mean of the three within-period species diversities is

$$H'_P(S) = \sum_{j=1}^{3} q_j H'_j(S).$$

Analogously, we can measure the period diversity for each species. That of the ith species is

$$H'_i(P) = -\sum_{j=1}^{3} u_{ij} \log u_{ij} \quad \text{for } i = 1, 2, 3, 4.$$

The mean over all four species is

$$H'_S(P) = \sum_{i=1}^{4} p_i H'_i(P).$$

Finally, taking account of both classifications simultaneously, each individual mosquito is seen to belong to one of twelve mutually exclusive classes corresponding to the twelve cells in the 4×3 table. The diversity of the community in respect of this double classification may be written

$$H'(SP) = -\sum_i \sum_j \frac{n_{ij}}{N} \log \frac{n_{ij}}{N}.$$

Then, since

$$\frac{n_{ij}}{N} = \frac{n_{ij}}{n_{i.}} \times \frac{n_{i.}}{N} = u_{ij} p_i$$

and

$$\frac{n_{ij}}{N} = \frac{n_{ij}}{n_{.j}} \times \frac{n_{.j}}{N} = v_{ij} q_j,$$

it is also seen that

$$H'(SP) = -\sum_i \sum_j p_i u_{ij} \log p_i u_{ij}$$

$$= -\sum_i \sum_j q_j v_{ij} \log q_j v_{ij}.$$

The way in which these various diversities are interrelated can be summarized by the following pair of equations

$$\left. \begin{array}{l} H'(SP) = H'(S) + H'_S(P) \\ H'(SP) = H'(P) + H'_P(S). \end{array} \right\} \tag{12.1}$$

In words, the first equation states that the total diversity (i.e., that based on the double classification) is the sum of the species-diversity and the mean of the within-species period-diversities. The second equation states that the total diversity is also the sum of the period-diversity and the mean of the within-period species-diversities.

The first relation is proved as follows.

$$H'(SP) = -\sum_i \sum_j p_i u_{ij} \log p_i u_{ij}$$

$$= -\sum_i \sum_j p_i u_{ij} (\log p_i + \log u_{ij})$$

$$= \sum_j u_{ij} \left\{ -\sum_i p_i \log p_i \right\} + \sum_i p_i \left\{ -\sum_j u_{ij} \log u_{ij} \right\}$$

$$= \left\{ -\sum_i p_i \log p_i \right\} + \sum_i p_i H'_i(P)$$

$$= H'(S) + H'_S(P).$$

The second equation is proved analogously starting from

$$H'(SP) = -\sum_i \sum_j q_j v_{ij} \log q_j v_{ij}.$$

Table 12.1 gives a numerical example (artifical).

The existence of the relationships shown in equations (12.1) is the third of the three properties of the Shannon-Wiener Diversity Index which, together, make it so useful in ecology. It is worth reiterating what was said

TABLE 12.1   (Artifical data). The species, and periods of capture, of a
sample of 500,000 mosquitoes

	Feeding period			
	6 p.m. to 2 a.m.	2 a.m. to 10 a.m.	10 a.m. to 6 p.m.	Totals
Species 1	170,000	15,000	15,000	200,000
Species 2	60,000	50,000	40,000	150,000
Species 3	20,000	70,000	10,000	100,000
Species 4	10,000	10,000	30,000	50,000
Totals	260,000	145,000	95,000	500,000

$$H'(S) = 1.2799 \qquad H'(P) = 1.0145$$
$$H'_S(P) = 0.7916 \qquad H'_P(S) = 1.0570$$

$$H'(SP) = 2.0715$$

All diversities are in nats.

above, namely that the index $-\Sigma p_i \log p_i$ is the *only* function of the p_i values
that has these properties. The usefulness of the third property is that it
allows the diversity of a community to be split into components whenever
individuals are classified in more than one way. The classifications may be
wholly independent. They may be partly dependent (as in the example in
Table 12.1). Or one classification may be entirely dependent on another as is
true, for instance, when organisms are classified first into genera and then
into species within each genus; (all the members of one species necessarily
belong to the same genus). Thus one may split a community's overall diversity into hierarchical components corresponding to the hierarchical levels of
taxonomic classification.

Given a large, highly diversified community one may treat its total diversity as being made up of separate species-diversities within each of the several genera, of separate genus-diversities within each of the families, of family diversities within the orders, and so on up the taxonomic hierarchy (Pielou,
1967b, 1969).

As an example, imagine an extensive forest with trees of seven species
belonging to four genera; suppose their relative proportions are as shown
in the 4 × 7 table in Table 12.2. These data have been invented for illustration, and for simplicity we shall here consider two taxonomic levels only,
genus and species. The argument is easily extended to take account of as
many taxonomic levels as desired though the manipulations become considerably more cumbersome; whereas a two-dimensional table suffices to
portray the data when only two taxonomic levels are considered, a three-

TABLE 12.2   The relative proportions of the trees in an imaginary forest

Genus	*P. strobus*	*P. resinosa*	*P. banksiana*	*A. saccharum*	*A. rubrum*	*B. papyrifera*	*T. occidentalis*	
Pinus	0.300	0.050	0.025	–	–	–	–	0.375
Acer	–	–	–	0.400	0.125	–	–	0.525
Betula	–	–	–	–	–	0.090	–	0.090
Thuja	–	–	–	–	–	–	0.010	0.010
	0.300	0.050	0.025	0.400	0.125	0.090	0.010	1.000

dimensional table would be needed if the individuals were classified in three ways and, in general, an n-dimensional table if they were classified in n ways.

The total species-diversity in the imaginary forest will be denoted by $H'(SG)$ since the proportions used in its calculation are arrived at by a double classification (into genera and species) of all the trees. Therefore, writing w_{ij} for the proportion of the total forest made up of individuals in the jth species of the ith genus,

$$H'(SG) = -\sum_i \sum_j w_{ij} \ln w_{ij} \quad \text{nats.}$$

In the example

$$H'(SG) = (0.300 \ln 0.300 + \cdots + 0.010 \ln 0.010) \quad \text{nats}$$

$$= 1.4925 \text{ nats.}$$

Next, let p_i denote the proportion of trees in the ith genus for $i = 1, 2, 3, 4$. In the table these proportions are shown in the column of row totals at the right. The genus-diversity of the forest, $H'(G)$, is

$$H'(G) = -\sum_i p_i \ln p_i = 0.9689 \text{ nats in the example.}$$

The species-diversities within each of the four genera are as follows. In *Pinus* (genus # 1)' the three species are present in proportions 0.300/0.375, 0.050/0.375 and 0.025/0.375; or 0.800, 0.133 and 0.067. Therefore in this genus the species-diversity is

$$H_1'(S) = 0.6277 \text{ nats.}$$

In *Acer* (genus # 2) the two species are present in proportions 0.762 and 0.238 and therefore $H_2'(S) = 0.5489$ nats.

$H_3'(S)$ and $H_4'(S)$, the species-diversities within *Betula* and *Thuja*, are both zero since these genera are represented by only one species each and $1 \ln 1 = 0$.

Thus $H'_G(S)$, the mean species-diversity within the four genera is

$$H'_G(S) = \sum_{i=1}^{4} p_i H'_i(S) = 0.5236 \text{ nats.}$$

It is easily confirmed that

$$H'(SG) = H'(G) + H'_G(S). \tag{12.2}$$

This formula is analogous to one of the pair in equations (12.1). Indeed, the splitting of total species-diversity into hierarchical components in this way is a special application of the third property of the Shannon-Wiener index. It is special because the second taxonomic classification (into species) is completely dependent on the first (into genera). Therefore the "genus-diversity within a species" is clearly zero by definition: in symbols, $H'_S(G) = 0$. Thus the equation to add to (12.2), to complete the analogy with the pair in equations (12.1), is the identity $H'(SG) = H'(S)$. Both these terms symbolize the same thing; the *total* species-diversity in the forest.

We return now to the more general case and the example of mosquitoes feeding in different periods. It is worth considering what the various diversity indices would be if the two classifications in the example were completely independent, that is, if the period during which an individual mosquito was active was independent of its species. We should then have

$$n_{ij} = \frac{n_{i.}n_{.j}}{N}.$$

Also, since $n_{ij}/n_{i.} = n_{.j}/N$ it follows that $u_{ij} = q_j$; and since $n_{ij}/n_{.j} = n_{i.}/N$ it follows that $v_{ij} = p_i$. Then $H'_i(P) = H'(P)$ for all i. (In words: the period-diversity within each species is the same for all species, and equal to the period-diversity for all the mosquitoes taken together.) Likewise $H'_j(S) = H'(S)$ for all j. (In words: the species-diversity within each period is the same for all periods, and equal to the species-diversity of all the mosquitoes taken together.)

Then both equations in (12.1) reduce to the same thing, namely

$$H'(SP) = H'(S) + H'(P). \tag{12.3}$$

That is, when individuals are subject to two *independent* classifications the diversity of the doubly-classified individuals is the sum of the two separate diversities pertaining to the separate classifications.

The second diversity index to be discussed here will, for convenience, be called the *Simpson Diversity Index;* this name has also been applied to other indices derived, like it, from the "measure of concentration" first proposed by Simpson (1949).

Writing, as before, p_i for the proportion of a community that belongs to the ith species, Simpson's measure of concentration is

$$\lambda = \sum_i p_i^2.$$

Clearly λ measures something that is the inverse of diversity. For a given number of species, the greater a community's evenness the smaller its λ value, and λ is a minimum when $p_i = 1/s$ for all i.

Two interpretations can be put upon λ. In the first place, it is the probability that if two individuals are drawn at random from the community they will belong to the same species; the more diverse the community, the less likely this is to happen. In the second place, λ measures the "expected probability of an event" (Weaver, 1948). This concept is most easily explained in terms of a numerical example. Imagine a community with only three species, the first being common ($p_1 = 0.6$), the second moderately common ($p_2 = 0.3$) and the third rare ($p_3 = 0.1$). Let each species have associated with it a "commonness value" (actually its probability of occurrence). If an individual is drawn at random from the community the probability that it will belong to the common species is p_1, and if it does its commonness value is also p_1. Likewise the probability is p_2 that its commonness value will be p_2, and p_3 that it will be p_3. Thus the *expected* commonness value (equivalently, the average of the commonness values that would be obtained if the experiment of drawing an individual at random were repeated many times) is $\lambda = \Sigma p_i^2$ (which is 0.46 in the numerical example). It is seen that if a community is made up of a very large number of species, all fairly evenly represented, then *all* the species are rare and there is nothing surprising (since no other outcome is possible) in obtaining a rare species when an individual is drawn from the community at random. In this case the expected probability (or expected commonness value) is low and the diversity is high. Conversely, in a community with one or two very common species, each with a high commonness value, the most likely outcome of a random draw is that one of these common species will be picked; then λ is high.

To measure diversity we need a function of λ that increases as λ decreases. There are various possibilities of which the best* is to take $-\log \lambda$; the reason for choosing this function is that it and the Shannon-Wiener function are both special cases of a much more general mathematical function of a set of proportions (or probabilities) known as the "entropy of order α" of

---

* I am indebted to Dr. C.D.Kemp for pointing this out

the set (Renyi, 1961). This function, here* denoted by the symbol $H^{(\alpha)}$ is defined as

$$H^{(\alpha)} = \frac{\log \sum p_i^{\alpha}}{1 - \alpha}.$$

It is found (as will be shown below) that when $\alpha = 1$, the function reduces to the Shannon-Wiener index and that when $\alpha = 2$ it reduces to $-\log \lambda$.

To find $H^{(\alpha)}$ when $\alpha = 1$ it is necessary to use l'Hospital's rule (see page 394) since merely to substitute 1 for α in the formula for $H^{(\alpha)}$ yields the indeterminate fraction $0/0$. But using the rule and the fact that

$$\frac{d}{d\alpha} [\log \sum p_i^{\alpha}] = \frac{1}{\sum p_i^{\alpha}} [\sum p_i^{\alpha} \log p_i]$$

we have

$$H^{(1)} = \lim_{\alpha \to 1} H^{(\alpha)} = \frac{\lim\limits_{\alpha \to 1} \dfrac{d}{d\alpha} [\log \sum p_i^{\alpha}]}{\lim\limits_{\alpha \to 1} \dfrac{d}{d\alpha} [1 - \alpha]} = \frac{\sum p_i \log p_i}{-1} = H',$$

the Shannon-Wiener index.

To find $H^{(\alpha)}$ when $\alpha = 2$ is straightforward. Simply substituting 2 for α shows that

$$H^{(2)} = -\log \left(\sum p_i^2 \right) = -\log \lambda.$$

Consequently $-\log \lambda$ is the most appropriate function of λ to use as a measure of diversity, and since λ itself was first introduced by Simpson (1949) to measure "concentration" (the opposite of diversity), $-\log \lambda$ may be called Simpson's diversity index.

Since H' and $-\log \lambda$ are both special cases of the more general formula $H^{(\alpha)}$, they have many similarities. Both are zero for a community with only one species and both have the same maximum value, $\log s$, in a completely even s-species community. In all other cases $-\log \lambda < H'$. Names for the units of diversity measured by $-\log \lambda$ have not yet been invented; as with H', of course, various units are possible corresponding to different choices of base for the logarithm. The similarity of H' and $-\log \lambda$ is demonstrated graphically in Figure 12.1. Imagine a community with three species. Then every point in an equilateral triangle represents a possible species-composition for the community. This follows from the fact that the sum of the perpendicular distances, p_1, p_2 and p_3, say, from any point in the triangle to its sides

---

* The customary symbol is H_{α} but $H^{(\alpha)}$ is used here since H with an unbracketed subscript is needed in another context.

is a constant and if the constant is set equal to unity, p_1, p_2 and p_3 represent proportions summing to unity (see Figure 12.1a). Therefore, to every point in the triangle there corresponds a value of H', and also a value of $-\log \lambda$, and the way in which these diversity indices vary with community composition may be shown by contours as in Figure 12.1b. The left half of the triangle in the figure shows the H' contours and the right half the $-\log \lambda$ contours, which have been juxtaposed to facilitate comparison. Natural logarithms were used to obtain the numerical values and the contours are at intervals of 0.1 units. At the center of the triangle, where $p_1 = p_2 = p_3 = 1/3$, both indices have equal maximum values given by

$$H'_{max} = (-\ln \lambda)_{max} = \ln 3 = 1.0986 \text{ units.}$$

Another property that the Simpson index shares with the Shannon-Wiener index is that given by equation (12.3) (page 296); namely, if the individuals in a community are subjected to two *independent* classifications

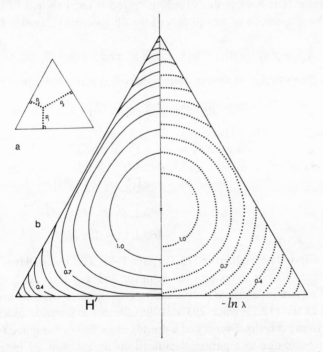

FIGURE 12.1 A comparison of the Shannon–Wiener Diversity Index, H', and the Simpson Index, $-\log \lambda$, for all possible three-species communities. Every point in an equilateral triangle represents a possible composition for a three-species community (see (a): for every point in the triangle, $p_1 + p_2 + p_3 = 1$). The contours in (b) show how the diversity indices vary with community composition. H', solid contours; $-\log \lambda$, dotted contours

the Simpson index based on the double classification is the sum of the indices based on the separate classifications. Thus, recall the earlier example in which it was supposed that mosquitoes had been classified according to species and also according to feeding period. For generality, we now suppose there are s species, present in proportions $p_1, p_2, ..., p_s$; and that the day is divided into t periods in which the proportions of mosquitoes that are active are $q_1, q_2, ..., q_t$; (any individual mosquito feeds in only one period). Then the Simpson index for the species-diversity may be written

$$-\log \lambda(S) = -\log \sum_{i=1}^{s} p_i^2 ;$$

the analogous formula for the period-diversity is

$$-\log \lambda(P) = -\log \sum_{j=1}^{t} q_j^2 .$$

Now assume that a mosquito's feeding period is independent of its species so that the proportion of mosquitoes of the ith species that feed in the jth period is

$$r_{ij} = p_i q_j \quad \text{with} \quad i = 1, 2, ..., s \quad \text{and} \quad j = 1, 2, ..., t.$$

Then the Simpson index for the doubly classified community may be written

$$-\log \lambda(SP) = -\log \left[\sum_i \sum_j r_{ij}^2 \right] .$$

Therefore

$$-\log \lambda(SP) = -\log \left[\sum_i \sum_j p_i^2 q_j^2 \right]$$

$$= -\log \left[\left(\sum p_i^2 \right) \left(\sum q_j^2 \right) \right]$$

$$= -\log \sum p_i^2 - \log \sum q_j^2$$

$$= -\log \lambda(S) - \log \lambda(P).$$

This corresponds to equation (12.3) (page 296). The resemblance between the Shannon-Wiener index and the Simpson index is thus fairly close. But for the Simpson index there is no relation analogous to that shown in equations (12.1) and (12.2) (pages 293 and 296). With the Simpson index it is *not* possible to express the diversity of a doubly-classified community as the sum of components due to separate classifications unless they are independent; in particular the Simpson index cannot be split into hierarchical taxonomic components. Consequently it seems fair to say that the Shannon-Wiener index is much more useful in ecology.

Before going into practical matters (in the next section) it remains to consider how the *evenness* of a community may be measured. It has already been pointed out that the diversity of a community depends on both the number of species and the evenness of their representation. If these two properties of a community are to be measured separately we need a measure of evenness. A useful one (Pielou, 1969) is to take the ratio of the observed diversity as measured by H' to the maximum value H' could possibly have in a community with the same number of species. This value, H'_{max}, would be found in a completely even community in which all s species were present in the same proportion, $1/s$.

Thus

$$H'_{max} = -\sum \frac{1}{s} \log \frac{1}{s} = -\log \frac{1}{s} = \log s.$$

The evenness, J' say, is given by the ratio

$$J' = \frac{H'}{H'_{max}} = \frac{-\sum p_i \log p_i}{\log s}. \tag{12.4}$$

J' is a ratio and has no dimensions; since H' and H'_{max} must, of course, be measured in the same units, the value of J' is unaffected by what these units are, i.e., by the base of the logarithms. But it is interesting to notice that the evenness of a community as measured by J' is numerically identical to its diversity if the latter be calculated using logs to base s. To see this, put

$$p_1^{p_1} p_2^{p_2} \ldots p_s^{p_s} = 1/A, \quad \text{say}$$

so that $H' = \log_x A$, where we have used the arbitrary number x as base for the logarithms.

Then

$$J' = \frac{\log_x A}{\log_x s}$$

whence

$$J' \log_x s = \log_x A, \quad \text{or} \quad s^{J'} = A.$$

Taking logarithms to base s of both sides of the last equation (and noting that $\log_s s = 1$) gives $J' = \log_s A$ which is the value of H' when logarithms to base s are used.

2 MEASURING DIVERSITY IN PRACTICE

In the preceding section it was argued that the most useful measure of ecological diversity is the Shannon-Wiener index H'. The discussion was theoretical and we now come to practical problems.

The values of p_i to be substituted in the formula for H' are the relative amounts of the several species in the community under study and the way in which the amounts are best measured depends on the type of organisms concerned. If, for instance, the community consists of all the warblers in a forest, or all the pollen grains in a section of peat core (which would be a legitimate "community" in this context), the most obvious measure of the relative amount of each species is the relative number of the community's individuals belonging to that species; in these examples the individuals are distinct, countable, and of comparable size. But when the individuals of the different species are very unequal in size (as is true of the fishes in a large lake, for example) or when "individuals" as such do not exist and the species are present as clumps or patches with indefinite boundaries (as is true of vegetatively reproducing herbs), it is best to set p_i equal to the relative weight (fresh or dry) of species i; or to the relative "cover" in the case of such organisms as cushion plants, mosses, lichens and crustose bryozoa. Deciding what the p_i values are to represent is the first decision to make when a diversity value is required.

One next needs to discover, by field observation, the number of species in the community and the numerical values of the p_i's. These data are often hard to acquire and there are two chief difficulties.

First, estimates of the proportional amounts of the rarer species will usually have large sampling errors and are likely to be unreliable.

Second, if only a small sample is taken from a large community it is likely that several of the rarer species in the taxocene defined as comprising the community will be missing from the sample. Thus one may even be ignorant of s, the number of species, let alone the s different values of p_i required to evaluate H'.

A method of coping with these difficulties will be discussed below, but first it must be pointed out that there is a way of evading them. Up to now it has been assumed (tacitly) that the community whose diversity was to be measured was too large to be censused in its entirety and that H' therefore had to be estimated from a sample. Such an estimate is of no value unless one can also estimate its sampling error, and hence specify a confidence interval which has known probability of bracketing the true, unknown, value of H' (cf. page 381). Because of the practical difficulties of obtaining satisfactory estimates, it is often simpler to redefine the community under investigation in such a way that it can be censused fully. Put another way, this amounts to saying that the collection of organisms actually examined is to be treated as *itself* the entity whose diversity is to be determined; it is not to be regarded merely as a sample from a larger population whose diversity is to be inferred from it, with, of course, inevitable sampling error. Whether this is a reasonable way of dealing with the difficulties of estimating H' must be decided in the light of the reasons for which diversity measurements were wanted in the first place. It is well to recall that a diversity index is of no ecological interest in itself; it is the value of only one of a community's many measurable properties. To arrive at ecological conclusions, measurements of at least two of these properties are usually required to discover whether, and if so how, they are interrelated. Thus a decision on how large a community must be for measurements on it to give useful results depends in part on what other properties (besides diversity) are to be observed. Some examples will clarify the argument.

Suppose one wishes to judge whether there is a relationship between the diversity of a second-growth forest (one that has grown up after fire) and its age (Pielou, 1966b). The most direct method is to carry out a complete census of the trees on various plots where the fire history is known. Any one plot thus yields two measurements, its diversity and its age, both determinable without sampling error. The measurements are regarded as pertaining to the plot itself, not to some larger area of forest which it is vaguely hoped the plot "represents". And conclusions on the relationship, if any, between age and diversity can only be drawn when data are available from a number of plots.

A second example: one might wish to determine whether there were a relationship between the area of open water in a prairie slough and the diversity of duck communities pausing there during the spring migration. It should be possible to discover the areas of several bodies of water, and the composition of the duck communities temporarily feeding on them, by direct observation. Thus for each slough the two properties required could be determined rather than estimated, and the interest of the results would reside in the evidence they gave as to the relationship between the properties measured.

In the two examples just described, there is no need to infer the diversity of a large community from observations on a sample since the sample measurements themselves are relevant to the problem concerned. As a contrast, we consider two other examples in which useful results can only be obtained from comparatively large communities. Thus if a comparison was to be made between the Atlantic and Pacific coasts of North America as to the diversities of the infauna of sandy beaches (in a given latitude belt), many large communities would have to be compared; censusing them would be impracticable and it would be necessary to estimate their diversities from samples. Similar remarks apply if the diversities to be compared are those of the breeding bird communities of decidous woodlands in, say, England and France. In general, estimation methods are needed when communities are so defined that they occupy areas of "geographic" extent.

For practical purposes we must therefore consider two distinct problems: (i) how to determine the diversity of a collection (i.e., a sample) that can be completely censused; and (ii) how to estimate the diversity of a large community.

The diversity of a fully-censused collection

Suppose a collection containing s species consists of N_1 individuals of the first species, N_2 of the second, ... and N_s of the sth. The total size of the collection is $N = \Sigma N_i$.

(If the organisms are such that the abundances of the species are better measured in terms of weights, say, than as numbers of individuals, then the N_i values denote numbers of units of weight; and correspondingly for any other method of measuring species-abundances.)

Ecologists often use $-\sum \dfrac{N_i}{N} \log \dfrac{N_i}{N}$ as a measure of such a collection's diversity but this formula has a serious drawback. It is true that it is the same as that for H' if we put $N_i/N = p_i$. However the Shannon-Wiener formula is *not* appropriate for finite collections, only for conceptually infinite ones.

In mathematical information theory, which is the context in which the Shannon-Wiener formula was originally devised, H' measures the information content of a language, or code, in which one can write many messages. There is no end to the number of messages that could be written in the code, one can continue indefinitely. Analogously, use of H' as a measure of the diversity of a community tacitly assumes that indefinitely many particular samples (equivalent to messages) can be taken from the community without depleting it.

Now suppose one has a collection that is to be treated as an entity in itself, not as a sample from something bigger. It is analogous to a particular message in some code rather than to the code itself. The appropriate formula for measuring such a community's diversity is that used in information theory to measure the information content of a particular message. It is

$$H = \frac{1}{N} \log \frac{N!}{N_1! \, N_2! \, ... \, N_s!} \quad \text{units,}$$

where

$$N! = N(N - 1)(N - 2) ... 2.1.$$

This formula was first used to measure ecological diversity by Margalef (1958). It has become known as the *Brillouin index* (see Brillouin, 1962). The units are the same for this index as for the Shannon-Wiener index and depend on the base of the logarithms (see page 290). It is a useful convention

(Pielou, 1966a) to use H' (note the prime) for the Shannon-Wiener index and H (without a prime) for Brillouin's. It is easy to prove (see, for instance, Pielou, 1969) that if *all* the N_i are very large, then

$$\frac{1}{N} \log \frac{N!}{N_1! \, N_2! \, \ldots \, N_s!} \rightarrow -\sum \frac{N_i}{N} \log \frac{N_i}{N}.$$

(This justifies the use of the Shannon-Wiener formula to calculate the diversities shown in Table 12.1 on page 294.) However, collections of which this is true are extremely uncommon. It is far more often found, when a community is censused, that at least a few of the species are poorly represented, perhaps by no more than one or two individuals.

This is another reason, in addition to that of strict mathematical appropriateness, for using H rather than $-\Sigma \, (N_i/N) \log (N_i/N)$ to measure the diversity of a fully censused collection. Its use emphasizes that the collection at hand is the thing whose diversity is determined; and that one is not trustfully assuming that N_i/N gives a precise estimate of p_i, the proportion of the ith species in a large community from which a random sample has been drawn with proper statistical precautions.

It should be noticed that H is always determined, never estimated. Therefore it has no standard error.

By analogy with equation (12.4) (page 301), the evenness of a fully censused collection is given by

$$J = H/H_{max}$$

where the Brillouin index is used in both numerator and denominator. H_{max} is the maximum possible diversity of a collection having both the same number of species, s, and the same number of individuals, N, as the observed collection. A finite collection has maximum evenness (and hence maximum diversity) when the individuals are divided up among the species as evenly as possible, but when N is not a multiple of s the species cannot all have the same number of individuals. Then, suppose

$$N = s \left[\frac{N}{s} \right] + r$$

where $[N/s]$ is the integer part of the quotient N/s and r the remainder. Then H_{max} is the diversity of a collection in which $(s - r)$ species contain $[N/s]$ individuals and the remaining r species contain $[N/s] + 1$ individuals.

The diversity of a fully censused collection can also be split into hierarchical components since, using symbols analogous to those in equation (12.2) (page 296)

$$H(SG) = H(G) + H_G(S).$$

The derivation and use of this formula are most easily demonstrated with an example. We shall use data given by Lloyd, Inger and King (1968) on the amphibians in a rain forest in Borneo; to keep the example small the members of only one family in their collection, the Pelobatidae, will be considered. The collection contained $N = 198$ individual frogs of this family, belonging to three genera and five species, as follows. (In parentheses after each name is shown the symbol that will be used for the number of individuals in the taxon concerned, and the number actually observed.)

Genus	Species
Leptobrachium $(l = 169)$	$\begin{cases} L.\ gracilis & (l_1 = 151) \\ L.\ hasselti & (l_2 = 18) \end{cases}$
Megophrys $(m = 8)$	$\begin{cases} M.\ baluensis & (m_1 = 3) \\ M.\ monticola & (m_2 = 5) \end{cases}$
Nesobia $(n = 21)$	$N.\ mjobergi\ (n_1 = 21).$

We now calculate the total species-diversity, $H(SG)$, of the collection. Many logarithms of factorials are needed and they can be obtained from statistical tables.*

Using natural logarithms,

$$H(SG) = \frac{1}{N} \ln \frac{N!}{l_1!\,l_2!\,m_1!\,m_2!\,n_1!}$$

$$= \frac{1}{N} \{\ln N! - \ln l_1! - \ln l_2! - \cdots - \ln n_1!\}$$

$$= 0.7790 \text{ nats.}$$

Alternatively, this may be written

$$H(SG) = \frac{1}{N} \ln \left\{ \frac{N!}{l!\,m!\,n!} \cdot \frac{l!}{l_1!\,l_2!} \cdot \frac{m!}{m_1!\,m_2!} \cdot \frac{n!}{n_1!} \right\}$$

$$= \frac{1}{N} \ln \frac{N!}{l!\,m!\,n!} + \frac{l}{N} \left(\frac{1}{l} \ln \frac{l!}{l_1!\,l_2!} \right)$$

$$+ \frac{m}{N} \left(\frac{1}{m} \ln \frac{m!}{m_1!\,m_2!} \right) + \frac{n}{N} \left(\frac{1}{n} \ln \frac{n!}{n_1!} \right)$$

$$= 0.4809 + \frac{169}{198} \times 0.3255 + \frac{8}{198} \times 0.5032 + \frac{21}{198} \times 0.$$

---

* For example, *Biometrika Tables for Statisticians, Volume I* (1954), Cambridge University Press.

It is seen that the first of these four components is the genus-diversity of the collection, $H(G)$; and that the other three components are the species-diversities withih each genus multiplied by the proportion of the total collection in that genus. Thus, in symbols again,

$$H(SG) = H(G) + \frac{l}{N} H_l(S) + \frac{m}{N} H_m(S) + \frac{n}{N} H_n(S);$$

(the subscript letters in $H_l(S)$, $H_m(S)$ and $H_n(S)$ are the initials of the genera). Finally, therefore,

$$H(SG) = H(G) + H_G(S).$$

In the example,

$$H(G) = 0.4809 \text{ nats} \quad \text{and} \quad H_G(S) = 0.2981 \text{ nats.}$$

Estimating diversity from a sample

If a community is too large to be censused its diversity must be estimated from a sample. In what follows we shall assume that the community is to be regarded as "indefinitely large" so that what is estimated is a value of the Shannon-Wiener index H'. And for conciseness, the sampling units will be called quadrats; it is assumed that they are located at random.

Estimating H' for a large community is not straightforward. In the first place, it is not satisfactory to treat N_i/N, the observed proportion of species i in the whole sample (i.e. all quadrats together), as a substitute for p_i in the formula for H'. To do so yields a biased estimate of H' (see Pielou, 1966c), and furthermore, one has no way of finding the standard error of the estimate.

In the second place, it obviously will not do to calculate a diversity value for each quadrat taken separately and average the results over all quadrats. This would nearly always yield much too small a value since (except in very depauperate communities) the contents of a quadrat usually fall far short of representing the community as a whole. This is especially true of a patchy community whose pattern forms a coarse mosaic; a single quadrat can then contain only a small portion of the pattern and a small fraction of the total number of species.

A way around these difficulties will now be outlined; the derivations are given in Pielou (1966c). Suppose the observations on the quadrats are arranged, quadrat by quadrat, in random order. Let $\mathscr{H}_1$ denote the diversity, as given by the Brillouin index, of the first. Now pool the data from the second quadrat with those of the first and calculate the diversity of the combined pair of quadrats; denote it by $\mathscr{H}_2$. Similarly, add the data from the third quadrat to those of the pooled pair and calculate $\mathscr{H}_3$. Continue in this

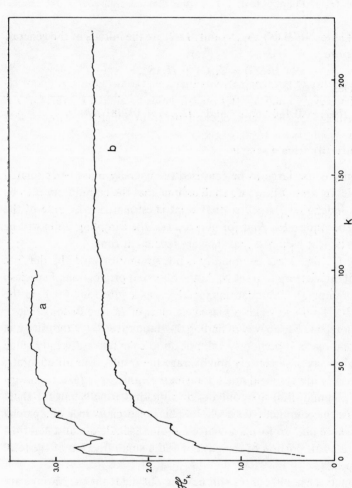

FIGURE 12.2 Graphs of $\mathscr{H}_k$ versus k. $\mathscr{H}_k$ is the diversity, measured by the Brillouin index, of the pooled contents of k quadrats. (*a*) Data are weights of ground vegetation in meter-square quadrats; Place: mixed woodland in Québec Province. (Redrawn from Pielou, 1966c). (*b*) Data are numbers of amphibians and reptiles caught in 25-foot square quadrats. Place: lowland rain forest in Borneo. (Redrawn from Lloyd, Inger and King, 1968)

manner adding quadrats to the pool and calculating the Brillouin index of the newly expanded sample each time.

If we now plot the successive $\mathscr{H}$ values against the number of quadrats accumulated the result is likely to be a curve that rises steeply at first and then levels off. Figure 12.2 shows two examples. Suppose the curve becomes horizontal (except for a slight "roughness") when the accumulated sample contains t quadrats. Then, for all $k \geqq t$, one may calculate

$$\frac{M_{k+1}\mathscr{H}_{k+1} - M_k\mathscr{H}_k}{M_{k+1} - M_k} = h, \quad \text{say}, \tag{12.5}$$

where M_k is the total number of individuals in the accumulated pool of k quadrats. Thus every quadrat after the tth yields a value of h. An estimate of H', the diversity of the whole community from which the sample was drawn, is now given by the mean of the observed values of h, that is,

$$\hat{H}' = \frac{1}{n}\Sigma h$$

where n is the number of h's calculated; (equivalently, n is the number of quadrats added to the pool after the within-pool diversity $\mathscr{H}$ had reached a constant level). The sampling variance of this estimate is

$$\text{var}(\hat{H}') = \frac{1}{n(n-1)}\left\{\Sigma h^2 - \frac{(\Sigma h)^2}{n}\right\}$$

and hence a 95% confidence interval for the diversity of the community is given by

$$\hat{H}' \pm 1.96\sqrt{\text{var}(\hat{H}')}.$$

Two other points should be noticed. The first concerns the problem of determining t, i.e., of judging at what point the curve of $\mathscr{H}$ against k levels off; recall that equation (12.5) can be used only when $k \geqq t$. The best way is to judge the value of t subjectively from the appearance of the curve; and then to test whether the calculated values of h are serially correlated (see Appendix 5.1, page 381). Provided there is no significant serial correlation in the sequence of h's, equation (12.5) may be used.

The second point is this: it is clear that the estimate of H' must be affected by the order in which the quadrats are accumulated, for obviously the whole estimation procedure can be repeated many times with different random orderings. If this is done, one may take, as the final estimate of H', the median of the several calculated estimates (Lloyd, Inger and King, 1968).

The obvious way to estimate J', the evenness of a large community, is to use equation (12.4), that is, to put

$$J' = \frac{\hat{H}'}{\log s}.$$

The sampling variance of J' is

$$\operatorname{var}(J') = \frac{\operatorname{var}(\hat{H}')}{(\log s)^2}.$$

But to obtain these estimates requires that s, the total number of species in the whole community, be known. An estimate of evenness should *not* be based on the number of species in the sample, which (especially in a very uneven community) may fall considerably short of the total. Indeed, unless s is known, it is impossible to estimate a community's evenness. This is hardly surprising; a community that contains a number of rare species whose presence, though perhaps suspected, has not been proved, clearly has far lower evenness than a sample would lead one to suppose.

3 BETA DIVERSITY

The discussion on diversity measurements in preceding sections presupposes throughout that one is treating the area delimited for study as a single "community". There is no objection to doing this even though one cannot be sure (without doing elaborate tests) that a subjectively defined community is internally homogeneous. Thus the methods for determining or estimating diversity described in section 2 give the diversity of a chosen taxocene within a chosen area regardless of whether the community so defined is "natural" in any sense (cf. page 288).

Whittaker (1960, 1972) has proposed that a distinction be made between two contrasted kinds of diversity: that in a true "natural community", which he calls *alpha diversity;* and that in a large heterogeneous region, which he calls *gamma diversity*. Both are measured in the same way and in the same units. The distinction is useful and it will certainly make for greater clarity in the literature when ecologists make a practice of specifying whether the diversities they are measuring are alpha or gamma. An alpha diversity pertains to a small area and is a property of a particular community; even though recognition of such an entity is nearly always subjective the risk of being seriously mistaken is negligible. A gamma diversity pertains to a large and almost certainly heterogeneous collection of organisms. Thus when one measures the diversity of a taxonomic group in a whole island, or in a geographic region, the result is a gamma diversity. Likewise the diversity of the insects collected by a light trap that has operated all night is a gamma diversity; it will contain samples from a number of distinct communities, which are active in different periods, and which each has its own alpha diversity.

To summarize: alpha and gamma diversities measure the same thing and differ only in that the first applies to homogeneous and the second to heterogeneous collections.

Whittaker uses the name *beta diversity* for a wholly different concept, that of spatial variability in community composition. Thus, suppose a given community, defined as all the members of a designated taxocene occurring in a designated area, is spatially homogeneous; in other words, the contents of any sampling quadrat are independent of the quadrat's location in the area. Then the beta diversity in the area is zero. Conversely, if community composition varies, so that the contents of sampling quadrats vary with their location, then the community is spatially heterogeneous and has positive beta diversity. Spatial heterogeneity in an area may result from the presence of two or more internally homogeneous communities, or from gradual (clinal) variation in species-composition. In either case it gives rise to what Whittaker (1972) variously calls beta diversity, within-habitat diversity, biotic change, or species-replacement. Whittaker's method of measuring beta diversity presupposes that change in a community's composition occurs along an environmental gradient; the data needed for calculating it are obtained by observing the contents of sampling quadrats ranged in a row along the gradient. One then determines the relationship between the "similarity" of two quadrats and their distance apart (details are given below); and hence the distance that must separate a pair of quadrats to ensure that, on average, their similarity shall be half that of a pair so close that the distance between them is effectively zero. The more rapidly the similarity between quadrats falls off with distance, the greater the beta diversity.

To carry out the procedure outlined above, one must first have a method of measuring the similarity between two quadrats. Many similarity measures have been devised and two that are useful in the present context are the *coefficient of community*, *CC*, and the *percentage similarity*, *PS*. They will be described first and we can then consider which is more useful in the measurement of beta diversity.

Let the quadrats whose similarity is to be measured be labelled X and Y.

To find their coefficient of community one needs to know s_X and s_Y, the numbers of species in quadrats X and Y; and s_{XY} the number of species common to both quadrats.

Then

$$CC = \frac{200s_{XY}}{s_X + s_Y}.$$

Clearly, if the quadrats have identical species lists, $s_X = s_Y = s_{XY}$ and $CC = 100$. Conversely, if the quadrats have no species in common $CC = 0$. (The coefficient of community is here expressed as a percentage; it is sometimes given as a fraction).

To find the percentage similarity of two quadrats one needs to know the quantity of each species in each quadrat; depending on the kinds of organisms, this is measured by counting individuals or by weighing biomass.

Now suppose there are s species altogether in the two quadrats.

Let Z be the total quantity of all species in both quadrats.

Let Z_j be the quantity of species j in both quadrats and let X_j and Y_j be the quantities of species j in quadrats X and Y respectively.

Thus

$$\sum_{j=1}^{s} (X_j + Y_j) = \sum Z_j = Z.$$

Then the percentage similarly is

$$PS = 200 \sum_j \min \left(\frac{X_j}{Z}, \frac{Y_j}{Z} \right).$$

It is clear that only if all species are evenly divided between the two quadrats (so that $X_j = Y_j = \frac{1}{2}Z_j$ for all j) shall we have

$$PS = 200 \sum \frac{Z_j}{2Z} = 100.$$

At the other extreme, if each species is present in only one quadrat, then $\min [(X_j/Z), (Y_j/Z)] = 0$ for all j and $PS = 0$.

In all other cases $0 < PS < 100$.

In choosing between CC and PS as similarity indices, one needs to consider whether the difference between the quadrats is slight or pronounced. If slight, that is if the species lists for the two quadrats are much the same, then PS is the preferred similarity measure. But if the difference is more marked, that is if each quadrat of the pair contains many species that are not present in the other, CC is the better measure. In fact CC measures the similarity between species lists and PS measures the similarity between species quantities. In many contexts both indices seem equally suitable and the choice becomes subjective. In the discussion below we assume, for concreteness, that CC is used.

Returning now to the measurement of beta diversity, it will be recalled that we wish to find the relationship between the similarity of two quadrats and their distance apart, assuming the quadrats to have been ranged at equal intervals in a row along an environmental gradient. Let this interval be taken as the unit of distance. To examine the observed relationship, let the quadrats be labelled A, B, C, D, E, F, etc. Then the observed mean similarity of quadrats separated by one distance unit, CC_1 say, is the mean of the similarities of the pairs (AB), (BC), (CD), etc. Likewise, the mean similarity between quadrats separated by two units, CC_2, is the mean of the similarities of pairs (AC), (BD), (CE), etc. The mean similarities CC_3, CC_4, ... are defined analogously.

In nearly all the clinally varying communities he investigated, Whittaker (1972) found that if log CC_r were plotted against r, the first few points were colinear (except for negligible departures presumably due to sampling variation). Figure 12.3 shows an example.

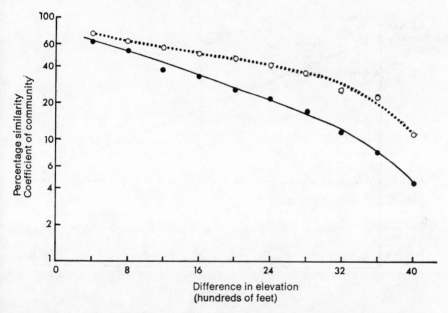

FIGURE 12.3 The relationship between the similarities of pairs of sampling units and their distance apart. The data are from a community of forest trees in the Great Smoky Mountains. Similarities are measured by the Coefficient of Community, ○···○; and by the Percent Similarity, ●——●. Note the logarithmic scale on the ordinate. The "distance" between two sampling units is measured by the difference in their elevations. (Redrawn from Whittaker, 1972)

This empirical finding leads to a convenient method of measuring beta diversity in the strip of ground (a transect) sampled by the row of quadrats. Let a straight line be fitted by eye to the colinear points. Denote by log CC_0 the value of the ordinate where the fitted line cuts it. Thus CC_0 is taken as an estimate of the expected similarity of two quadrats lying side by side with respect to the gradient (equivalently, at zero distance apart along the gradient). Observe that $CC_0 \neq 100$ even though we are envisaging quadrats at sites with identical conditions; owing to sampling variation two quadrats in a homogeneous community will hardly ever have identical contents and therefore we expect to find $CC_0 < 100$.

Next suppose that the total length of the row of quadrats is n units; CC_n is thus the similarity of the first and last quadrats of the row and it is usually found (cf. Figure 12.3) that the point representing this value lies below the line fitted to the first few points. We therefore write $\log CC_n'$ for the ordinate of the point *on the fitted line* corresponding to a distance of n units.

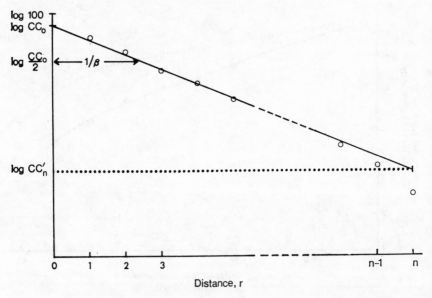

FIGURE 12.4 Graphical determination of beta diversity. The sloping line is fitted, by eye, to the first few, apparently co-linear, data points. Then $1/\beta$, the reciprocal of the beta diversity, is the distance between pairs of quadrats whose average coefficient of community is $(CC_0)/2$

Now consider Figure 12.4. To measure the beta diversity of the transect under investigation, we need to know the abscissa of the point whose ordinate is $\log(CC_0/2)$; in other words, we require the expected distance apart of quadrats whose similarity is one-half the expected similarity, CC_0, of quadrats at zero distance apart. Let this distance be denoted by $1/\beta$; it is the distance required for a *half-change* in community composition (Whittaker, 1960).

From Figure 12.4 it is seen that

$$\frac{1/\beta}{n} = \frac{\log CC_0 - \log (CC_0/2)}{\log CC_0 - \log CC'_n}$$

whence

$$\frac{1}{\beta n} = \frac{\log 2}{\log CC_0 - \log CC'_n}$$

or

$$\beta = \frac{\log CC_0 - \log CC'_n}{n \log 2}.$$

Thus the index β is the reciprocal of the length (measured as a proportion of the total length of the transect) required for a half-change. If, as we have supposed above, similarity is measured by the coefficient of community, the half-change is in the complement of species in the quadrats; if percentage similarity is used, the half-change relates to changes in the quantities of the species.

We have also supposed, up to this point, that "distance" along a transect meant distance in the ordinary sense. Depending on the nature of the environmental gradient under study, it is often more appropriate to measure the distance between quadrats in terms of change in the environmental factor, whatever it is, that varies along the gradient; as examples the differences among the spaced quadrats might be in respect of elevation, aspect, depth of water table or duration of snow cover. The sampling quadrats should be equidistant with respect to the environmental factor concerned. Thus in the example shown in Figure 12.3, which relates to forest composition in the Great Smoky Mountains, this factor was elevation; the difference in elevation between adjacent quadrats was always 400 feet and the distance between them as measured along the ground was therefore variable.

CHAPTER 13

The Ecological Niche

TWO OF THE most remarkable properties of an ecological community are these: the fact that in most communities a number (sometimes a very large number) of species coexist without excluding one another although groups of them make fairly similar demands on the environment and presumably compete to some extent at least; and the fact that some of the species in a community are represented by an abundance of individuals whereas other species, no less successful to judge from the way their members persist and thrive, are rare. The desire to explain these facts has led to the theory of the niche, a central topic in modern ecology.

1 DEFINITION OF THE NICHE

As a preliminary to discussing the topic, the first requirement is a definition of the word *niche*. Different authors have used it in different senses. To Grinell (1917) the niche of a species was made up of all the sites where it could live and these were determined by habitat conditions. To Elton (1927), the niche is the function performed by the species in the community of which it is a member. One of the several examples he gives is the following: in arctic Canada in winter the remains of seals killed by polar bears are eaten by arctic foxes; in tropical Africa the remains of antelopes and zebras killed by lions are eaten by hyenas. The arctic fox and the hyena can therefore be said *ƒo* fill the same niche in their different communities, a niche defined by the *tunction* (that of eating carrion left by the largest carnivore in the community) that both species perform.

Miller (1967) has summarized these two contrasting concepts by remarking that to Grinnell an animal's niche is its "address", and to Elton its "profession". Alternatively, using Elton's concept, the niche of a species may be termed, metaphorically, its "role" and the words are often used interchangeably. This is an objectionable practice, however, as it seems to imply that natural communities and their members exist to fulfil some undefined purpose. So teleological a notion has no place in ecology.

The word *niche* is also used occasionally to connote the "address" of a particular individual animal. For example R.B.Miller (1958) described the territory of a trout, which it occupies and defends fiercely throughout its life, as its niche. In this context, the word "territory" seems more useful because it is less likely to be misunderstood. It should be noticed that the meaning attached to "niche" by Grinell and by R.B.Miller are not the same. Grinnell's niche is a collective name for the set of *all* territories that are, or could be, occupied by a species.

A formal definition of the niche was proposed by Hutchinson (1957) in a famous paper enigmatically titled "Concluding Remarks"; in it he set forth what is now widely regarded as the clearest conceptual model of the niche and the one most likely to be useful in ecological research. Consider the totality of environmental variables that affect the welfare of a species. These include all the abiotic factors (temperature, water supply, etc.) and also all the biotic factors (food, predators, diseases, etc.) that impinge on the species in any way. Suppose there are n such factors; (n is always very large). It is clear that for a given species to succeed in a given region the numerical values of all the factors must lie between certain limits peculiar to the species. If any factor, temperature say, goes permanently outside the tolerance limits for the species, then the species will inevitably disappear from the region. Therefore we may define the niche of the species as the totality of sets of conditions that are compatible with its persistence and success. Conceptually, these can be thought of as forming a "solid figure" of n dimensions. For suppose each variable factor were measured along one of the axes of an n-dimensional coordinate frame and that it were known, for a particular species, the precise limits of its tolerance in respect of each factor and every combination of factors. We could then (only conceptually of course) portray the niche of the species as the *n-dimensional hypervolume* bounded by these limits. Hutchinson describes this hypervolume as the *fundamental niche* of the species.

It is not supposed that a species is equally successful at every point of its fundamental niche, that is, in every set of conditions in which it can occur. In conditions close to the limits of its tolerance it may be sparse, and near the center of its niche, where all conditions are close to optimal and none borderline, it may be abundant. By definition, the fundamental niche includes *all* sets of conditions where the species can manage to persist even in small numbers.

In saying that the fundamental niche of a species is the totality of sets of environmental conditions compatible with its persistence, we are not asserting that it does in fact persist wherever an acceptable set of conditions occurs. If a competing species requires the same set of conditions and is a strong enough competitor to oust the first species, then the niche actually occupied by the first species will be reduced. Thus a species' fundamental niche is made up of two disjunct subsets: one is the subset of all conditions in which it could succeed but for the fact that it is excluded by competitors; the other is the subset of all conditions in which it does succeed in fact. The latter is termed the *realized niche* of the species.

Many illustrative examples of the difference between a realized and a fundamental niche could be given and some have already been described in Chapter 10 where competitive exclusion was discussed (see page 234 et seq.). Thus Miller's (1967) study on the exclusion of redwinged blackbirds by yellowheaded blackbirds (page 235) should be recalled. Where yellowheaded blackbirds occur, redwinged blackbirds are excluded from part of their fundamental niche (conditions characterized by deep-water marshes with emergent cattails, bulrushes and reeds) and confined to the remaining part (conditions characterized by shallower marsh and weedy or shrubby marsh borders) which forms the realized niche.

It is important to notice the distinct meanings of the words "niche" and "range" (or "occupied habitat"). The word niche (in Hutchinson's usage) connotes a mental concept; it is a hyperspace, or hypervolume, in an n-dimensional coordinate frame; equivalently, it is the set of all points within the hypervolume and every point represents one acceptable combination of values of the n factors that determine a species' success. By contrast, the range (or occupied habitat) of a species is the actual space where the species is found; it is concrete as opposed to abstract. Confusion is hard to avoid since often the words used to describe a point in the hypervolume can equally well be used to describe a place in the real world. In fact at those sites in the real world where the species actually occurs three things are true; habitat conditions correspond in all respects with those at some point inside the species' fundamental niche; competitiors have not preempted the site; and the species has migrated into the site if it evolved somewhere else.

We can now make a formal statement of the theory of the niche as follows: *It is impossible for two (or more) species to coexist indefinitely in the same region if their niches are identical.*

It is seen that the theory is simply a restatement of the theory of competitive exclusion or, equivalently, of the coexistence principle (see page 234) and it suffers from the same difficulties of interpretation. It is true that the theory is often invoked to expain why there are not more than the observed number of species in a given community. The conventional explanation would be "all the niches are filled", but this statement merely begs another question: how dissimilar must two species be (or, which comes to the same thing, how dissimilar must two niches be) for both to occur in the same area?

Equally, the theory is sometimes pressed into service to explain why there are so many species in a given community; but the answer that there are as many species as there are niches is, of course, no answer at all. It merely raises the following questions. Is a niche created by a species; that is, does a species become adapted until it can avail itself of some hitherto unused resource in an area? Or do niches exist independently of species so that the concept of a "vacant niche" waiting to be filled is meaningful? It is easy to envisage niches that "ought to" be filled—for instance, why are there no freshwater flatfish?—but such speculation is unlikely to produce any worthwhile results. The recognition of vacant niches does not usually come until after they have been filled, by the establishment in what had hitherto been thought to be a "full" community of a successful immigrant.

The theory of the niche is the starting point for more than idle speculation, however. It is the chief unifying theory that underlies research in nearly all of community ecology and in the following pages we shall consider some of the major topics that can profitably be treated as offshoots from it. Before that, in a mainly anecdotal section, some examples of *niche diversification*, as it is called, will be described. They illustrate some of the ways in which a number of apparently similar species can differ among themselves sufficiently to ensure that the resources of the environment are partitioned among them and none is excluded by the others.

2 EXAMPLES OF NICHE DIVERSIFICATION

The ecological literature contains numerous accounts of niche diversification and a short list of illustrative examples cannot begin to do justice to the topic. The seven examples given below, in short, numbered paragraphs, have been chosen to demonstrate the wide array of taxonomic groups which have been studied in this context; and to show also the contrasting ways in which a group of similar or related species can differ among themselves sufficiently to ensure that each obtains an adequate share of the resources it needs.

1   Heatwole and Davis (1965) have described a remarkable case of niche diversification in three ichneumonid wasp species of the genus *Megarhyssa*, namely *M. atrata*, *M. macrurus* and *M. greenei*. All three species occur together in beech-maple forests in Michigan and all are parasitoids of a single host insect, the wood wasp *Tremex columba*, in whose larvae (grubs) they

lay their eggs; an ichneumonid larva then develops within each successfully parasitized wood wasp grub, consuming its host as it grows. The host grubs are wood borers and are found in dead logs and stumps. The three parasite species differ among themselves in the lengths of their ovipositors, and a female will lay eggs only in a grub which her ovipositor just reaches when it is inserted at right angles to the wood surface. Since the grubs are at various depths in the wood, any individual grub is a target for only one of the ichneumonid species and they therefore parasitize three different segments of the host population.

2   An example (one of many) of the ways in which animal species partition the space they inhabit has been given by Schoener (1968). He observed four species of the lizard genus *Anolis* in South Bimini, a small island in the Bahamas, and found that the members of each species spent most of their time in somewhat different microhabitats: *A. sagrei* on the ground or on low perches; *A. distichus* on the trunks and larger branches of trees; *A. angusticeps* on small, high twigs; and *A. carolinensis* on or near leaves.

3   Any group of sympatric species of similar life styles may exhibit niche diversification; the phenomenon is not limited to congeneric species. For example, O'Neill (1967) has described it in seven species of diplopods (millipedes) which, though they are all in different genera, seem to compete directly; they all feed on leaf litter and decaying wood in maple-oak forests in Illinois. It was found, however, that each species had its specialized microhabitat in, on, or near fallen logs. These were: the heartwood of logs; the superficial wood of logs; just under the bark; beneath but on the surfaces of logs; beneath logs on the ground; among the leaves of the litter; and on the ground under the litter.

It is interesting to consider the concept of vacant niches in the context of this example. It is hardly credible that any ecologist, on examining a log and its surroundings from which all millipedes had been removed, could recognize and describe the seven distinct niches (using the word non-rigorously) that we now know, by hindsight, to be occupiable by millipedes. Possibly, of course, there are still some vacant niches, as yet undiscerned, but nevertheless available for future millipedes. It seems more reasonable to suppose that the observed diversification has evolved, in the sense that natural selection has favored those millipedes in each species whose habits are such that they experience the least interference from other species.

4   Sea birds provide another good example of habitat diversification. At big sea bird colonies many species nest simultaneously, often in enormous numbers, and the different species usually have their preferred nesting sites. Thus in colonies on the coast of Newfoundland (Tuck, 1960; Godfrey, 1966)

common puffins and Leach's petrels nest in burrows in the sloping soil at the tops of cliffs; kittiwakes, common murres and thick-billed murres on narrow ledges on vertical cliffs; gannets and murres on the level tops of inaccessible stacks; black guillemots on talus slopes; and razorbills in rock crevices and under overhangs. There is not a perfect one-to-one correspondence between bird species and nesting sites; for example, a few common murres nest among the gannets on the gannet stack at Cape St. Mary's, Newfoundland (Tuck, 1960). Even so, the degree of niche diversification is notable considering that competition for nesting sites is presumably slight; though the density of the nests within a colony is extremely high, there is no shortage of suitable habitat nearby and plenty of space for the colonies to expand.

5   When several rather similar species of animals live in a confined space, it may be diversification of their diets that makes coexistence possible. Many cases are known. For instance, Keast (1966) found that different species of small freshwater fishes had strikingly different diets even though their sizes were similar. Figure 13.1 shows part of his results; the pie diagrams show the compositions of the diets (in the month of May) of the four fish species found in a shallow, weedy pond formed by the widening of a small stream flowing through a meadow in southern Ontario. *Pimephales* (5.5 to 7.5 cm long) and *Chrosomus* (6 to 8 cm in both species) are minnows (Cyprinidae); *Eucalia* (3.5 to 5 cm) is a stickle-back (Gasterosteidae).

6   Populations of closely related species that succeed in the same area are often found to differ from one another in the time at which they make their heaviest demands on the available resources. Each species could be said to have its own temporal niche. Examples have been given by Coleman and Hynes (1970) who studied the life histories of stoneflies (Plecoptera) and mayflies (Ephemeroptera) whose aquatic immature stages occured together in a shallow, stony, well-oxygenated stream flowing through agricultural land in southern Ontario. Their observations on four congeneric pairs of species are particularly striking and are shown diagramatically in Figure 13.2. Every species has its own period of most active growth and, as is clear from the diagrams, the growth periods within each congeneric pair did not overlap.

7   Niche diversification is far more often found among animals than among plants. This is not surprising since most vascular plants, being autotrophic and sessile, have no means of partitioning resources of food or of living space. They can, however, have "temporal niches" as have the insects described in the preceding example. This has been demonstrated by Harper (1968) who describes experiments in which fiber flax and oil seed flax (two varieties of the species *Linum usitatissimum*) were grown together. It was found that the different varieties grow at different rates and can thus make much better

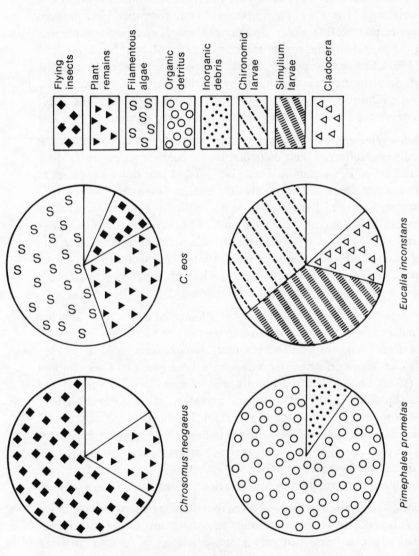

Flying insects
Plant remains
Filamentous algae
Organic detritus
Inorganic debris
Chironomid larvae
Simulium larvae
Cladocera

C. eos

Eucalia inconstans

Chrosomus neogaeus

Pimephales promelas

FIGURE 13.1   Diverse diets of the four species of fish occurring in a shallow pond in Ontario. The diagrams show the relative proportions of the different components of their diets during May. Unshaded sectors represent unclassified components each forming less than 10% of the diet. (Adapted from Keast, 1966)

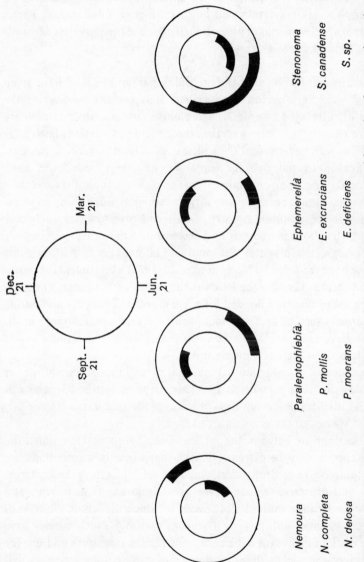

FIGURE 13.2 The periods of most active growth of the aquatic immature stages of four pairs of congeneric species of stoneflies (of the genus *Nemoura*) and mayflies (the other three genera). For every pair, the periods of most active growth are shown by black arcs, with the outer arc pertaining to the upper of the named species pair. A third species of *Stenonema*, *S. vicarium*, grew steadily for most of the year and is not shown in the figure. (Data from Coleman and Hynes, 1970)

use of the incident light than they could if their growth cycles were synchron-ized. One result of this phenomenon is that a mixed sowing of the two varie-ties yields more dry weight per unit area of soil surface than does either variety growing alone. It is clear that the annual rate of energy fixation by natural many-species vegetation will be much higher if the growth cycles of the different species are out of phase than if all make their greatest demands for incident light at the same time.

To dispel any impression that the preceding paragraphs may have given that species' niches are narrowly prescribed, it is worth remembering the great versatility displayed by animals and plants in nature. Most animals are compelled to change their diets with the seasons and observations lasting for a month or two can never reveal the wide range of foods a species uses dur-ing a year. Seasonal variation in the availability of certain foods forces many animals to make local seasonal migrations; as a result, their diets and their habitats change simultaneously. The large herbivores offer some good ex-amples. Thus, in the mountain country of western North America, mule deer graze at high elevations in summer and descend into the valleys, where they browse on evergreens, in winter. Seasonal habitat changes in African game animals have been recorded by Vesey-Fizgerald (1960) who studied the move-ments of the various herds in the Rukwa Rift Valley of Tanzania. The grass plains of the valley floor are flooded for some months of the year and during this time animals such as buffalo, eland, topi and zebra, which graze in the plains in the dry season, move into the surrounding woodlands which provide a different habitat as well as a different diet.

Besides changes due to the annual seasonal cycle, in most animal species an individual's diet changes with its age. This is obviously true of mammals, whose diet is milk for the first few days or weeks of life, and it is also obviously true of most species of insects, fishes and birds.

Indeed, the contrast between the niches of the adults and the immature stages of a species may be extreme. Thus winged insects whose immature stages are aquatic (e.g. dragonflies, mosquitoes, mayflies) and marine gastropods and crustacea whose larvae are planktonic (e.g. cowries and barnacles) offer striking examples of species in which the whole life style of an individual changes abruptly as it reaches maturity. Such species have two wholly distinct niches (or subniches), one for the immature and one for the adult stage; the niches differ in respect of diet, habitat, enemies and diseases. This raises the question of whether a niche is something that per-tains to a species or to a "life style". It could be argued that a species of dragonfly, say, has a single niche which is a property of the species, but that this niche, thought of as a many-dimensional hypervolume, consists of two disjunct parts. Alternatively, the disjunct parts can each be regarded as separate niches. This is merely a matter of terminology and every ecologist is free to use the word as he likes. The point will not be discussed further here. To avoid irrelevant complications in subsequent discussion it will be assumed, tacitly, that any one species has only one life style.

Of far more profound significance to the theory of the niche is the problem of delimiting a species. The matter is so crucial to all our subsequent discussions that fairly lengthy consideration must be given to it. This is done immediately below in Section 3.

3 POLYMORPHIC SPECIES

If a collection of plants and animals, gathered in the course of an ecological investigation, were handed to several taxonomists and they were asked to list, independently, the species in the collection, it is most unlikely that their lists would all contain the same number of names. "Splitters" and "lumpers" would disagree; the former would distinguish as species many groups of specimens that the latter would class together as distinct forms, or *morphs*, of a single *polymorphic species*. The term *polymorphic species* is used here to denote any species with several recognizably different forms (or *morphs*) regardless of whether they occur in the same or different geographic regions, or the same or different environments. However, all taxonomists would agree that the species they recognize are of unequal rank. Some are internally homogeneous and clearly differentiated from related species; others are ill-defined, heterogeneous assemblages labelled "species" more as a matter of convenience than anything else. Thus the word "species" sometimes applies to what seems to be a truly natural and indivisible group of organisms; and at other times to an assortment of fairly similar organisms about which little is known concerning their interrelationships.

Disagreements among taxonomists are therefore the "fault", if it is one, of the material they work with, and ecologists who turn to taxonomists for help in identifying their material should never forget this. The desire of ecologists to make their subject precise, quantitative, and logically rigorous should not lead them to dismiss these uncertainties as trivial simply because they are inconvenient. Fuzzy thinking is rightly abhorrent, but "fuzzy" material must often be endured. Thus arguments based on niche theory are often used to explain why the number of species of a particular taxocene in a particular area is what it is instead of being larger or smaller (an easy exercise), or to predict the number of species that could occur in an area (a difficult, if not impossible task). But such theorizing is a waste of time unless due allowance is made for the fact that species are not all of equal rank; and also, for this is not (necessarily) the same thing, that species are not all equal in the "widths" (more strictly, the hypervolumes) of their niches.

The difference just remarked should be emphasized. Conceptually, the niche of a species could be big because all members of the species have the same widely spaced tolerance limits in respect of the factors that determine its niche. Alternatively, a niche could be big because it consists of the union of a number of slightly different, partially overlapping subniches, each of which is the niche of a distinct subpopulation of the species. The evidence to date strongly supports the latter explanation in most cases that have been investigated. If this be true in general, certain theoretical consequences follow and we shall explore them later. First a few examples will be given of polymorphism in nature. The literature on the subject is enormous and the choice of examples to be discussed here is unavoidably arbitrary; they have been picked to illustrate the great range of animals and plants that have been studied in this connexion.

We begin with botanical examples. Many plant species consist of a number of variants, that is, groups of individuals that are in some way distinctive. The distinguishing characters may be morphological, physiological, or both and they may or may not be genetically determined. A rather complicated terminology has grown up for labelling the different kinds of *variants* (itself a non-committal term) depending on whether they occur in the same or different habitats, and in the same or different geographical regions. Moreover a new terminology is supplanting the old, which adds to the complications. The whole matter has been clearly discussed by Briggs and Walters (1969) and we shall adopt their system insofar as special terms are needed here. The fact that in many species of plants, genetic variants have evolved that are adapted to different environments was first discovered by Turesson (a short account of whose work is given by Briggs and Walters) who described the variants as *ecotypes*. In the newer terminology* they are called *genoecodemes*.

---

* This is the *-deme* terminology due to Gilmour and Gregor (1939) (and see Davis and Heywood, 1963; and Briggs and Walters, 1969). Every term ends with the suffix -deme which merely connotes a group of individuals within a species *sensu lato*). One or more explanatory prefixes are then attached to this "neutral" suffix. For instance, the members of a:

topodeme   occur in a single, specified region;
ecodeme   occur in a specified kind of habitat (not necessarily confined to one continuous region);
genodeme   are a genetically related group known to differ genetically from other such groups;

For example, McNeilly (1968) has described two genoecodemes of the pasture grass *Agrostis tenuis*. They were found near Drws y Coed in Wales, in a valley containing a small, derelict copper mine. The ground was a mosaic of normal soil and soil contaminated with ore. Distinct genoecodemes of *A. tenuis* grew in each habitat, one being tolerant and the other intolerant of the heavy metal contamination. On contaminated soil, only tolerant seedlings can establish themselves and they therefore experience no competition. But both the tolerant and the intolerant genoecodemes can grow on normal soil; there they do compete, though the tolerant form, being a somewhat weaker competitor, is gradually eliminated. The valley could therefore be said to contain two niches for *A. tenuis*, and two genoecodemes to occupy the niches. All the same, *A. tenuis* would almost certainly be treated as "one species" by anyone unaware of these detailed studies who was compiling a list of plant species found in the region.

A somewhat similar example is due to Jones (1966). In studies on *Lotus corniculatus* (bird's foot trefoil) in different parts of England, he found that the species is dimorphic in that some plants contain, and others do not, a cyanogenic glucoside. The two morphs are different genodemes. The cyanogenic plants are selectively favored in virtue of their toxicity to such herbivores as slugs, snails and voles; and the acyanogenic genodeme is believed to persist because it is faster-growing and more fertile, and hence has a selective advantage where the herbivores are sparse. It would be incorrect to say that *L. corniculatus* has two distinct niches since places with different herbivore densities differ only in degree, not in kind. Thus the system consists of one "species" with two genodemes, and a niche whose dimension representing the density of enemies varies continuously.

In the examples above the contrasted groups of plants occurred together in small areas. More often the different groups within a polymorphic species occur in different geographic regions. Variation may be clinal (continuous) or stepwise, and the different local populations can be called,

---

gamodeme   can interbreed and are close enough to do so;
topoecodeme   occur in a specified habitat in a specified region;
genoecodeme   form a genetically distinct ecodeme;
hologamodeme   are capable of interbreeding though not necessarily close enough in space to do so;
clinodeme   belong to one of a series of related groups exhibiting clinal (continuous) variation.

Observe that each prefix has one meaning only. Thus one need never use a term freighted with hidden, unintended implications; it is always possible to devise a term conveying exactly what is meant, neither more nor less.

respectively, clinodemes or topodemes. If the differences are known to be genetically determined the prefix geno- can be attached to these terms. When genetic differences account for observed geographic differences, the fact may be demonstrated by transplant experiments in which specimens of a species collected from many different sources are all grown together under uniform conditions in an experimental garden. If the differences are genetic the distinguishing characteristics of the local groups of plants will, of course, be retained in cultivation. Experiments of this kind are often of economic importance; for instance, foresters investigate the properties of trees from different geographic sources by growing seedlings from them in a single nursery. Members of different genotopodemes or genoclinodemes usually vary in a number or morphological and physiological traits. For example Mergen (1963) has described the variability in three-year old seedlings of *Pinus strobus* (Eastern white pine) grown together in Connecticut from seeds collected in eight different regions; the source regions ranged in latitude from 35°N (in North Carolina) to 57°N (in New Brunswick, Canada). It was found that "southern" seedlings had longer needles with more stomata and fewer resin ducts than "northern" ones. Also, that pines grown from seed collected in areas with extremely cold winters were better able than others to withstand frost damage; the needles of experimental plants were exposed to low temperatures in the laboratory and the extent of damage resulting was judged by the amount of discoloration shown by the treated needles after a three-week recovery period. Thus, as these detailed studies show, *Pinus strobus* as a whole has a "large niche", as indeed it must considering the huge extent of its geographic range. But the individual trees, and the local populations to which they belong, have narrower niches than the species as a whole.

McMillan (1960) has shown how whole sets of species may show parallel variation. That is, the several species that together dominate a particular kind of vegetation may each be represented in a given local community by a genoecodeme adapted to the local conditions. It will be recalled that the niche of a species is sometimes figuratively described as its "role"; extending the metaphor, one could say of such a set of species that it represents a whole play in which the parts are performed by a local cast of actors. Ward (quoted in McMillan, 1960) has described this phenomenon in sagebrush communities in western North America. McMillan's own experiments, on grass species typical of middle western grasslands, provide another illustration. In an experimental garden in Lincoln, Nebraska, he grew eight different grass species using material from two sources; the southern source (Manhattan, Kansas) is at 39°N latitude and the northern source (Watertown, South Dakota) at 45°N. To provide replicates, vegetative shoots were taken from three clones

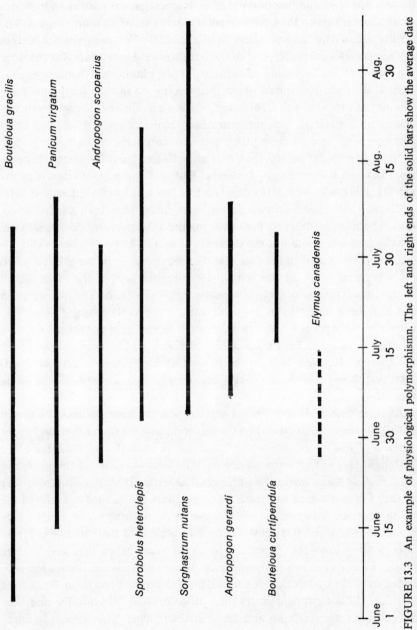

FIGURE 13.3   An example of physiological polymorphismn. The left and right ends of the solid bars show the average date of anthesis of transplanted grass clones from, respectively, a northern and a southern source. The order was reversed in *Elymus canadensis* (broken bar). (Adapted from McMillan, 1960)

at each source. The date of the beginning of anthesis (the period in which flowers are open and functioning) of each specimen was used as an indicator of its adaptedness to the environmental conditions of its source region. The results are shown schematically in Figure 13.3. For each species (except *Elymus canadensis*) the left end of the bar shows the mean date (for the three replicates) of the beginning of anthesis for the plants from Watertown; the right end of the bar shows the corresponding mean date for plants from Manhattan. It is seen that for seven of the eight species (*E. canadensis* was anomalous) material from more northern genoecodemes, where fall frosts come earlier, flowered before their southern counterparts.

Next we consider polymorphism in animals and the influence its existence must have on the concept of the niche. Out of a vast field of interesting examples that might be described only a few can be picked for mention here.

Clarke (1969) has discussed dimorphism in the land snail *Cepaea nemoralis*. The morphs differ in shell pattern, and wild populations of snails are usually found to contain some members of each. The persistence of the two morphs is attributed to the fact that thrushes preying on the snails tend to eat whichever is, at the moment, the commoner morph; the rarer morph at any time is thus selectively favored because of its rarity. The phenomenon has also been observed in another snail of the same genus, *C. hortensis*. Clarke calls the process *apostatic selection*. It seems likely that this mechanism often accounts, also, for the successful coexistence of two competing species (cf. Chapter 10, page 240). In the present example, however, the competitors that are enabled to coexist are merely two morphs of the same species.

Another case of dimorphism in a gastropod has been discussed by Giesel (1970). In studies carried out on the Oregon coast, he found that the limpet *Acmaea digitalis* occurred in two forms, one with dark shells and the other with light shells. Furthermore, the dark limpets were always found on the dark rock surfaces where their veligers had settled from the plankton; that is, they were on a dark substrate. The light limpets inhabited colonies of the white goose-neck barnacle *Pollicipes polymerus* which provide a light substrate; they reached this substrate by migrating into the barnacle colonies after settling onto rock from the plankton. Thus when they are in their proper habitats the two morphs are cryptically colored and somewhat protected from the oystercatchers and surfbirds that prey upon them. The forms therefore differ morphologically (in shell color) and behaviorally, and they have distinct microhabitats and hence distinct niches. They appear to interbreed freely, however, and belong to a single "good" species.

A common form of dimorphism in animals is that between the sexes. Selander (1966) has discussed sexual dimorphism in birds and the niche diversification that results from it. Thus he quotes observations by Norris that in the Brown-headed nuthatch (*Sitta pusilla*) the males forage lower down on the trunks of trees than do the females; and observations by Pitelka that in the Long-billed dowitcher (*Limnodromus scolapaceus*), a wader in shallow freshwater pools, the males can forage in deeper water than the females because they have longer legs and bills. From these examples and many others it is clear that sexual dimorphism may bring about an enlargement of the area in which a species can successfully search for food. The difference between the sexes in a dimorphic species is often as great as that between two different monomorphic species and the niche of a dimorphic species can presumably be twice as large.

What might be called chemical polymorphism, a genetically determined polymorphism in protein structure, is also known to occur in many animals. Powell (1971) has reported it in *Drosophila willistoni*, and found that he could control the genetic variability of experimental populations by changing the amount of environmental variability they were exposed to. In his experimental cages he arranged for spatial heterogeneity by putting different foods in the foodcups in each cage; and for temporal heterogeneity by keeping the cages warm and cool in alternate weeks. After 45 weeks (about 15 generations) he measured the amount of genetic polymorphism in the experimental populations by determining the mean number of alleles at each of 22 protein loci. It was found that populations in heterogeneous environments maintained greater genetic variability; that this was true both for environments that varied only spatially and for environments that varied only temporally; and that the effect was still more pronounced in environments that were heterogeneous in both respects. In sum, what may be called "multiple niche polymorphism" was brought about experimentally and at the same time it was shown that the protein differences among the flies were not selectively neutral.

Protein polymorphism has also been studied in wild populations. For example Doyle (1972) observed the phenomenon in the brittle star *Ophiomusium lymani* which is an abundant component of the benthos at depths between 1500 and 3000 meters off the North Carolina coast. He found the species to be polymorphic in respect of its proteins, with different morphs predominating in collections from different depths. Another example: Koehn (1969) has described dimorphism in respect of serum esterase in a species of freshwater fish, the Gila mountain sucker *Catastomus clarkii*. Populations of the fish from different regions (extending from southern Arizona to northern Nevada) were found to contain different relative proportions of the two morphs. One morph, in which serum esterase activity increases with increasing temperature, predominates in southern populations; in more northern populations the prevalent morph has serum esterase whose activity decreases with increasing temperature.

Polymorphism manifests itself in still another form in species that show *character displacement*. This name was proposed by Brown and Wilson (1956), who give many examples, to describe the following phenomenon. Members of two closely related species are markedly different from each other in regions where they occur together (i.e., where their geographical ranges overlap); but outside the region of overlap, where one or other species occurs alone, the differences between them are much less pronounced, so much so that specimens from these areas may be almost indistinguishable. This can be put the other way round by saying that where the two species occur together they have diverged so that the differences between them are exaggerated. Thus, in the words of Brown and Wilson (1956), "species populations show displacement where they occur together, and convergence where they do not".

Brief mention of only two examples will be made here. Wilson (1955), in a description of the ant genus *Lasius*, reported that in the eastern United States, where the two species *L. flavus* and *L. nearcticus* occur together, several characters distinguish them. However, *flavus* populations in the west, where this species occurs alone, exhibit far more intraspecific variability and many of the diagnostic differences that separate it from *nearcticus* disappear. Amadon (1947) has described character displacement in Hawaiian birds. For example in Kauai (the westernmost island of the chain) the genus *Phaeornis* (of the thrush family, Turdidae) is represented by a pair of species whose members differ conspicuously in bill size and behavior; in the remaining islands, where one or other of the species occurs alone, they have bills of intermediate size. The same phenomenon was reported for the genus *Loxops*, of the Hawaiian honey-creepers (Drepanidae).

Other examples of character displacement have been given by Hutchinson (1959) and Mayr (1963).

The reason for the phenomenon seems clear. To quote Mayr (1963), "competition in an area of geographic overlap should exert strong selection pressure. Those individuals of two overlapping species should by most favored by selection that have the least need for the resources jointly utilized by both species."

If this explanation is correct (and there is no reason to doubt it) it has important implications for the concept of the niche. The argument is as follows. Where two competing species occur together, and undergo character displacement, each of the co-occurring species-populations exhibits less ecological and morphological variability than either species does when it lives alone (Mayr, 1963). Each species' niche is thus "narrower" in the region of overlap than elsewhere. This leads to the conclusion that the niche of a species is not wholly determined by the species' own properties, but by species and environment acting jointly. This should not be taken to mean, however, that an individual organism (animal or plant) can narrow its requirements and "move along" (like a guest at a crowded table) to make room for others. It means, rather, that the population with the narrower niche is less variable genetically or, equivalently, that it has a smaller gene pool.

This brings us to what seems to me to be the crux of the matter. A niche is not the posession of a "species", which may consist of an enormous number of individuals spread over a large geographical area, but the possession of a *local intrabreeding population* which is usually quite small both in number of members and in geographic extent. It is these local populations of interrelated individuals, sharing a common gene pool, that should be treated as the "ecological units" among which environmental resources are partitioned, and among which competition takes place.

In the same way Ehrlich and Raven (1969) have argued that it is these local intrabreeding populations that function as the "units" in evolution. These authors have assembled, from a number of sources, a wealth of varied examples to show that in nature gene exchange takes place within, and is restricted to, local populations that are quite small. The fact that spatially distant populations often resemble one another closely does not mean that they share the same gene pool. On the contrary, a "good" taxonomic species may be made up of a number of completely isolated populations that cannot possibly have exchanged genes for many generations, for morphological differentiation is *not* an inevitable consequence of isolation. Further, even where gene exchange is not demonstrably impossible (as it is, for example, between populations of obligate cave-dwelling animals living in different caves) it may still be exceedingly uncommon. For instance, though the pollen of wind-pollinated plants can, in theory, be carried enormous distances, in fact it is not. Thus Colwell (1951) showed that for *Pinus coulteri* (Coulter pine) very little pollen was carried more than 50 meters from the tree that produced it. Similarly in insect-pollinated species, the distances across which fertilizations commonly occur may be very short. For species of *Clarkia* and *Delphinium* it was shown by Roberts and Lewis (1955) that groups of plants as little as 15 meters apart could be reproductively isolated. The butterfly species *Euphydryas editha* occurs as scattered local colonies and Ehrlich (1965) found that a gap between colonies of only 100 meters can be an effective barrier to gene flow.

Although some details of their arguments have been challenged (Shapiro, 1970; Baker, 1970), these examples and many others which have been brought together by Ehrlich and Raven (1969) demonstrate convincingly that in those species that have been carefully investigated, interbreeding is restricted to small local populations. It will be recalled that in botanical contexts such a population is called a *gamodeme* (see page 326), the term for a more or less isolated intrabreeding community. When Gilmour and Gregor (1939) introduced this term they wrote "these 'breeding communities' are likely to become increasingly important in the intensive study of evolutionary problems and we propose to name them *gamodemes*". I consider that these "breeding communities" are the units ecologists should be concerned with no less than students of evolution; and also that the comment of Gilmour and Gregor just quoted applies to animals no less than to plants. In what follows, for brevity and convenience, the term "gamodeme" will be used to connote a local intrabreeding population whether the organisms concerned are animals or plants.

In the foregoing discussion it was been tacitly assumed that the members of a gamodeme are morphologically and genetically similar. This need not always be so. Even when there is a free exchange of genes among a group of individuals, they may still be differentiated into clearly distinct morphs owing to differential selection. Maynard Smith (1970), in a theoretical discussion, has shown that such morphs can maintain their genetic distinctness, in spite of the fact that they interbreed freely, provided each has its own niche. Or, which comes to the same thing, provided each morph constitutes a population whose size is regulated by a separate limiting resource specific to the morph.

As a conclusion to the foregoing discussion (and to this section) the following statement therefore seems reasonable. The population of organisms that "owns" a niche may be either a homogeneous gamodeme, or else one of the morphs of a differentiated gamodeme.* Such a population (which may be difficult to recognize in the field) is the basic ecological unit from the point of view of niche investigations.

4 THE FACTORS CONTROLLING DIVERSITY

One of the most interesting problems to engage community ecologists is that of accounting for the fact that some habitats contain large numbers of species and others only a few. In terms of the concept of the niche, the state of affairs to be explained is that plants and animals have been able to create

---

* Or a single stage in the life cycle of one of these entities if the organism has different life styles in different stages (cf. page 324).

a far larger number of niches for themselves in some circumstances than in others. In this section we shall consider some of the many theories that have been put forward in this context.

It should be realised from the outset, however, that the theories are crude and the arguments underlying them lacking in precision. This is unavoidable in the present state of our knowledge but should not be overlooked. In debates on such subjects as why there are so many more species of trees or birds or bats in low latitude forests than in high, or why there are so many more mollusk species on tropical than on temperate shores, the words "diversity" and "niche" are bandied about freely. But in these discussions "diversity" is most often used as a synonym for "number of species" without regard to their relative abundances (cf. page 289); and a "niche" is most often thought of as the property of a whole species rather than of a local intrabreeding population or gamodeme (see the end of section 3 of this chapter, above). I believe it is because of these crudities that no satisfyingly convincing theory has yet been put forward to explain the way in which the products of biological evolution are *ultimately* determined by the abiotic properties of the environment. Theorizing about the concepts of "diversity" and "the niche" when the words are so loosely used is like performing a chemical analysis with impure reagents and a rusty balance. However, to refrain from theorizing until all the detailed knowledge we wish for has been gathered would call for an inhuman degree of restraint; it seems better to err on the side of rashness rather than timidity and theorize as best we may with what we have. In the rest of this chapter, therefore, the words *diversity* and *niche* will be used in the rather vague senses given above.

Numerous hypotheses, some testable and some not, have been advanced to explain why the flora and fauna are much more diverse in some habitats, or geographic regions, than in others. In an attempt to make the discussion as systematic as possible these theories are discussed below under four headings.

1 Structural complexity of the environment hypothesis

The idea that more species of animals (of a single taxocene) should be able to establish themselves in a given habitat if that habitat has a complicated "structure" (in the geometrical sense) has been suggested by many authors. In colloquial terms, such a habitat is rich in nooks and crannies for animals to occupy. Some examples: MacArthur (1964, 1968) has shown that the number of nesting bird species in a woodland is sometimes very closely related to the structural complexity of the vegetation; the more niches (in the nook-and-cranny sense) there are available in the different vegetation layers (tree canopies, shrubs and ground vegetation), the more species of nesting birds there are likely to be. Kohn (1968) has decribed the ecology of marine gastropods of the genus *Conus* in tropical islands in the Indian Ocean. He found 28 species altogether in the "complicated" habitats provided by subtidal reef platforms which offer a mixture of substrates including living and dead corals and boulders. A simpler type of habitat made up of smooth intertidal rock benches covered by a thin layer of sand yielded only 11 of these species (plus one other found in the simpler habitat only). Kohn ascribed the difference in species-richness to the difference in topographic complexity of the habitats, but presumably it is partly due also to the fact that the simple habitat is intertidal and the complex habitat subtidal.

Thus there is evidence to support the hypothesis that a complicated habitat (i.e., one offering a mosaic of different substrates) can support a more varied fauna than a uniform habitat. But some astonishingly diverse faunas have been found in seemingly simple habitats. For instance Pielou and Verma (1968) found 257 species of arthropods (insects and mites) in a collection of fruiting bodies of the annual bracket fungus *Polyporus betulinus* and a less structured, more uniform substrate is hard to envisage. Similarly Matthewman and Pielou (1971) found 182 arthropod species in the perennial bracket fungus *Fomes fomentarius* collected in the same area. Clearly, a straightforward relationship between species numbers and environmental complexity, though occasionally found, cannot be taken for granted.

The next three theories have been devised to explain the well-known fact that nearly all kinds of ecological communities are far richer in species in the tropics than in temperate latitudes. In some areas, and in respect of some taxocenes, there is a steady latitudinal gradient in species diversity (Pianka, 1966) in the sense that communities of a particular kind (e.g. forest trees, forest birds, littoral invertebrates) become steadily richer in species as one goes from high latitudes to low.

2 The Community-Age Hypothesis

The low diversity of high latitude communities is ascribed by some workers (e.g. Fischer, 1960) to the effects of the last (Pleistocene) ice age. It is argued that there has not been enough time since the melting of the last ice sheet (about 10,000 to 12,000 years ago in southern Canada) for stable communities to develop on the newly uncovered ground. According to this theory existing communities have yet to reach their full quota of species. Immigrant species are still entering them from more southerly areas that were never glaciated, and new species are still evolving in them. However, the theory fails to account for the fact that the steepest gradients in species numbers do not occur in recently glaciated regions; they are found, instead, in warm temperate regions that were never under ice (Simpson, 1964). Further, there is paleontological evidence to show that the steady increase in diversity from the north pole to the equator has existed since long before the Pleistocene glaciation which cannot, therefore, be responsible for it. According to Stehli *et al.* (1969) this gradient in diversity has existed at least since the Permian period. Fossil planktonic foraminifera of the Cretaceous period (70 to 80 million years ago), when the climate over the whole earth was much more equable than it is now, show a clear latitudinal gradient in species numbers in the northern hemisphere. Likewise fossil brachiopods of Permian age (270 million years ago) tend to contain progressively more families in samples obtained nearer the equator. Stehli *et al.* hypothesize that evolution proceeds faster in the warmer climates of low latitudes; as a result there is a more rapid turnover of species; that is, new species evolve and existing ones become extinct at a faster rate. Consequently more species are extant at any one time.

3 The Productivity Hypothesis

According to this hypothesis, which was developed in its entirety by Connell and Orias (1964), species numbers are highest where the annual rate of production of organic matter is highest. Thus the most species-rich terrestrial communities known are those in equatorial rain forests where the annual net primary productivity is greater than anywhere else on earth and where, also, the rate of nutrient cycling is probably fastest. Briefly, the hypothesis states that high productivity enables every species to produce a large population; and that the larger the population, the larger the species' gene pool and hence the greater its intraspecific variability. This is itself amounts to great diversity if we take as a measure of diversity the number of local intrabreeding populations (or gamodemes) in a region (see page 334). And a greater diversity in the conventional sense, that of a greater number of taxonomically distinguishable "species", is also likely to result. This will happen because any large, genetically variable species-population is liable to break up into a number of smaller, somewhat isolated local populations which may undergo evolutionary divergence until they rank as separate species. In highly productive environments, where the amount of energy available per unit area is great, this process is facilitated by the fact that animals need not travel far to obtain food; thus the members of local populations will rarely mingle and reproductive isolation will be enhanced.

There has been much discussion (see, for example, Pianka, 1966) on whether the great productivity of the wet tropics should lead to a higher or lower degree of competition among populations requiring fairly similar resources; and also, whether the exploitation of prey populations by predators is higher or lower. It has also been debated whether the coexistence of many fairly similar species in highly diverse communities results from the fact that (because of intense competition) each is highly specialized and occupies a "narrow niche"; or from the fact that (because of mild competition) considerable "niche overlap" can be tolerated. It is easy to invent any number of plausible chains of argument to lead from any starting point to any desired conclusion, and a sufficiently diligent search will usually yield circumstantial evidence to support at least some of the links of the theorist's favorite chain. For example, two such chains are as follows:

$$(i) \ \{\text{Intense competition}\} \rightarrow \left\{ \begin{array}{l} \text{Many narrow niches} \\ \\ \text{(Possible where} \\ \text{productivity is high)} \end{array} \right\} \rightarrow \{\text{Many species}\};$$

$$(ii) \ \left\{ \begin{array}{l} \text{Intense} \\ \text{predation} \end{array} \right\} \rightarrow \left\{ \begin{array}{l} \text{Mild competition} \\ \\ \text{(because species} \\ \text{populations kept} \\ \text{small)} \end{array} \right\} \rightarrow \left\{ \begin{array}{l} \text{Great niche} \\ \text{overlap} \\ \text{possible} \end{array} \right\} \rightarrow \{\text{Many species}\}$$

(Terms such as "niche width" and "niche overlap" are sufficiently self-explanatory not to need a rigorous definition in a qualitative discussion such as this one.)

What should be clear from the remarks above is that many of the different "explanations" of latitudinal diversity gradients are not truly comparable since the chains of reasoning comprising them start at different points. Thus some theories offer proximate causes, and some ultimate causes, for high diversity. Klopfer and MacArthur (1961), for example, have argued that the high diversities of tropical bird communities result from the fact that tropical species can tolerate a greater degree of niche overlap than can birds of temperate latitudes. Their evidence in support of this view is that the range of bill sizes (as measured by length of culmen) in groups of sympatric (and hence coexisting) congeneric species is much less in the tropics than it is farther north. If one accepts this finding, it leads to the inference that competition among birds must be less intense in tropical conditions and hence that reduced competition is the "explanation" of high tropical diversity. But it provides only a proximate cause and leaves unexplained *why* (if it be true) competition should be less intense in the tropics.

One could as well argue, as Savile (1960) has, that competition reaches its minimum intensity in the excessive harshness of arctic environments. Writing of plant communities, Savile argues that in the Canadian Arctic Archipelago plant populations are regulated entirely by abiotic environmental factors. The growing season is short, water is deficient, temperatures are low and the winds are fierce; as a result, the proportion of ground covered by vegetation is often less than 10% and such plants as do grow are too spaced out to compete. Competitive exclusion does not, therefore, occur and nothing prevents the coexistence in a single habitat of two or more closely related species within each of such genera as *Saxifraga*, *Draba*, *Potentilla* and *Cerastium*.

Looking back at the foregoing discussion it is seen that we could argue that:

i) Diversity is higher in the tropics (than in the arctic) *because* competition is mild (Klopfer and MacArthur). This in turn is (presumably) because productivity is high.

ii) Diversity is lower in the arctic (than in the tropics) *in spite* of the fact that competition is mild which is because productivity is low (Savile).

Thus even if one accepts that high productivity is an ultimate cause of high diversity, there is plenty of room for disagreement about the chain of proximate cause-and-effect relations connecting them.

In any case, many ecologists do not accept that high productivity leads to high diversity since cases are known in which the opposite appears to be true. Most of the contrary evidence concerns freshwater plankton (chiefly diatoms). The number of diatom species in a sample from a body of water is often found to be inversely related to the nutrient content of the water. Indeed, this finding is the basis for using counts of diatom species as indicators of water quality. A full account of the subject has been given by Williams (1964) in a report on the work of the National Water Quality Network of the U.S. Public Health Service. Large-scale investigations have been carried out in many major waterways (both lakes and rivers) of the U.S. to determine how the qualitative and quantitative characteristics of plankton communities are related to the degree and kind of pollution. It has been found that at "eutrophic" stations, where unnaturally enriched waters support a dense standing crop of plankton, the number of species is considerably lower than at "clean" (oligotrophic) stations where productivity and standing crop are low.

The fact that studies on the plankton communities of inadvertently enriched waters in highly developed countries on the one hand, and on the bird communities of tropical forests on the other, lead to diametrically opposed conclusions about the relationship between productivity and diversity is not very surprising. One obvious contrast between the two cases is that the forest communities have evolved over long periods whereas the dense, species-poor plankton communities grow in response to sudden environmental change. On a more general level, however, it is unreasonable to

22*

postulate that high productivity is ever the sole cause of high diversity. Admittedly it may well be a contributory cause, and no doubt debate will continue as to its importance relative to other contributory causes. It seems likely, however, that high productivity and high diversity are the joint outcome of an overriding cause. This brings us to the fourth of the hypotheses devised to account for latitudinal gradients in species diversity.

4 The Environmental Stability Hypothesis

It is probably fair to say that this is the most widely accepted hypothesis to account for latitudinal trends in species diversity. In brief, the argument is that in stable tropical environments a great many small populations of highly specialized animals and plants can thrive and perpetuate themselves with negligible risk of being wholly destroyed by catastrophic changes in the weather. Conversely, in high latitudes, where the coming of winter kills off all but a small fraction of all poikilothermic animals, specialization cannot be attained. If a poikilothermic species is to survive at all it must be extremely fertile and produce numerous offspring each breeding season to ensure that a few survivors, at least, shall live to breed in the succeeding season. Only a comparatively small number of different species (in a given taxocene) can therefore coexist. Homiothermic animals and perennial plants are not, of course, subject to regular winter kill. But they, too, especially when immature, can be the victims of unpredictable extremes of climate and these are more likely to occur in high latitudes than in low.

The stability theory predicts not only that there will be a greater number of species in warm, stable, wet-tropical environments than in the variable and unpredictable environments of higher latitudes, but also that natural selection will have favored different kinds of organisms. Pianka (1970) has summarized the contrasts between the sorts of organisms that flourish in circumstances associated with low diversity communities and those that flourish in circumstances associated with high diversity communities. To use terms introduced by MacArthur and Wilson (1967), the former are the product of *r-selection* and the latter of *K-selection*. Briefly summarized, the distinguishing characteristics of these two classes of organisms are as follows.

In environments where *r-selection* predominates, those species are favored that are highly productive in the sense of producing very large numbers of offspring. They live in environments characterized by rapidly changing physical conditions in which catastrophic mortality occurs from time to time leaving few survivors. Populations are therefore generally below their saturation levels and there is little competition. The individual organisms tend to be short-lived, often with lifetimes of one year or less. And they tend to be small, fast-growing, early-maturing, and to be what Cole (1954) has called *semelparous*; this term describes species in which an individual reproduces only once in its lifetime as do, for instance, univoltine insects and annual plants. Selection thus favors species having high values of r, the intrinsic rate of natural increase. This is because high fertility is a necessity if a species is to produce enough offspring each year to ensure its survival in spite of the sporadic occurrence of calamities. Among poikilotherms the tremendous die off caused by the onset of winter leaves few survivors and these survive more by chance than because of the possession of superior qualities.

In environments where K-selection predominates, those species are favored that are efficient, non-wasteful converters of food (or, more generally, of energy obtained from the environment) into offspring. They live where environmental conditions are fairly constant, and form populations that are at their saturation levels most of the time, under the control of density-dependent regulation. Therefore intra- and interspecific competition are intense. The individuals tend to be long-lived, large, slow-growing, late-maturing, and to be what Cole (1954) has called *iteroparous* (i.e., each female bears young several times in a lifetime). Selection favors species that produce a comparatively small number of high-quality offspring and survival is less a matter of chance than of competitive superiority in a particular habitat.

In the two contrasted environments there are thus contrasting kinds of ecological community. The first kind, which is characteristic of environments where conditions are harsh and unstable, is typically make up of comparatively few, non-specialized, r-selected species. The members of such a community are often exposed to severe physiological stresses and the fate of the community as a whole is governed by the physical properties of the environment. Sanders (1968) has proposed the description *physically controlled* for these communities. The second kind of community, which Sanders calls *biologically accommodated*, occurs where physical conditions are relatively uniform. Such a community typically contains a great many, fairly specialized, K-selected species. Their fate is far more dependent on biological than on physical stresses, and community composition is determined by biological processes such as competition and predator-prey interactions.

Janzen (1967) has suggested yet another reason (in addition to those above) why the stability of tropical climates should lead to the development of high diversity. His argument is that animals living in regions with continuously high temperatures never become adapted to cold. Therefore, if they live at the bottoms of valleys in mountainous country, they can never migrate into neighboring valleys because they are unable to withstand the cold of the intervening mountain ridges. The ridges thus serve as effective barriers between populations in adjacent valleys; the populations kept apart by them are prevented from interbreeding, and their reproductive isolation leads to divergence and speciation. This state of affairs is to be contrasted with that in the mountainous regions of temperate latitudes. The inhabitants of valleys in these regions are acclimatized to cold since they endure it every winter; for them, the ridges between neighboring valleys are no barrier to migration and neighboring populations are not reproductively isolated.

It should be noticed that this mechanism can explain the occurrence of a large number of species (of a given taxocene) in a tropical mountainous region of large area, containing several valleys; what is predicted is that, in tropical conditions, different valleys will tend to contain different species. But the theory has nothing to say concerning the diversity of a community within any one valley. This raises a problem that should be kept in mind in all discussions of the factors controlling species diversity. A large area may contain numerous species either because it supports a single, uniform, species-rich community; or because the organisms living there belong to a number of separate communities, each no richer in species than an average temperate-climate community, which form a spatial mosaic. This recalls the distinction emphasized by Whittaker (1960) between alpha diversity and gamma diversity (page 310). It has also been discussed by MacArthur (1965) who uses the self-explanatory terms "within-habitat diversity" and "between-habitat diversity". The distinction is often overlooked and it should not be if our understanding of ecological diversity is to advance.

Yet another reason for expecting high diversity to be associated with high environmental stability has been given by Connell and Orias (1964). They comment that the energy any organism assimilates from its environment must be channelled into two pathways. Part is needed to keep the organism's homeostatic processes functioning. These are the internal processes whereby the organism buffers the effects of fluctuating external conditions; (for example, in mammals sweating is the homeostatic process which ameliorates the effect of excessively high external temperature). The other part of the assimilated energy is needed for reproduction. In short, the two functions to which assimilated energy must be allocated may be called, respectively, "maintenance" and "productivity". Clearly, the more stable the environment, the less the maintenance energy required and hence the more there is available for productivity. The reasons for supposing high productivity to lead to high diversity have been discussed above (page 337). We now see how the stability of the tropical environment can reinforce its high productivity to give great ecological diversity.

The final reason to be mentioned here (no doubt there are others) for high species numbers in the tropics is that the long growing season, at least in regions without a prolonged period of drought, permits greater tempora niche diversification (see page 321). It is assumed that a greater number of species can be "fitted in". Whittaker (1960), however, starting from the same premise reached the opposite conclusion. In the context of forest vegetation, he argued that species are more likely to evolve life histories with non-over-lapping periods of maximum growth (in other words, temporal niches) where the climate is markedly seasonal than where it is uniform. It may be recalled (see page 321) that according to Coleman and Hynes (1970) the aquatic immature stages of some insect species showed peak growth in an Ontario stream in midwinter. Thus, for some organisms, winter cold need not make a temporal niche unusable.

This completes our discussion of the reasons that have been offered to ex-plain the latitudinal gradients in diversity shown by nearly all groups of terrestrial plants and animals. The "environmental stability hypothesis" seems (to me) the most persuasive. However, one should not overestimate the stability of conditions in the wet tropics. For example, Lloyd *et al.* (1968), in an account of the diversity of reptiles and amphibians in rain forest country in Borneo, comment that the environment is not at all stable for amphibians. It is true that temperatures remain extremely steady; daily maxima show an annual range of 6 °C and daily minima of only 2 °C. But the rain comes in torrential downpours at unpredictable intervals and during these storms most forest streams become raging torrents in which tadpoles cannot possibly sur-vive. Conditions that seem stable to an adult human may be fraught with disasters for a very much smaller animal.

Even if climatic stability is among the more important causes of high diversity in "ordinary" circumstances, it cannot ensure high diversity. Thus, consider limestone caves; in the parts of a cave above the highest level ever reached by floods physical conditions are probably more constant than any-where else on earth (except for the deep sea); yet the fauna is usually very depauperate, no doubt because of the extreme shortage of food (Poulson and Culver, 1969). Now consider the deep sea; Hessler and Sanders (1967) have shown that, contrary to the long-held belief that the deep-sea fauna was depauperate, it is in fact richer in species than that of any other habitat known. Presumably the extremely stable physical conditions, together with the perpetual rain of nutrient material settling down through the overlying layers of water, are the joint causes of this great diversity. In any case, the contrast between cave faunas and those of the deep-sea floor show convincingly that diversity is not governed by environmental stability alone.

To conclude this section it should be pointed out that species-counts made in different habitats may not be easily comparable. Thus in the preceding para-graph, cave faunas were compared with those of the deep sea benthos. It is worth describing how the data on which these comments are based were obtained.

The study of cave communities by Poulson and Culver (1969) was carried out in a cave system in the Mammoth Cave National Park, Kentucky. The authors collected specimens at 22 sites at each of which a trap (a cup baited with a piece of rotten liver) was left in place for a week. Besides counting specimens (chiefly insects) caught in the traps, they also censused, by eye, the area around each trap. The greatest number of species recorded at any one site was four.

The deep-sea benthic samples analysed by Hessler and Sanders (1967) were obtained in the western Atlantic between Massachusetts and Bermuda with a specially designed epifaunal sampler. It consists of a flattened plankton net mounted on runners to prevent its sinking into the substrate as it is dragged along the bottom. The whole device, called by the authors an epi-benthic sled, is about $4\frac{1}{2}$ feet wide and weighs 350 pounds. It is towed along the sea floor, at speeds of 1 to $1\frac{1}{2}$ knots, for a period of one hour before being brought to the surface for sorting and identification of the catch. In terms of numbers of species, the richest haul was obtained from a depth of 1400 meters on the continental slope; there were 365 species in the sample. Even the poorest sample, from 4700 meters depth in the Sargasso Sea, contained 196 species.

Thus though it is reasonable to conclude that there is an enormous difference in species richness between these two extreme kinds of communities, it is not reasonable to attempt a direct comparison between the actual numerical results. The difference between the observational methods is too great for such a comparison to be legitimate.

5 COMMUNITY DIVERSITY AND COMMUNITY STABILITY

There has been much speculation, and a certain amount of research, on the subject of whether there is any relation between the diversity of an ecological community and its "stability" *. It is prompted by the notion (it is scarcely more than that) that a diverse community "ought" to be stable. The argument,

---

* This section is concerned with *community stability* (the tendency of a community's composition to remain constant) as opposed to *environmental stability* whose ecological effects were explored in Section 4. The distinction is important and should be borne in mind. The much-used term *ecosystem stability* will be avoided because of its ambiguity. It is commonly used to mean what is here called *community stability*, but such use is unfortunate since the word *ecosystem* itself connotes both a community and its physical setting.

a plausible one, is that if there are several prey species for every predator species, the predator populations are in no danger of sudden "crashes" because if one prey population becomes scarce there is always another to turn to. Likewise if there are several predator species attacking every prey species, no prey population ever has a chance to "explode" because if one predator does not keep it in check another will.

If this phenomenon actually occurs, and is at all widespread, it should cause ecological communities as a whole to be fairly clearly divisible into two classes, those that are of high diversity and very stable, and those that are of low diversity and very unstable. This is because the mechanism just described would be self-reinforcing; in a species-poor community instability would soon lead to the chance extinction of one or another of the unstable component populations, leaving the community still more depauperate and still more unstable. The fact that natural communities in general are not separable into obviously distinct classes, plus the fact that many exceptions are known to the postulated positive correlation between diversity and stability, show that these characteristics are not so simply related after all. The mechanism described above does, in any case, hinge on the assumption that where numerous species are present, belonging to several trophic levels, the food relationships constitute an anastomosing "food web" rather than a number of unbranched "food chains". For, if the food chains were unbranched, that is, if every predator species relied for food on only one prey species and every prey species was attacked by only one predator species, a species-rich community would be no more stable than a species-poor one.

These remarks make it clear that any relationship between the diversity and the stability of ecological communities is not something to be taken for granted. Research on the topic needs to be done without preconceptions. It seems likely that high community stability is no more the result of a single cause (namely high diversity) than high diversity itself is the result of a single cause (such as environmental stability). It was emphasized in the last section that the diversity of a community is usually controlled by several causes whose relative importance differs in different circumstances. Presumably the same may be said of a community's stability. For example both may, on occasion, be determined (in part) by environmental stability.

If "community stability" is to be investigated, one must obviously decide how it should be defined and measured. The problem is at least as difficult as the measurement of diversity but it has not received nearly so much attention. Before discussing it, it is necessary to point out the contrast between the stability of a one-species population and that of a many-species community. Even for one-species populations the concept of stability is open to various interpretations and no useful discussion is possible until a definition of population stability and a way of measuring it have been decided upon. Murdoch (1970) suggests that a population be called stable if its size exhibits no trend, and fluctuations around the mean are of small amplitude. This is a useful definition as far as it goes but it provides no means of measuring stability quantitatively. One might (in principle) do this by equating the stability of a population to the fraction of time during which its size is between, say, 90% and 110% of its long-term average. (For species that inevitably fluctuate annually, the measurement of long-term stability would be based on estimates of population size made at the same stage in the life cycle each year). Then stability would be low (as it should be) both for populations exhibiting trends and for populations undergoing fluctuations of large amplitude. There are, of course, great practical difficulties in measuring the stability of a natural population by this method, but the same could be said of any method.

Now consider the stability of a many-species community. The several species will usually be of very unequal abundance and of very different generation lengths. Should the community be described as "highly stable" if *most* of its species are extremely stable and only a few unstable; or if *all* its species are moderately stable? Either of these definitions may be acceptable and one must make a choice. As a quantitative measure of community stability one might take the proportion of individuals in the community belonging to species whose population-stability exceeded some specified numerical value, for example a value of 0.5 measured in the way suggested in the preceding paragraph. Alternatively, community stability could be equated to the proportion of species (as opposed to individuals) whose populations meet the same requirement. Again, a decision must be made on how terms are to be defined and community properties measured.

The greatest difficulty in attaching a clear meaning to the concept of community stability arises from the fact that in many-species communities the different species are likely to have different generation times. The difficulty does not arise when the stabilities of separate species-populations are measured. But in a mixed community the population-stabilities of the short-lived species is bound to be affected by the mere "persistence" of the long-lived ones. For example, the community formed by all the vascular plants in a forest will contain trees whose average life spans are measured in decades or centuries, and, in chance clearings, annual weeds with life spans of only one year. The fluctuations, from one generation to the next, of the weed populations are much less affected by analogous fluctuations in the tree populations than by the long-continued presence of individual trees.

Margalef (1969) has suggested a measure of stability-plus-persistence that makes no attempt to separate the two things. He defines the measure, symbolized by S, in two ways. One may put

$$S \propto \sum b_i/m_i; \quad \text{or else} \quad S \propto \sum b_i t_i.$$

Here b_i is the biomass of the ith species; m_i is the instantaneous death rate of the ith species; and t_i is the half-life of the biomass, that is, the time required for a unit of biomass (of the ith species), present at any moment, to be reduced by one half. The reduction may be by death or by what Margalef calls "attrition of biomass" which includes such losses of living material from an area as leaf-fall from trees and emigration of animals.

Another way of describing S is to call it the "average residence time" of a unit of biomass. Also, S can be said to give an inverse measure of the rate of turnover; the higher the value of S the slower the turnover.

The relationship between S and stability in the usual sense (of uniformity in the size of a population over many generations) is not at all direct. Thus the effect on S of population *in*stability (in one of a community's component species-populations) depends on the extent to which this instability stems from changes in the birth rate or changes in the death rate of the fluctuating species. If the birth rate alone were varying, the population's size could fluctuate wildly but the effect on S would be nil.

We now consider some actual observations on population- and community-stability.

The question of whether the stabilities of one-species insect populations depend on the numbers of species of food plant eaten by the insects was investigated by Watt (1964). From the records of the Canadian Forest Insect Survey he obtained data on the yearly abundances of all the important pest caterpillers (Macrolepidoptera) occurring in Canadian forests; he then devised a measure of population-stability and calculated its value for each species. The number of species of host tree attacked by each pest species is also known. Data were thus available for estimating the correlation between population-stability, on the one hand, and number of tree species eaten, on the other. Watt subdivided these data into ten batches, one from each of the ten classes of Canadian forests he distinguished; the forests were classified on the basis of their geographical region and on whether they were coniferous, deciduous, or mixed. Of the ten correlations calculated, it was found that six implied no significant correlation between population-stability and the number of tree species eaten; three showed significant negative correlation, implying that species with a wide range of foods available to them were *less* stable than others; and only one showed the positive correlation predicted by theory. So fas as it goes, therefore, this investigation does not support the view that a population's size is less likely to fluctuate if the species has many foods to choose from.

Evidence relating to the same problem, but of an entirely different kind, has been given by Murdoch (1969). He carried out experiments to determine whether predatory marine snails would switch to alternative prey when their customary prey was in short supply. Two species of drilling whelks, *Thais emarginata* and *Acanthina spirata* were the predators in these experiments. The prey species offered to *Thais* were the mussels *Mytilus edulis* and *M. californianus*; the prey offered to *Acanthina* were *M. edulis* and the barnacle *Balanus glandula*. The results showed that the predators did *not* switch, and this led Murdoch to conclude that "switching in nature probably is rare, and that at least invertebrate predators in general do not stabilize the numbers of their prey by this mechanism".

The results of these experiments contrast with those of Clarke (1969) (see page 330) in his studies of apostatic selection. However, in Clarke's work the predators were vertebrates (thrushes preying on land snails). If Murdoch's conclusion is generally applicable to *in*vertebrates it casts doubt on the common assumption that high diversity makes for great stability. It should not be taken for granted that where many species of herbivores and carnivores occur together the "food web" must necessarily be a complicated network. That is, one must not assume that each carnivorous predator attacks a wide range of herbivorous prey species and that each prey species is the victim of a wide range of attacking predators. The variety of foods eaten by an animal may be much less than the human observer is apt to suppose from a glance at the riches apparently available. The fact that a complicated food web can be envisaged does not necessarily mean that it exists; it seems quite likely that even in very diverse communities the food chains may be, for the most part, unbranched and unconnected. In every case the true state of affairs must be discovered by observation, not guesswork.

To study the stability and diversity of whole, undifferentiated communities may lead to the neglect of interesting within-community relationships. In any community with several trophic levels, the primary producers (the autotrophic plants comprising the vegetation) constitute part of the environment for the next higher trophic level (the herbivores). Similarly, the herbivores are part of the environment for the primary carnivores, and so on up the trophic pyramid. Thus if *environmental* stability is an important determinant of diversity, it follows that *community* stability in the vegetation, say, which is equivalent to *environmental* stability for the herbivores, must promote diversity among the herbivores. Thus stability and diversity at one trophic level must affect, and be affected by, diversity and stability at another.

Hurd *et al.* (1971) investigated these relationships experimentally and took account, also, of the effect of ecological succession on diversity and stability. They define the stability of a community, or system, as "the ability of a system to maintain or return to its ground state after an external perturbation." They therefore judged the stabilities of the systems they studied by observing the effects on them of experimental perturbations.

The experimental communities were two abandoned hayfields in central New York. One field (the young field) had gone out of cultivation seven years before the experiment and the other (the old field) 16 years before; the fields thus differed in their successional stages. The experimental treatment consisted in the addition of fertilizer to experimental plots within the fields. The communities in the fields were made up of sub-communities belonging to three trophic levels; primary producers (plants); herbivores (various arthropods) feeding on the plants; and carnivores (also arthropods) feeding on the herbivores. The experiment enabled diversities and productivities to be compared between treated and control plots, between the old and the young fields, and among the three trophic levels. Productivities were measured in dry weights of biomass produced; and diversities were equated to the numbers of species found in samples.

So far as the primary producers (the plants) were concerned, the results accorded with long-accepted theory. The vegetation was less productive, more diverse, and less responsive to the experimental perturbation (the fertilizer application) in the old field than in the young. The magnitude of the response to perturbation is an inverse measure of stability; it can therefore be said that the primary producers were more stable in the old field than in the young. Thus high stability, high diversity, and low productivity went together.

It was at the herbivore and carnivore level that results were obtained that were at variance with established theory. Although diversity was greater in the old field than in the young, stability was less; that is, the response to perturbation was greater. Also, among herbivores, productivity was greater. Thus at the two "consumer" levels (herbivores and carnivores) high diversity was associated with low stability. Looking at relationships between, rather than within, trophic levels it appears that low stability in the consumers was linked to high diversity in the primary producers. Hurd *et al.* hypothesize that herbivorous insects (and through them the carnivores that feed on them) are normally food-limited. Enriching the soil for the food plants enabled more insects to live upon them and hence led to the higher productivity at the herbivore level. It also enabled insect species to survive that in a less productive environment would have been excluded by stronger competitors; hence the higher diversity.

To conclude this section it should be pointed out that studies on the diversity and stability of natural communities are of more than academic interest. If the deleterious effects of growing human populations and spreading technological influences on natural communities are to be minimized, we obviously need to know which communities are most sensitive, and which most resistant, to the perturbations caused by pollution. The knowledge is needed if delicate communities are to be adequately protected, and it may well turn out that the stability of a community can be most easily gauged by measuring some other community property such as diversity. The relationship between stability and diversity is not straightforward, however, as this section has shown, and more research is needed.

6   HOW MANY SPECIES CAN COEXIST IN A COMMUNITY?

At the beginning of this chapter it was remarked that in most communities a number of species coexist without excluding one another in spite of the fact that many of them make fairly similar demands on environmental resources. This leads to the question: how many species *can* coexist in a given environment?

In terms of the niche concept the question can be reformulated so that one asks: how many niches can a given environment contain? The researcher is then (overtly, at any rate) investigating niches rather than species. It is arguable whether this helps. Instead of considering niches it might be better, as Mayr (1966) has commented, to liken the resources of the environment to "a gigantic warehouse from which each species makes withdrawals".

However the question is formulated, there are various ways of attempting to answer it, including that of dismissing it as inherently unanswerable. Those who attempt to find an answer start from the assumption that in any area two opposing processes are at work. One process, competitive exclusion, tends to reduce the number of species. The other process, the continual addition (through immigration and evolution) of new species tends to increase it. When the processes are in equilibrium the number of species is expected to remain constant, and this is the number theorists attempt to derive.

In terms of niches, one might say that equilibrium is attained when niche differentiation has reached a limit. This is only another way of saying that there is a limit to the degree to which a species can become specialized in its requirements and still persist. Too high a degree of specialization is obviously incompatible with survival, for if the number of individuals in a species-population falls too low (as a result of overspecialization), extinction of the population is inevitable. The mathematical basis for this assertion need not detain us here; its truth is intuitively obvious.

Attempts to determine the number of species that a given environment will contain when the two processes are in balance are of great theoretical interest and the fact that the assumptions on which they are based are unrealistically simple does not detract from their value. A "simple" model is as valuable to ecologists as an "ideal gas" is to physicists, as Mead (1971) has pointed out. To derive, by mathematical reasoning, the logically inevitable outcome of chosen assumptions is always rewarding. When the results predicted are intuitively surprising, they may suggest appropriate modifications to intuition-based predictions concerning natural situations. Also, a simple model is often useful as a "limiting case" to a battery of more realistic models which are too complicated for all their mathematical consequences to be derived.

When simple models are discussed, however, one must never lose sight of the ways in which the true state of affairs *probably* departs from the chosen assumptions. Thus models designed to predict the number of species a particular environment will contain overlook (deliberately) three important factors.

These are, first, that in natural communities equilibrium between competitive exclusion on the one hand and immigration plus evolution on the other may never be attained. Gradual changes occur in nearly all environments (whether they should be thought of as slow fluctuations or as trends makes no difference to the present argument) and are presumably accompanied by at least minor evolutionary changes in the species-populations present. Evolution is not necessarily a slow process. Perceptible heritable changes in populations can occur in a very few generations; examples have been given by Sammeta and Levins (1970) who remark that "demographic and at least microevolutionary time are commensurate." Thus to assume that a community has reached equilibrium in regard to the acquisition of new species and the loss of old ones is probably seldom justified.

The second factor ignored by the simple models relates to sessile species or those whose range of movement is small. It is that a species may be competing with one set of competitors in one part of the area under discussion and with another set of competitors in another part (McIntosh, 1970). Even in a wholly homogeneous area (if such a thing exists) the sedentary and sessile members of a community are likely to have patchy patterns because of their modes of reproduction. As a result, the intensity of the competition facing any individual organism, and its chances of success, must be affected by its location.

The third complication that simple models cannot (at present) allow for is that the competing units that should be counted are not "species" in the ordinary sense but gamodemes (see page 334). One can, of course, alter the wording of all discussions on "species numbers" to accord with this stipulation. However, because of the difficulty of recognizing, and hence of coun-

ting, gamodemes in nature it is difficult to compare the observed and predicted numbers of them in any area.

Now let us consider an example of the kind of mathematical argument
used in connection with species-richness problems. MacArthur and Levins
(1967) (and see also Levins, 1968) set themselves the task of deciding under
what conditions it would be possible for an immigrant species to invade an
area successfully it its requirements came "between" those of two species already established there. Let the two established species be labelled N_1 and N_2
and denote their numbers of individuals by the same symbols. Also, suppose
that the growth rates of the two species-populations can be described by
Gause's equations (see equations (9.3), page 206) which in this case are
taken to be

$$\frac{1}{N_1} \frac{dN_1}{dt} = r - sN_1 - \beta sN_2$$

$$\frac{1}{N_2} \frac{dN_2}{dt} = r - sN_2 - \beta sN_1.$$

Notice that the constants r, s and β are equal for the two species. That is,
both have the same intrinsic rates of natural increase, r. Both experience the
same degree of *intra*specific competition, which is measured by s. And the
two species are evenly matched in *inter*specific competition as shown by the
presence in both equations of the same factor βs. The coefficient βs shows
that the degree to which the growth of each species is reduced by *inter*specific
competition is β times the degree to which it is reduced by intraspecific competition; (in the present discussion βs takes the place of the factors denoted
by u_M and u_N in Chapter 9, page 207).

The two species are assumed to coexist in stable equilibrium and for this
to be possible we must have $\beta < 1$ (see the conditions for stable equilibrium
given on page 211). It is also easy to see that if a population of either species
were growing without the other, it would cease to grow when

$$\frac{1}{N} \frac{dN}{dt} = r - sN = 0, \quad \text{that is, when} \quad N = \frac{r}{s}.$$

Thus $K = r/s$ is the saturation level for each species considered separately;
(subscripts need not be used here since species N_1 and species N_2 are assumed
to behave identically). When the two species-populations are together, both
populations stop growing when

$$\frac{1}{N} \frac{dN}{dt} = r - sN - s\beta N = r - sN(1 + \beta) = 0;$$

or, equivalently, when

$$N = \frac{r}{s(1 + \beta)}.$$

Therefore, writing K^* for the saturation size of each population when the two populations are competing, we see that

$$K^* = \frac{r/s}{1 + \beta} = \frac{K}{1 + \beta}.$$

Now suppose that a new species, species M, whose resource needs are intermediate between those of the two established species, is introduced to the area. Its so-called "logistic parameters", the constants r and s, are assumed to be identical with those of species N_1 and N_2. Thus if it were alone, species M's growth would accord with the logistic law

$$\frac{1}{M} \frac{dM}{dt} = r - sM;$$

and its saturation level (when alone) would be $K_M = r/s$. In the presence of the established species, however, when their population sizes are N_1 and N_2, we have

$$\frac{1}{M} \frac{dM}{dt} = r - sM - \alpha s N_1 - \alpha s N_2$$
$$= r - sM - \alpha s(N_1 + N_2).$$

That is, the growth rate per individual of the intruding species is reduced by the amount $\alpha s(N_1 + N_2)$ when $N_1 + N_2$ competitors of the other two species are present.

We now come to the crucial question which is: under what conditions will species M contrive to grow in spite of the presence of species N_1 and N_2? For this to happen we must have

$$\frac{1}{M} \frac{dM}{dt} > 0 \quad \text{when } M \text{ is very small and when} \quad N_1 = N_2 = K^*.$$

In words, species M's growth rate must exceed zero even when species N_1 and N_2 are both at their (joint) saturation levels K^* each.

The required condition is therefore given by

$$\frac{1}{M} \frac{dM}{dt} = r - sM - \alpha s(K^* + K^*) > 0.$$

When M is very small, the term sM is negligible and the condition becomes

$$\frac{1}{M} \frac{dM}{dt} = r - 2\alpha s K^* > 0$$

or
$$(r/s) > 2\alpha K^*;$$

equivalently, since $(r/s) = K_M$ and $K^* = K/(1 + \beta)$, the condition is

$$K_M > \frac{2\alpha K}{1 + \beta}.$$

Thus for species M to establish itself successfully in the face of competition from saturated populations of species N_1 and N_2, it is necessary that

$$\frac{K_M}{K} > \frac{2\alpha}{1 + \beta}, \quad \text{or} \quad 1 + \beta > 2\alpha \quad \text{(since } K_M = K\text{)}.$$

The inequality relates the ratio of the saturation sizes of the species-populations when alone to coefficients in their growth equations, and hence to the intensity of intra- and inter-specific competition within and between their populations.

It is easier to derive this inequality than it is to interpret it in concrete terms. As MacArthur and Levins (1967) state, if species N_1 and N_2 do *not* compete (so that $\beta = 0$) the condition for species M to immigrate successfully becomes, simply, $\alpha < \frac{1}{2}$. But when N_1 and N_2 compete with each other (i.e., when $\beta > 0$) their ability to compete with species M is reduced and a somewhat larger value of α, which measures the extent to which N_1 and N_2 together reduce M's growth rate, will not preclude M's successful invasion. But still, if $1 + \beta < 2\alpha$, invasion is impossible.

We cannot here pursue these theoretical studies further. The foregoing paragraphs demonstrate the sort of reasoning involved. They demonstrate, also, the difficulty of linking theoretical arguments to real-life situations. The existence of difficulties is not, of course, a deterrent; work should continue until they are overcome. In a more recent development, MacArthur (1969) has devised a model to predict how many species of "consumer" animals can maintain themselves on a set of food resources when due allowance is made for the growth rates of the resources themselves (which are living) and the efficiency with which the consumers use what they eat for further growth.

An entirely different approach to the problem of species numbers is due to MacArthur and Wilson (1963) (and see also MacArthur and Wilson, 1967). They started from the fact that the number of species in any area is the resultant of the number gained (by immigration and local evolution) and the number lost (by emigration and local extinction). Then, considering oceanic islands, they discussed the factors that control the rates of gain, and of loss, of species. Thus the rate of immigration of new species (new, that is, to a particular island) must be greatest into islands near to a mainland coast, and least into islands a long distance away. Immigration and evolution rates will also depend on the species-richness of the mainland from which the immigrants come, and on the variety of habitat conditions available in the island itself. But for islands stocked from a given mainland source, distance from this source is the overriding factor determining the rate of gain of new species, which is chiefly through immigration. The rate at which an island loses species, that is, the rate at which they become locally extinct, depends on the island's size and is obviously more rapid for small islands than for large. The foregoing statements are most easily summarized graphically, as in Figure 13.4.

In the figure, the abscissa shows the number of species already present; the solid lines show the rate at which species are gained and the dashed lines the rate at which they are lost. Consider the solid lines first; clearly the rate at which new species are gained must decrease as the number already there goes up, since the probability that a newly-arrived immigrant belongs to a species not already in the island will get steadily less as the island's roster

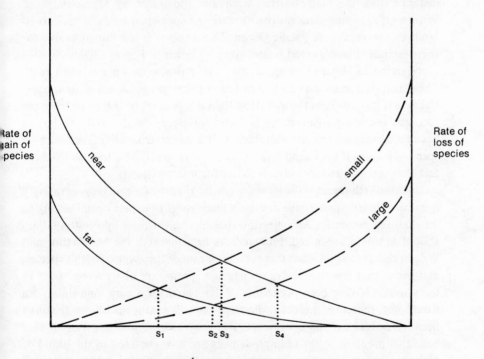

FIGURE 13.4 The rates of gain (solid line) and of loss (broken line) of species from islands of different sizes and different distances from the mainland. The equilibrium number of species is given by the value of s at which the island's gain-curve and loss-curve intersect. (Adapted from MacArthur and Wilson, 1963)

of species grows; therefore, the solid curves descend from left to right. Also, the curve relating to a near island, which acquires new species rapidly, is above that relating to a far island, which acquires them more slowly. Next, consider the dashed lines; these descend from right to left since the rate at which species are lost must obviously be greater when there are many species than when there are few. Also, the curve relating to a small island, which loses species rapidly, is above that relating to a large island, which loses them more slowly.

The equilibrium number of species in an island of given size at a given distance from the mainland is thus, in principle, obtainable. It can be read off as the abscissa of the point where the island's species-gain curve and its species-loss curve intersect. Four such values are shown in the figure; thus s_1 represents the equilibrium number of species on a far, small island; s_2 the number on a far, large island, and so on.

The dependence of an island's species-richness on both its size and its distance from the mainland has been well illustrated by MacArthur and Wilson (1963) using data on the numbers of species of birds in various islands in the southwest Pacific Ocean. New Guinea is the faunal source for these islands. The observed relationship is shown in Figure 13.5.

Referring to Figure 13.4 again, two other predictions are implicit in it. These are: (i) that an island's size will have a stronger influence on its species-richness if it is far from the mainland than if it is near; (this follows from the fact that the species-gain curve is flatter for a far island). And (ii) that an island's distance from the mainland will have a stronger influence on its species-richness if the island is large than if it is small; (this follows from the fact that the species-loss curve is flatter for a large island).

The MacArthur and Wilson theory (their "theory of island biogeography") is extraordinarily persuasive and even lends itself to experimental testing as we shall see presently. An objection that has been raised by Preston (1968) should not pass unnoticed, however. As he remarked, the MacArthur and Wilson theory predicts that there is a high rate of turnover of island species, and hence that few or none of an island's species are likely to be "old" in the sense of having been occupants of the island for a very long time. One would not, therefore, expect to find species still extant on islands that had become extinct on the continents. But several examples are known that belie this prediction; for example lemurs are now confined to the island of Madagascar, whereas they were widespread in the Eocene. Madagascar is so large, however, that one would not expect a theory concerning small islands to apply to it.

MacArthur and Wilson's theory of island biogeography has been tested in the most direct manner possible, by experiment (Wilson and Simberloff,

1969; Simberloff and Wilson, 1969; Simberloff, 1969; Wilson, 1969). Seven small islands in Florida Bay were used as the experimental islands; they were very similar to one another and each consisted of one or a few red mangrove trees (*Rhizophora mangle*) rooted in sand or mud that (except for one of the islands) was covered by the sea at high tide. The islands ranged in diameter from 11 to 25 meters; their distances from the nearest shore ranged from 2 to 1188 meters. The resident fauna of the islands consisted of arthropods, chiefly insects and spiders but also some scorpions, pseudoscorpions, centipedes, millipedes and arboreal isopods. All the animals lived somewhere on the mangrove trees, in living or dead wood, under dead bark, and in tree holes, leaves, flowers, shoots and fruits.

Careful sampling of the fauna present before the experiment began gave estimates of the initial number of species on each island; the numbers ranged from 25 to 41. Further, it was known from studies of control islands that the number of species in an island resulted from the existence of a state of dynamic equilibrium. This was shown by the fact that the number of species on any particular island remained fairly constant even though the actual species composition varied considerably.

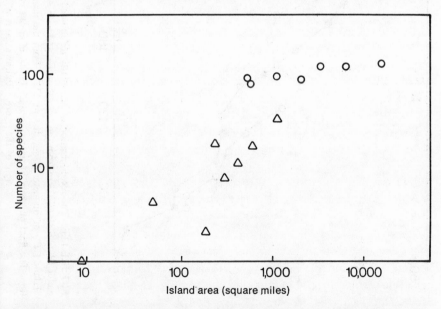

FIGURE 13.5   The numbers of species of birds on islands of different sizes in the southwest Pacific. The source region for most species is New Guinea. "Near" islands (less than 500 miles from the source) are shown by triangles; "far" islands (more than 2000 miles from the source) are shown by circles. Both axes are scaled logarithmically. (Adapted from MacArthur and Wilson, 1963)

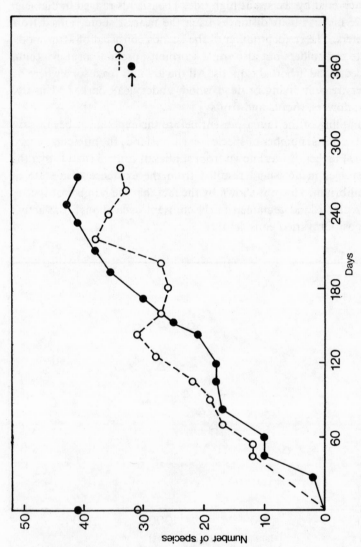

FIGURE 13.6   The recolonization of artificially defaunated mangrove islands. ●——●, "Island E9", of diameter 18 m, distant 379 m from the nearest source and with 41 arthropod species before defaunation. ○——○, "Island E3", of diameter 12 m, distant 172 m from the nearest source and with 31 species before defaunation. The final numbers of species in the islands, after one year, are shown by the unconnected symbols in the right of the figure. (Redrawn from Wilson and Simberloff, 1969)

The experimental treatment of the islands consisted in their artificial defaunation. Each island was covered with a plastic tent and fumigated with methyl bromide at a concentration which killed all the arthropods but caused no more than minor, temporary damage to the plants. Thus at the end of the treatment each island was, momentarily, devoid of all arthropod life. Observations were then begun to record their recolonization. All the islands were monitored at 18 day intervals for a year; the monitoring process, which took about 20 man-hours, consisted in a careful examination of large samples of all possible arthropod habitats so that an estimate of the number of species present on any island could be made. In this way the recolonization process was followed and it was possible to judge its speed and how it was affected by each island's size and distance from the land.

It was found that on all islands (except the most distant one) the number of species first rose to a level slightly above the original, pre-experiment level; this occurred in about 250 days. But after reaching a peak, the number of species on each island then fell slightly until, after a year, it was approximately at its predefaunation value. Figure 13.6 gives two representative graphs each showing how a peak value was reached but not maintained.

Wilson and Simberloff interpret their results as follows. There appear to be two equilibria, the first *non-interactive*, and the second *interactive*. Non-interactive equilibrium is attained when the rates at which an island gains and loses species first become equal. At this stage the species-populations are fairly small, and they do not interfere with one another; since the populations have not had time to grow large, they are not regulated by such processes as predation and competition, and the extinctions that occur result from chance or from density-independent causes. This equilibrium does not last long. The species-populations continue to increase in size and, as a result, begin to interact appreciably; density-dependent regulatory processes come into operation, including competitive exclusion. Consequently, there is a slump in the number of species to a new, somewhat lower, equilibrium level, that of interactive equilibrium. In the experimental islands this is known to have persisted for at least two years following the experiment (Wilson, 1969).

The final section of this book constituted one more illustration of the way in which theory, field observation, and experiment can all be brought to bear on one topic; and how the results of the three kinds of approach can reinforce one another and lead to a greater degree of ecological understanding than could otherwise be achieved. I believe that only by using all these methods can ecology advance; and advance it must, if our environment is to be enjoyed rather than endured in the future.

Exponential Functions

Using the Binominal Theorem, $(1 + a)^n$ may be expanded as follows:

$$(1 + a)^n = 1 + n \cdot a + \frac{n(n - 1)}{2!} \cdot a^2 + \frac{n(n - 1)(n - 2)}{3!} \cdot a^3 + \cdots$$

where

$$2! = 2 \times 1, \quad 3! = 3 \times 2 \times 1, \quad \ldots \quad \text{and in general,}$$

$$c! = c \times (c - 1) \times (c - 2) \times (c - 3) \times \cdots \times 3 \times 2 \times 1.$$

$$(c! \text{ is called "}c \text{ factorial"}).$$

Similarly

$$\left(1 + \frac{r}{n}\right)^n = 1 + n \cdot \frac{r}{n} + \frac{n(n - 1)}{2!} \cdot \frac{r^2}{n^2} + \frac{n(n - 1)(n - 2)}{3!} \cdot \frac{r^3}{n^3} + \cdots$$

$$= 1 + r + \frac{1\left(1 - \dfrac{1}{n}\right)}{2!} \cdot r^2 + \frac{1\left(1 - \dfrac{1}{n}\right)\left(1 - \dfrac{2}{n}\right)}{3!} \cdot r^3 + \cdots$$

$$\therefore \quad \text{as} \quad n \to \infty \quad \text{so that} \quad \frac{1}{n}, \quad \frac{2}{n} \quad \text{etc.} \to 0$$

we have

$$\lim_{n \to \infty} \left(1 + \frac{r}{n}\right)^n = 1 + r + \frac{r^2}{2!} + \frac{r^3}{3!} + \frac{r^4}{4!} + \cdots.$$

Now

$$\lim_{n \to \infty} \left(1 + \frac{r}{n}\right)^n = e^r \quad \text{by definition.}$$

$$\therefore \quad e^r = 1 + r + \frac{r^2}{2!} + \frac{r^3}{3!} + \frac{r^4}{4!} + \cdots.$$

Notice, also, that

$$\lim_{n \to \infty} \left(1 + \frac{r}{n}\right)^{nt} = \lim_{n \to \infty} \left[\left(1 + \frac{r}{n}\right)^n\right]^t = [e^r]^t = e^{rt}$$

as required in equation (1.1) of Chapter 1.

The numerical value of e, or e^1, is obtained from

$$e = \lim_{n \to \infty} \left(1 + \frac{1}{n}\right)^n$$

$$= 1 + 1 + \frac{1}{2!} + \frac{1}{3!} + \frac{1}{4!} + \cdots$$

$$= 1 + 1 + \frac{1}{2} + \frac{1}{6} + \frac{1}{24} + \frac{1}{120} + \cdots$$

$$= 2.71828\ldots$$

Confidence Limits

The probability distribution of N_t is approximately normal with expectation, or mean, $\mathscr{E}(N_t) = N_0 e^{rt}$, and variance

$$\text{var}\,(N_t) = N_0(b + d)\,e^{rt}(e^{rt} - 1)/r,$$

as given in equation (1.2). (The derivation of this equation will be found in Pielou (1969)). The variance (which is the square of the standard deviation) is a measure of the "spread" of the probability distribution.

We now inquire into the implications of this statement, using the numerical example described in Chapter 1. This population had birth rate $b = 0.2434$ and death rate $d = 0.0611$ so that $r = b - d = 0.1823$. The population's size was $N_0 = 1000$ in 1970 and we wished to make predictions about N_{12}, its size in 1982.

The probability distribution of N_{12} is shown graphically in Figures 1.4a and 1.4b. First consider Figure 1.4a and suppose we wished to find the probability that the size of the population in 1982 would fall between *chosen* lower and upper limits (denoted by x_L and x_U) with, for instance, $x_L = 8800$ and $x_U = 9100$. The required probability is equal to the area shown shaded in Figure 1.4a expressed as a fraction of the total area under the curve. In this example the probability is 0.091.

Now consider the converse problem. We are given a probability, 0.95 or 95% for example, and asked to state what are the limits between which we expect to find the size of the population in 1982 with *this* probability. Whereas, before, we were told the limits and asked to find the probability, now we are given the probability and asked to find the limits (with the proviso that they are to be symmetrical about the mean). The answer is shown graphically in Figure 1.4b. The area of the shaded portion is 95% of the total area under the curve and it can be proved that for this to be true the lower and upper limits (the 95% confidence limits) are

$$x_L = \mathscr{E}(N_{12}) - 1.96\,\sqrt{\text{var}\,(N_{12})} \quad \text{and} \quad x_U = \mathscr{E}(N_{12}) + 1.96\,\sqrt{\text{var}\,(N_{12})}$$

or 8242 and 9586 as already given.

To obtain 99% confidence limits all that is necessary is to replace the numerical coefficient 1.96 by 2.58. Then the 99% confidence limits are

$$\mathscr{E}(N_{12}) \pm 2.58 \sqrt{\mathrm{var}\,(N_{12})} \quad \text{or} \quad 8028 \text{ to } 9800 \text{ individuals.}$$

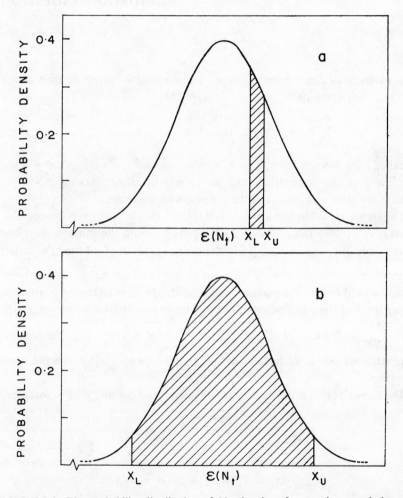

FIGURE 1.4   The probability distribution of N_t, the size of a growing population at some future time t. $\mathscr{E}(N_t)$ is the expected value. In both graphs the area of the hatched part (as a proportion of the total area under the curve) is the probability that N_t will lie between the values X_L and X_U. In (b) this area is 0.95 of the total and X_L and X_U are thus the lower and upper 95% confidence limits

Summation Notation

THE EXPRESSION Σc_x means the sum of a sequence of symbols each of the form c_x. For example

$$\sum_{x=0}^{6} c_x = c_0 + c_1 + c_2 + c_3 + c_4 + c_5 + c_6$$

is called "the sum of the c_x's from $x = 0$ to $x = 6$"; $x = 0$ and $x = 6$ are the limits of the summation and the sum is of all c_x's having subscripts (which are whole numbers) in the range 0 to 6 inclusive.

The limits are often not given in full if they are obvious from the context. Thus if c_x denotes the proportion of a population in the xth age class one could write $\sum_x c_x$ or simply $\sum c_x$ to denote summation of *all* the c_x's. Either expression denotes the sum of the proportions in all age classes. These forms could be used even if the maximum attainable age were unknown so that the upper limit of the summation could not be given. Alternatively, one might write $\sum_{x=0}^{\infty} c_x$ for the sum of all the c_x's. For values of x greater than the greatest attainable age one must have $c_x = 0$ so that $\sum_{x=0}^{\infty} c_x$ and $\sum c_x$ denote the same thing.

The subscript that takes a sequence of values need not be "x". For instance,

$$\sum_{j=2}^{5} n_j = n_2 + n_3 + n_4 + n_5$$

and

$$\sum_{a=3}^{7} x_a = x_3 + x_4 + x_5 + x_6 + x_7.$$

Sometimes the terms to be summed have two subscripts; one or the other or both may take a sequence of values. Thus

$$\sum_{x=0}^{3} n_{xt} = n_{0t} + n_{1t} + n_{2t} + n_{3t};$$

$$\sum_{t=1}^{5} n_{xt} = n_{x1} + n_{x2} + n_{x3} + n_{x4} + n_{x5};$$

$$\sum_{x=1}^{3} \sum_{t=2}^{4} n_{xt} = \sum_{x=1}^{3} (n_{x2} + n_{x3} + n_{x4})$$

$$= (n_{12} + n_{13} + n_{14}) + (n_{22} + n_{23} + n_{24})$$

$$+ (n_{32} + n_{33} + n_{34}).$$

It is easy to see that

$$\sum_{x=1}^{3} \sum_{t=2}^{4} n_{xt} = \sum_{t=2}^{4} \sum_{x=1}^{3} n_{xt}.$$

That is, the order of the summation signs does not matter.

It sometimes happens that the term after a summation sign is a constant (and therefore has no subscript). Thus let k be a constant.

Then

$$\sum_{j=1}^{r} k = k + k + \cdots + k \quad \text{with } r \text{ terms}$$
$$= rk.$$

The terms summed can be functions of simpler terms as shown in the following three examples which illustrate further points to notice.

1.
$$\sum_{j=1}^{4} 6x_j^3 = 6x_1^3 + 6x_2^3 + 6x_3^3 + 6x_4^3.$$

It is also seen that

$$\sum_{j=1}^{4} 6x_j^3 = 6 \sum_{j=1}^{4} x_j^3$$

and in general that

$$\sum_j cy_j = c \sum_j y_j.$$

2.
$$\sum_{i=2}^{3} x_i y_i = x_1 y_1 + x_2 y_2 + x_3 y_3.$$

Observe that

$$\sum_i x_i y_i \neq \left(\sum_i x_i \right) \left(\sum_i y_i \right)$$

since

$$\left(\sum_{i=1}^{3} x_i \right) \left(\sum_{i=1}^{3} y_i \right) = (x_1 + x_2 + x_3)(y_1 + y_2 + y_3)$$

$$= x_1 y_1 + x_1 y_2 + x_1 y_3 + x_2 y_1 + x_2 y_2$$

$$+ x_2 y_3 + x_3 y_1 + x_3 y_2 + x_3 y_3.$$

Likewise

$$\sum_j (x_j - a)^2 \neq \left\{ \sum_j (x_j - a) \right\}^2$$

for

$$\sum_{j=1}^{3} (x_j - a)^2 = (x_1 - a)^2 + (x_2 - a)^2 + (x_3 - a)^2$$

$$= x_1^2 + x_2^2 + x_3^2 - 2a(x_1 + x_2 + x_3) + 3a^2$$

whereas

$$\left\{ \sum_{j=1}^{3} (x_j - a) \right\}^2 = (x_1 - a + x_2 - a + x_3 - a)^2$$

$$= (x_1 + x_2 + x_3 - 3a)^2$$

$$= x_1^2 + x_2^2 + x_3^2 + 9a^2 + 2(x_1 x_2 + x_1 x_3 + x_2 x_3)$$

$$- 6a (x_1 + x_2 + x_3).$$

Summarizing the above in the form of a general rule: The sum of a function is not equal to the function of the sum.

3. The only exception to the above rule that need concern us here is this:

$$\sum_i (x_i + y_i) = \sum x_i + \sum y_i.$$

Thus, if the summation were from $i = 0$ to $i = 2$ we have

$$\sum_{i=0}^{2} (x_i + y_i) = (x_0 + y_0) + (x_1 + y_1) + (x_2 + y_2)$$

$$= (x_0 + x_1 + x_2) + (y_0 + y_1 + y_2)$$

$$= \sum_{i=0}^{2} x_i + \sum_{i=0}^{2} y_i.$$

Similarly

$$\sum_{t=0}^{4} (b_t + c) = \sum_{t=0}^{4} b_t + \sum_{t=0}^{4} c = \sum_{t=0}^{4} b_t + 5c.$$

Matrix Multiplication

An $r \times c$ matrix is an array of rc "elements" (numbers or symbols) arranged in r rows and c columns. For example

$$\mathbf{A} = \begin{pmatrix} a_{11} & a_{12} & a_{13} \\ a_{21} & a_{22} & a_{23} \end{pmatrix} \quad \text{is a } 2 \times 3 \text{ matrix.}$$

To represent a matrix by a single symbol it is customary to use a letter printed in boldface. When a matrix is written out with each element labelled with a pair of subscripts the first number of the pair denotes the row, and the second number the column, in which the element occurs. Thus a_{23} is the element in the 2nd row and 3rd column of $\mathbf{A}$.

If a matrix has equal numbers of rows and columns, i.e. if $r = c$, the matrix is square, $r \times r$.

If there are r rows and only one column $(c = 1)$ the $r \times 1$ "matrix" is known as an r-element column vector.*

If there is one row $(r = 1)$ and c columns the $1 \times c$ "matrix" is known as a c-element row vector.

Thus

$$\mathbf{B} = \begin{pmatrix} b_{11} & b_{12} \\ b_{21} & b_{22} \end{pmatrix} \quad \text{is a } 2 \times 2 \text{ square matrix,}$$

$$\mathbf{e} = \begin{pmatrix} e_1 \\ e_2 \\ \vdots \\ e_r \end{pmatrix} \quad \text{is an } r\text{-element column vector}$$

and $\mathbf{d} = (d_1\ d_2\ \cdots\ d_c)$ is a c-element row vector. For vectors, a single subscript suffices to label the elements. It is customary to use boldface capital letters for matrices with more than one row and column; and boldface lower case letters for vectors.

---

* The name "vector" is used for a single row or column of numbers for the following reason. Any n-element vector (a list of n numbers) can be regarded as the coordinates of a point in n-dimensional space; the directed line joining the origin of the coordinates to this point is a vector in the physicist's sense.

The transpose of a matrix is obtained by writing its rows as columns and vice versa. If $\mathbf{E}$ is a matrix, its transpose is written $\mathbf{E}'$ (note the prime). For example the transpose of the matrices shown above are as follows:

$$\mathbf{A}' = \begin{pmatrix} a_{11} & a_{21} \\ a_{12} & a_{22} \\ a_{13} & a_{23} \end{pmatrix}, \quad \mathbf{B}' = \begin{pmatrix} b_{11} & b_{21} \\ b_{12} & b_{22} \end{pmatrix}$$

$$\mathbf{e}' = (e_1 \; e_2 \; ... \; e_r)$$

$$\mathbf{d}' = \begin{pmatrix} d_1 \\ d_2 \\ \vdots \\ d_c \end{pmatrix}$$

When two matrices are multiplied together, the order in which they are written is important, since, in general, $\mathbf{GH} \neq \mathbf{HG}$.

The matrix product $\mathbf{GH}$ is called $\mathbf{G}$ "post-multiplied" by $\mathbf{H}$ or $\mathbf{H}$ "pre-multiplied" by $\mathbf{G}$. This product exists *only* if the number of columns of $\mathbf{G}$ is equal to the number of rows of $\mathbf{H}$. The product itself, say $\mathbf{GH} = \mathbf{K}$, is a matrix that has the same number of rows as $\mathbf{G}$ and the same number of columns as $\mathbf{H}$. These rules become self-evident when we consider how a product is formed. For example, suppose $\mathbf{G}$ is a 2×4 matrix and $\mathbf{H}$ a 4×3 matrix; then the product $\mathbf{GH}$ is a 2×3 matrix and it is obtained as follows:

$$\mathbf{GH} = \begin{pmatrix} g_{11} & g_{12} & g_{13} & g_{14} \\ g_{21} & g_{22} & g_{23} & g_{24} \end{pmatrix} \begin{pmatrix} h_{11} & h_{12} & h_{13} \\ h_{21} & h_{22} & h_{23} \\ h_{31} & h_{32} & h_{33} \\ h_{41} & h_{42} & h_{43} \end{pmatrix} = \begin{pmatrix} k_{11} & k_{12} & k_{13} \\ k_{21} & k_{22} & k_{23} \end{pmatrix} = \mathbf{K}$$

with

$$k_{11} = g_{11}h_{11} + g_{12}h_{21} + g_{13}h_{31} + g_{14}h_{41}$$

$$k_{12} = g_{11}h_{12} + g_{12}h_{22} + g_{13}h_{32} + g_{14}h_{42}$$

$$\cdots \cdots \cdots \cdots \cdots \cdots \cdots \cdots \cdots \cdots \cdots \cdots$$

$$k_{23} = g_{21}h_{13} + g_{22}h_{23} + g_{23}h_{33} + g_{24}h_{43}.$$

That is to say, the element k_{ij} of $\mathbf{K}$ is obtained by multiplying together successive pairs of elements from the ith row of $\mathbf{G}$ and jth column of $\mathbf{H}$, and adding these products together. Clearly, if the number of elements in the ith row of $\mathbf{G}$ did not correspond with the number of elements in the jth column of $\mathbf{H}$, the term k_{ij} would be undefined. Thus since $\mathbf{H}$ has 3 columns and $\mathbf{G}$ has 2 rows one cannot form the product $\mathbf{HG}$.

Another consequence of the method of matrix multiplication is the following: If $\mathbf{J}$ is an $r \times c$ matrix, its transpose $\mathbf{J}'$ is a $c \times r$ matrix. Then the product $\mathbf{JJ}'$ is an $r \times r$ square matrix; and the product $\mathbf{J}'\mathbf{J}$ is a $c \times c$ square matrix.

To generalize: If **M** is a $p \times q$ matrix and **N** is a $q \times s$ matrix the product **NM** does not exist; and the product **MN** is a $p \times s$ matrix calculated as follows:

$$
\begin{bmatrix} m_{11}\ m_{12}\ \ldots\ m_{1q} \\ m_{21}\ m_{22}\ \ldots\ m_{2q} \\ \cdots \cdots \cdots \\ m_{p1}\ m_{p2}\ \ldots\ m_{pq} \end{bmatrix}
\begin{bmatrix} n_{11}\ n_{12}\ \ldots\ n_{1s} \\ n_{21}\ n_{22}\ \ldots\ n_{2s} \\ \cdots \cdots \cdots \\ n_{q1}\ n_{q2}\ \ldots\ n_{qs} \end{bmatrix}
=
\begin{bmatrix}
\sum_{j=1}^{q} m_{1j}n_{j1} & \sum_{j=1}^{q} m_{1j}n_{j2} & \ldots & \sum_{j=1}^{q} m_{1j}n_{js} \\
\sum_{j=1}^{q} m_{2j}n_{j1} & \sum_{j=1}^{q} m_{2j}n_{j2} & \ldots & \sum_{j=1}^{q} m_{2j}n_{js} \\
\cdots \cdots \cdots \cdots \cdots \cdots \\
\sum_{j=1}^{q} m_{pj}n_{j1} & \sum_{j=1}^{q} m_{pj}n_{j2} & \ldots & \sum_{j=1}^{q} m_{pj}n_{js}
\end{bmatrix}
$$

These rules apply without modification to vectors. For example if **z** is an r-element column vector (i.e. an $r \times 1$ matrix) and **z**′ is its transpose (an r-element row vector or $1 \times r$ matrix), the product **zz**′ is an $r \times r$ square matrix, and the product **z**′**z** is a 1×1 matrix (simply an "ordinary" number or *scalar*).

Thus

$$
\mathbf{zz}' = \begin{pmatrix} z_1 \\ z_2 \\ \vdots \\ z_r \end{pmatrix} (z_1 z_2 \ldots z_r) = \begin{pmatrix} z_1^2 & z_1 z_2 & \ldots & z_1 z_r \\ z_2 z_1 & z_2^2 & \ldots & z_2 z_r \\ \cdots \cdots \cdots \cdots \cdots \\ z_r z_1 & z_r z_2 & \ldots & z_r^2 \end{pmatrix}
$$

and

$$
\mathbf{z}'\mathbf{z} = (z_1 z_2 \ldots z_r) \begin{pmatrix} z_1 \\ z_2 \\ \vdots \\ z_r \end{pmatrix} = z_1^2 + z_2^2 + \cdots + z_r^2 = \sum_{j=1}^{r} z_j^2 .
$$

Notice that the system of labelling matrix elements with subscripts employed in this appendix is appropriate to the examples here used in which the matrices have no concrete meaning and have been written down merely to illustrate the operation of multiplication.

For the projection matrices and age distribution vectors of Chapter 2 the *rules* described in this appendix apply, but the subscripts represent ages and times (as explained in the text) and not simply the position of an element within a matrix (as in this appendix).

Variances and Covariances

CONSIDER A SET of N numbers, $X_1, X_2, \ldots, X_N$. They might represent measurements on N different objects, for example the weights of N birds. The mean, or average, of the set is

$$\bar{X} = \frac{1}{N} \sum_j X_j.$$

The variance of the set, var (X), is the mean of the square of the deviation of each number from $\bar{X}$; in symbols:

$$\text{var}(X) = \frac{1}{N} \sum_j (X_j - \bar{X})^2.$$

Now let each of the original measurements be expressed as a deviation from the mean; that is, for the original set of numbers substitute the set $x_1, x_2, \ldots, x_N$, with $x_j = X_j - \bar{X}$. (Consequently some of the x_j are positive and some negative, and all together they sum to zero.)
Then

$$\text{var}(X) = \frac{1}{N} \sum_j x_j^2.$$

An alternative way of writing the foregoing that will be useful later is as follows. Let the set of numbers $x_1, x_2, \ldots, x_N$ be written as a row vector with N elements, thus

$$\mathbf{x} = (x_1 \ x_2 \ \cdots \ x_N).$$

Writing $\mathbf{x}'$ for the transpose of $\mathbf{x}$, so that $\mathbf{x}'$ is an N-element column vector (see Appendix 2.2, page 367), it is clear that

$$\mathbf{x}\mathbf{x}' = \sum_j x_j^2$$

and hence that

$$\text{var}(X) = \frac{1}{N} \mathbf{x}\mathbf{x}' \tag{1}$$

Imagine, next, a set of N *pairs* of numbers, (X_{11}, X_{21}), (X_{12}, X_{22}), (X_{13}, X_{23}), ..., (X_{1N}, X_{2N}). For instance, to use an ecological example again, each pair of numbers might represent the height and diameter, respectively, of one of the N trees in a forest. Each X has two subscripts. The first, which is 1 or 2, denotes the kind of quantity represented: 1 for height and 2 for diameter. The second subscript, which is 1, 2, ... or N, shows which of the N individual trees yielded that quantity. Thus X_{1i} is the height of the ith tree and X_{2j} is the diameter of the jth tree. Also $\bar{X}_1 = \dfrac{1}{N} \sum_j X_{1j}$ is the mean height of all N trees and $\bar{X}_2 = \dfrac{1}{N} \sum_j X_{2j}$ is their mean diameter.

It would reduce the number of subscripts to use X's and Y's (instead of X_1's and X_2's) for the heigths and diameters. The reason for using what seems a more complicated symbolism will appear later.

Now, given two different observations on each of N individuals, we wish to know not only the variances of the X_1's and the X_2's, but also the covariance of the the variables, cov (X_1, X_2). This is a measure of the extent to which two variables are related to each other and vary together (or "covary") as explained on page 71. The covariance of X_1 and X_2 is defined as

$$\text{cov}\,(X_1, X_2) = \frac{1}{N} \sum_j (X_{1j} - \bar{X}_1)(X_{2j} - \bar{X}_2).$$

Obviously

$$\text{cov}\,(X_1, X_2) = \text{cov}\,(X_2, X_1).$$

As before, it is convenient to express the original measurements as deviations from their respective means. That is, we put

$$x_{1j} = X_{1j} - \bar{X}_1 \quad \text{and} \quad x_{2j} = X_{2j} - \bar{X}_2.$$

Then

$$\text{cov}\,(X_1, X_2) = \frac{1}{N} \sum_j x_{1j} x_{2j}.$$

Now let us find an expression for covariance analogous to equation (1). That is, we require a formula for covariance that uses matrix notation.

The original N pairs of numbers, expressed as deviations from their respective means, are

$$(x_{11}, x_{21}), (x_{12}, x_{22}), (x_{13}, x_{23}), ..., (x_{1N}, x_{2N}).$$

Let us now write them as a $2 \times N$ "data matrix", $\mathbf{X}$.

That is

$$\mathbf{X} = \begin{pmatrix} x_{11} & x_{12} & x_{13} & ... & x_{1N} \\ x_{21} & x_{22} & x_{23} & ... & x_{2N} \end{pmatrix}.$$

Each of the N columns of $\mathbf{X}$ represents the pair of measurements made on a single individual. Postmultiplying $\mathbf{X}$ by its transpose gives

$$\mathbf{XX'} = \begin{pmatrix} \sum_j x_{1j}^2 & \sum_j x_{1j}x_{2j} \\ \sum_j x_{2j}x_{1j} & \sum_j x_{2j}^2 \end{pmatrix}$$

whence

$$\frac{1}{N}\mathbf{XX'} = \begin{pmatrix} \text{var }(X_1) & \text{cov }(X_1, X_2) \\ \text{cov }(X_2, X_1) & \text{var }(X_2) \end{pmatrix}.$$

The convenience and compactness of matrix notation becomes clear when we consider data consisting of *several*, say k, measurements (instead of merely two) on each of N individuals. Thus suppose, for each of N trees, we measure height, diameter, age, width of crown, ..., thickness of bark, for a total of k measurements on each tree. Write each variable as a deviation from its mean over all the N trees. The "data matrix" is a $k \times N$ matrix, $\mathbf{X}$, in which each column represents all the k measurements made on a single tree, and each row represents the measurements of a single variable, say height, on all N trees.

That is,

$$\mathbf{X} = \begin{pmatrix} x_{11} & x_{12} & ... & x_{1N} \\ x_{21} & x_{22} & ... & x_{2N} \\ x_{k1} & x_{k2} & ... & x_{kN} \end{pmatrix}.$$

It will now be found, as was found above for the particular case when $k = 2$, that

$$\frac{1}{N}\mathbf{XX'} = \begin{pmatrix} \text{var }(X_1) & \text{cov }(X_1, X_2) & ... & \text{cov }(X_1, X_k) \\ \text{cov }(X_2, X_1) & \text{var }(X_2) & ... & \text{cov }(X_2, X_k) \\ \cdot\cdot\cdot\cdot\cdot\cdot\cdot\cdot\cdot\cdot\cdot\cdot\cdot\cdot\cdot\cdot\cdot\cdot\cdot\cdot\cdot\cdot\cdot\cdot\cdot\cdot\cdot\cdot\cdot\cdot\cdot\cdot \\ \text{cov }(X_k, X_1) & \text{cov }(X_k, X_2) & ... & \text{var }(X_k) \end{pmatrix}.$$

This is the "covariance matrix" of the data.

Table 3.3 on page 43 is an actual example of a covariance matrix and Appendix 3.2 (below) describes how it was calculated. Before doing the calculations, however, it is necessary to derive an expression for the variance of the sum of several variables, as promised on page 41. It will be shown that if several variables can be added to form a new variable, then the variance of the new variable is equal to the sum of the elements in the covariance matrix of the component variables. It would be pointless, of course, to add together such variables as the height, diameter and age of a tree, for example, since the sum of these numbers would be meaningless. Three examples of cases in which the sum of a set of variables is itself a variable of interest are the following:

a)   The length of life of an insect is the sum of the durations of each of its successive stages.

b)   The weight of an animal is the sum of the weights of its component organs.

c)   The logarithm of the trend index is the sum of the logarithms of its factors as shown in equations (3.3) and (3.5).

First, consider the sum of two variables: thus, let us put

$$X_{1j} + X_{2j} = Y_j \quad \text{for all } j,$$

and find the variance of the Y's. As before, we shall measure the variables as deviations from their means by putting

$$x_{1j} = X_{1j} - \bar{X}_1 \quad \text{and} \quad x_{2j} = X_{2j} - \bar{X}_2.$$

Then

$$\text{var}(Y) = \text{var}(X_1 + X_2) = \frac{1}{N} \sum_j (x_{1j} + x_{2j})^2$$

whence

$$\text{var}(Y) = \frac{1}{N} \sum_j x_{1j}^2 + \frac{2}{N} \sum_j x_{1j} x_{2j} + \frac{1}{N} \sum_j x_{2j}^2.$$

Thus

$$\text{var}(Y) = \text{var}(X_1) + 2\,\text{cov}(X_1, X_2) + \text{var}(X_2).$$

It is a straightforward matter to generalize the above argument. Suppose each Y is the sum of k component variables, that is

$$Y_j = X_{1j} + X_{2j} + \cdots + X_{kj} \quad \text{for all } j.$$

Then

$$\text{var}(Y) = \frac{1}{N} \sum_j (x_{1j} + x_{2j} + \cdots + x_{kj})^2$$

$$= \frac{1}{N} \sum_j x_{1j}^2 + \frac{1}{N} \sum_j x_{2j}^2 + \cdots + \frac{1}{N} \sum_j x_{kj}^2$$

$$+ \frac{2}{N} \sum_j x_{1j} x_{2j} + \frac{2}{N} \sum_j x_{1j} x_{3j} + \cdots + \frac{2}{N} \sum_j x_{k-1,j} x_{kj}$$

$$= \text{var}(X_1) + \text{var}(X_2) + \cdots + \text{var}(X_k)$$

$$+ 2\,\text{cov}(X_1, X_2) + 2\,\text{cov}(X_1, X_3) + \cdots + 2\,\text{cov}(X_{k-1}, X_k).$$

Or

$$\text{var}(Y) = \sum_{i=1}^{k} \text{var}(X_i) + 2 \sum_{i=1}^{j} \sum_{j=i+1}^{k} \text{cov}(X_i, X_j).$$

In words, the variance of the sum is equal to the sum of the variances of the component variables plus twice the covariance of every pair of component variables.

The Analysis of Generation Survival
in the Oystershell Scale

THE PURPOSE of this appendix is to show how Table 3.3 (page 43) was arrived at. The raw data consisted of ten life tables, one for each tree sampled, of which Table 3.4 is an example. The numbers shown here have been rounded for brevity.

TABLE 3.4   Life table for the oystershell scale on an apple tree

Stage	Estimated number present	Trend index factors, T	log $100T$
Eggs (1963)	2971.1		
		$S_E = 0.2857$	1.4559
Small larvae	848.9		
		$S_S = 0.1034$	1.0144
Big larvae	87.8		
		$S_B = 0.3536$	1.5485
Adults	31.0		
		$F = 10.6839$	3.0287
Eggs (1964)	331.5		

The second column of Table 3.4 gives the estimated numbers present of each of the stages listed in the first column. The trend index, I, is the ratio of the number of eggs present in 1964 to the number in 1963 and thus $I = 331.5/2971.1 = 0.1115$. The third column gives the proportions surviving from each stage to the next, S_E, S_S and S_B, and the fertility factor, F.

For example,
$$S_E = 848.9/2971.1 = 0.2857.$$

The four proportions are the factors of the trend index. The last column gives the logarithm (to base 10) of 100 times the factor in the third column. For example, $\log (100 \times 0.2857) = 1.4559$. Multiplying each factor by 100 is equivalent to adding 2 to its logarithm and is done for convenience: it prevents any of the numbers in this column from being negative, but at the same time has no effect on the variances and covariances that are to be calculated. In what follows we shall call these numbers "the components of $\log I$" though in fact they sum to $\log (100S_E \times 100S_S \times 100S_B \times 100F)$ $= \log (10^8 \times I) = 7.046$.

As explained above, the total data consisted of ten tables of which Table 3.4 is a single example. Using the data from all ten trees, the mean value of each component of $\log I$ was calculated and the means are shown in Table 3.5.

TABLE 3.5   Means of the components of $\log I$

		Means of ten values of:		
$\log 100S_E$	$\log 100S_S$	$\log 100S_B$	$\log 100F$
1.1875	1.1283	1.5750	2.7978

We can now write down the 4×10 data matrix, $\mathbf{X}$, and its transpose, the 10×4 matrix $\mathbf{X}'$ (see Appendix 3.1, page 372). $\mathbf{X}'$ is shown in Table 3.6. The data in Table 3.4 refer to the ninth of the ten trees and are therefore to be found in the ninth row of $\mathbf{X}'$ (which is the ninth column of $\mathbf{X}$). It is seen that the elements of this row are, respectively

$$1.4559 - 1.1875 = 0.2684$$

$$1.0144 - 1.1283 = -0.1139$$

$$1.5485 - 1.5750 = -0.0265$$

$$3.0287 - 2.7978 = 0.2309.$$

That is, each element is the difference between a component of $\log I$ for the ninth tree (from Table 3.4) and the mean of this component over all trees (from Table 3.5). Note also that the elements in each column of $\mathbf{X}'$ sum to zero; this is because they are deviations from the column mean.

TABLE 3.6   The transpose, $\mathbf{X}'$, of the data matrix

0.0869	0.0152	−0.2648	−0.1656
0.1588	−0.1697	−0.1343	−0.0319
−0.1055	0.0118	0.1327	0.2322
−0.3251	0.0745	−0.0058	−0.3260
0.0329	0.2170	−0.4212	−0.2482
−0.1733	0.0103	0.2021	−0.4310
−0.1926	−0.4473	−0.0655	0.3300
0.1305	0.0256	0.0451	0.0479
0.2684	−0.1139	−0.0265	0.2309
0.1190	0.3766	0.5381	0.3617

It remains to compute the 4×4 covariance matrix $\dfrac{1}{N}\mathbf{XX'}$ with $N = 10$.

Since $\operatorname{cov}(X_1, X_2) = \operatorname{cov}(X_2, X_1)$ the matrix is symmetrical; i.e. the element in the ith row and jth column is equal to the element in the jth row and ith column.

TABLE 3.7   The covariance matrix

0.0321	0.0058	− 0.0030	0.0176
0.0058	0.0437	0.0167	−0.0113
−0.0030	0.0167	0.0621	0.0267
0.0176	−0.0113	0.0267	0.0731

The sum of all 16 elements is var $(\log I) = 0.3160$.

We are now in a position to derive the entries in Table 3.3 on page 43, which shows the variances and covariances of the components of $\log I$ as percentages of var $(\log I)$.

Thus

$$\operatorname{var}(\log S_E) = \frac{0.0321 \times 100}{0.3160} = 10.16\% \quad \text{of var } (\log I)$$

and

$$2\operatorname{cov}(\log S_E, \log S_S) = \frac{2 \times 0.0058 \times 100}{0.3160} = 3.67\% \quad \text{of var } (\log I).$$

Because of the symmetry of the covariance matrix there is no need to write $\operatorname{cov}(X_1, X_2)$ and $\operatorname{cov}(X_2, X_1)$ (for example) separately. They are added and the sum is entered in one cell of the table.

Solving some Differential Equations for Population Growth

THE EQUATIONS to be solved (see pages 47, 48) are

$$\frac{1}{N} \frac{dN}{dt} = r \tag{1}$$

$$\frac{1}{N} \frac{dN}{dt} = r - sN \tag{2}$$

$$\frac{1}{N} \frac{dN}{dt} = uN. \tag{3}$$

In what follows the A's (with different subscripts) are constants of integration; N_0 denotes population size at time $t = 0$.

To solve equation (1) put

$$\int \frac{dN}{N} = \int r\, dt.$$

Integrating gives

$$\ln N = rt + A_1$$

whence

$$N = A_2 e^{rt} \quad \text{where} \quad A_2 = \text{antilog}_e A_1.$$

When

$$t = 0, \quad N = N_0 = A_2.$$

Thus

$$N = N_0 e^{rt}.$$

This is the equation for *exponential growth*. Checking this result by differentiating it gives

$$\frac{dN}{dt} = r(N_0 e^{rt}) = rN$$

so that

$$\frac{1}{N} \frac{dN}{dt} = r$$

which is equation (1).

To solve equation (2) put

$$\int \frac{dN}{N(r - sN)} = \int dt.$$

The term on the left can be separated into partial fractions giving

$$\int \left\{ \frac{1}{rN} + \frac{s}{r(r - sN)} \right\} dN = \int dt$$

or

$$\int \frac{dN}{N} + \int \frac{s\, dN}{(r - sN)} = \int r\, dt.$$

Integrating gives

$$\ln N - \ln (r - sN) = rt + A_3.$$

Then

$$\frac{N}{r - sN} = A_4 e^{rt}$$

whence

$$N = \frac{r}{s + e^{-rt}/A_4} = \frac{(r/s)}{1 + e^{-rt}/(A_4 s)}.$$

But $r/s = K$ (see page 48) and we may put $1/(A_4 s) = A_5$, another constant; therefore

$$N = \frac{K}{1 + A_5 e^{-rt}}.$$

Now when $t = 0$,

$$N = N_0 = \frac{K}{1 + A_5}$$

whence,

$$A_5 = \frac{K - N_0}{N_0}$$

and finally

$$N = \frac{K}{1 + \left(\dfrac{K - N_0}{N_0} \right) e^{-rt}}$$

which is the *logistic equation*.

Checking this result by differentiating it gives

$$\frac{dN}{dt} = \frac{K}{1 + A_5 e^{-rt}} \cdot \frac{A_5 r e^{-rt}}{1 + A_5 e^{-rt}} = N \cdot \frac{A_5 r e^{-rt}}{1 + A_5 e^{-rt}}$$

Thus

$$\frac{1}{N}\frac{dN}{dt} = r \cdot \frac{A_5 e^{-rt}}{1 + A_5 e^{-rt}}.$$

Now, since $N = \dfrac{K}{1 + A_5 e^{-rt}}$ it is seen that

$$1 + A_5 e^{-rt} = K/N \quad \text{and} \quad A_5 e^{-rt} = (K - N)/N$$

$$\therefore \quad \frac{1}{N}\frac{dN}{dt} = \frac{r(K - N)/N}{K/N} = \frac{r(K - N)}{K}.$$

But

$$K = r/s$$

$$\therefore \quad \frac{1}{N}\frac{dN}{dt} = \frac{r\left(\dfrac{r}{s} - N\right)}{\dfrac{r}{s}}$$

$$= r - sN$$

which is equation (2).

To solve equation (3) put

$$\int \frac{dN}{N^2} = \int u \, dt.$$

Integrating gives

$$-\frac{1}{N} = ut + A_6.$$

Since, when

$$t = 0, \quad N = N_0,$$

$$-\frac{1}{N_0} = A_6.$$

$$\therefore \quad \frac{1}{N} = \frac{1}{N_0} - ut$$

and thus

$$N = \frac{N_0}{1 - uN_0 t}$$

is the required integral equation.

Checking this result by differentiating it gives

$$\frac{dN}{dt} = \frac{uN_0^2}{(1 - uN_0 t)^2} = uN^2$$

and thus

$$\frac{1}{N}\frac{dN}{dt} = uN$$

which is equation (3).

Calculation of a Serial Correlation Coefficient

CONSIDER A sequence of n observations $x_1, x_2, \ldots, x_n$. The sth order serial correlation coefficient, r_s, is the correlation coefficient between every observation and the sth subsequent one. It is calculated as follows. Let the data be written in two sets: Set 1 consists of the first $n - s$ observations and Set 2 consists of the last $n - s$ observations. The two sets, their cross-products and sums are:

Set 1	Set 2	Cross-products
x_1	x_{s+1}	$x_1 x_{s+1}$
x_2	x_{s+2}	$x_2 x_{s+2}$
$\vdots$	$\vdots$	$\vdots$
x_{n-s}	x_n	$x_{n-s} x_n$

Sums: $\displaystyle\sum_1^{n-s} x_i \qquad \sum_1^{n-s} x_{s+i} \qquad \sum_1^{n-s} x_i x_{s+i}$

The covariance of the two sets (see page 371) is

$$C = \frac{1}{(n-s)} \sum_{i=1}^{n-s} x_i x_{s+i} - \frac{1}{(n-s)^2} \left(\sum_1^{n-s} x_i \right) \left(\sum_1^{n-s} x_{s+i} \right).$$

The variance of the whole sequence of n observations is

$$V = \frac{1}{n} \sum_{i=1}^{n} x_i^2 - \frac{1}{n^2} \left(\sum_1^{n} x_i \right)^2$$

and the required serial correlation coefficient is $r_s = C/V$.

Confidence Intervals for Population Estimates

CONSIDER the quantity

$$\frac{\hat{Y} - Y}{s_{\hat{Y}}} = T, \quad \text{say,}$$

mentioned on page 102. Here Y is the true but unknown size of a natural population; $\hat{Y}$ is the estimate of Y based on observations on a sample; and $s_{\hat{Y}}$ is the calculated standard error of Y.

The quantity T has a probability distribution known as the "t distribution" whose shape depends *only* on ν, the number of degrees of freedom of $s_{\hat{Y}}$. Figure 6.2a shows the shape of the t distribution for $\nu = 3$ and 13. For large values of ν the distribution is indistinguishable from the standard normal distribution. Now assume that $\hat{Y}$, $s_{\hat{Y}}$ and ν are known and that the t distribution corresponding to this value of ν is as shown in Figure 6.2b. Let the value of T be calculated.

If the estimate $\hat{Y}$ were equal to the true population value Y, we should have $T = 0$. Obviously the actual value of T will most likely be some small positive or negative value. Indeed, the probability is 0.95 that T will be found to lie somewhere between the values $-t_{0.025}$ and $+t_{0.025}$ marked on the abscissa of Figure 6.2b, since the symmetrical area (shown shaded) under the

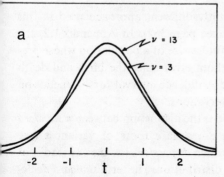
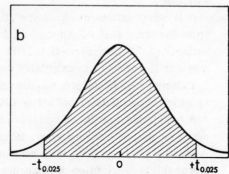

FIGURE 6.2 (*a*) The probability density of the t distribution for $\nu = 3$ and $\nu = 13$. (*b*) the hatched area, between $-t_{0.025}$ and $t_{0.025}$, is 95% of the total area under the curve

381

curve between these two values is 95% of the total area. This states in words what equation (6.1) states in symbols; equation (6.2) then follows.

Table 6.6 gives the values of $t_{0.025}$, required for calculating 95% confidence intervals. The tabulated numbers are known as percentage points (in this case 95%) of the t distribution.

TABLE 6.6   Percentage points of the t distribution

Degrees of freedom v	$t_{0.025}$	Degrees of freedom v	$t_{0.025}$
2	4.303	16	2.120
3	3.182	18	2.101
4	2.776	20	2.086
5	2.571	22	2.074
6	2.447	24	2.064
7	2.365	26	2.056
8	2.306	28	2.048
9	2.262	30	2.042
10	2.228	40	2.021
12	2.179	60	2.000
14	2.145	∞	1.960

Although when $v > 60$, $t_{0.025} < 2.000$ it is, nevertheless, common practice to use the value 2 instead of an exact value. That is, the 95% confidence interval is given as:

Calculated mean $\pm$ (2 × standard error).

This is as precise as one can hope to be, since $s_{\bar{y}}$ itself is subject to sampling variation.

It is worth commenting on the slightly different approach used in this appendix from that of Appendix 1.2 (see page 362). In Appendix 1.2, the variance of the number predicted (the future size of a population whose present size is known) was calculable without error since the birth and death rates were *given*. In the present case the confidence interval for Y depends on $s_{\bar{y}}$ which is calculated from actual observations.

A final point that should be explained is the distinction between a *standard deviation* and a *standard error*. Both are square roots of variances (see Appendix 3.1, page 370). Both measure the *dispersion* (i.e. the "spread") of a probability distribution or frequency distribution. The term *standard deviation* is used when the distribution is that of actual measurements or observations carried out on material objects; in other words, a standard deviation pertains to "raw data". The term *standard error* is used when the distribution is that of a "sample statistic", that is, some calculated value (such as the mean) *derived* from the raw data.

To discover empirically the standard error of the mean of n observations, say $\bar{x}_n$, it would be necessary to take many samples *each* of size n, calculate $\bar{x}_n$ for each separate sample, and then obtain the standard deviation of all these sample means. This would, by definition, be an estimate of the standard error of the sample means.

Such an empirical determination is never done in practice, however, since the standard error of the $\bar{x}_n$'s can be estimated from the standard deviation of the x's (the original observations or measurements). Thus the standard deviation of the x's is

$$s_x = \sqrt{\frac{1}{n-1}\Sigma x^2 - \frac{(\Sigma x)^2}{n(n-1)}}$$

and the standard error of $\bar{x}_n$, the mean of a sample of n observed x's, is estimated by

$$s_{\bar{x}} = \frac{s_x}{\sqrt{n}}.$$

Often it is more convenient to use variances rather than standard deviations and standard errors. Then $s_x^2 = \text{var}(x)$ is the variance of the original observations; and $s_{\bar{x}}^2 = \frac{s_x^2}{n} = \text{var}(\bar{x})$ is the sampling variance of the sample statistic $\bar{x}$, the mean of n observations.

Estimating the Coefficients of a Regression Line

SUPPOSE TWO observations are made on each of a sample of n units. Denote the pair of observations (which are counts or measurements of observable quantities) by $(x_1, y_1), (x_2, y_2), \ldots, (x_n, y_n)$. These n pairs of observations may be plotted as a scatter diagram (see Figure 6.3). A straight line is to be fitted to the points so that, for any given value of x, one may estimate the expected value of y corresponding to it. The line then represents the "regression of y on x". Let its equation be

$$y = \hat{a} + \hat{b}x.$$

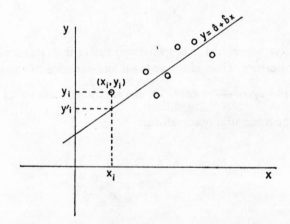

FIGURE 6.3   A regression line fitted to a scatter diagram

As explained on page 118, the line is to be that for which the sum of squares of the distances (parallel with the y-axis) of the points from the line is a minimum*. We must therefore determine what values $\hat{a}$ and $\hat{b}$ should have for the line to meet this requirement.

---

* Had we wished to estimate an expected x from a given y, we should have minimized the sum of squared distances parallel with the x-axis, to give the regression of x on y.

Let (x_i, y_i) be any observed point and let $y_i' = \hat{a} + \hat{b}x_i$ be the y-coordinate of the point on the line at which $x = x_i$ (see Figure 6.3).

The distance from the point (x_i, y_i) to the line is therefore

$$y_i - y_i' = y_i - \hat{a} - \hat{b}x_i.$$

Denote by U the sum of squares of all n of these distances.

That is

$$U = \sum_{i=1}^{n} (y_i - \hat{a} - \hat{b}x_i)^2.$$

To find the value of $\hat{a}$ for which U is a minimum, we must obtain $\dfrac{\partial U}{\partial \hat{a}}$, the partial derivative of U with respect to $\hat{a}$, and set it equal to zero. And likewise for $\hat{b}$.

Now

$$\frac{\partial U}{\partial \hat{a}} = -2 \sum (y_i - \hat{a} - \hat{b}x_i) \qquad (1)$$

and

$$\frac{\partial U}{\partial \hat{b}} = -2 \sum x_i (y_i - \hat{a} - \hat{b}x_i). \qquad (2)$$

Putting $\dfrac{\partial U}{\partial \hat{a}} = \dfrac{\partial U}{\partial \hat{b}} = 0$ yields equations whose solutions are the required values of $\hat{a}$ and $\hat{b}$.

From (1), $\Sigma (y_i - \hat{a} - \hat{b}x_i) = 0$ or $\Sigma y_i = n\hat{a} + \hat{b}\Sigma x_i$, whence, dividing through by n,

$$\bar{y} = \hat{a} + \hat{b}\bar{x}.$$

This shows that the regression line goes through the point $(\bar{x}, \bar{y})$. It also follows that

$$\hat{a} = \bar{y} - \hat{b}\bar{x},$$

and it remains to find $\hat{b}$.

From (2),

$$\Sigma x_i (y_i - \hat{a} - \hat{b}x_i) = 0$$

or

$$\Sigma x_i y_i - \hat{a}\Sigma x_i - \hat{b}\Sigma x_i^2 = 0.$$

In this equation, put $\hat{a} = \bar{y} - \hat{b}\bar{x} = \dfrac{1}{n}\Sigma y_i - \dfrac{\hat{b}}{n}\Sigma x_i$ to give

$$\Sigma x_i y_i - \frac{1}{n}\Sigma x_i (\Sigma y_i - \hat{b}\Sigma x_i) - \hat{b}\Sigma x_i^2 = 0.$$

Then

$$b \left\{ \Sigma x_i^2 - \frac{1}{n} (\Sigma x_i)^2 \right\} = \Sigma x_i y_i - \frac{1}{n} (\Sigma x_i) (\Sigma y_i)$$

whence

$$b = \frac{n \Sigma x_i y_i - (\Sigma x_i) (\Sigma y_i)}{n \Sigma x_i^2 - (\Sigma x_i)^2}$$

as in (6.10) on page 119.

The Proof that $\mathscr{E}(X) = \operatorname{Var}(X)$
for a Poisson Distribution

THE EXPECTED value of any variate that takes integer values only (i.e. a discrete variate) is

$$\mathscr{E}(x) = \sum_{\text{all } x} x p_x$$

where p_x is the probability that the variate takes the value x.

For a Poisson distribution, as shown on page 135,

$$p_x = \frac{m^x}{x!} e^{-m},$$

where m is a constant known as the parameter of the distribution.

Therefore, for the Poisson distribution, the expected value of the variate (or the "expectation") is

$$\mathscr{E}(x) = \sum_{x=0}^{\infty} x \frac{m^x}{x!} e^{-m}$$

$$= 0 + e^{-m} \sum_{x=1}^{\infty} \frac{m^x}{(x-1)!}.$$

Now $(x-1)! = 1$ when $x = 1$ since, by definition, $0! = 1$. Therefore, expanding the sum as a series, we see that

$$\sum_{x=1}^{\infty} \frac{m^x}{(x-1)!} = m\left(\frac{1}{0!} + \frac{m}{1!} + \frac{m^2}{2!} + \frac{m^3}{3!} \cdots\right) = m e^m,$$

(cf. Appendix 1.1, page 360).

Then

$$\mathscr{E}(x) = e^{-m}(m e^m) = m.$$

The variance of any variate is the expected value of the square of its deviation from the mean. Thus for a discrete variate we have, in symbols,

$$\operatorname{Var}(x) = \sum_{\text{all } x} [x - \mathscr{E}(x)]^2 p_x$$

$$= \sum_{x=0}^{\infty} (x - m)^2 p_x \quad \text{since } \mathscr{E}(x) = m.$$

Omitting the limits of summation for breviety, notice that

$$\Sigma (x - m)^2 p_x = \Sigma x^2 p_x - 2m \Sigma x p_x + m^2 \Sigma p_x.$$

But

$$\Sigma x p_x = m \quad \text{and} \quad \Sigma p_x = 1;$$

thus

$$\Sigma (x - m)^2 p_x = \Sigma x^2 p_x - m^2.$$

If we now substitute the appropriate formula for a Poisson probability in place of p_x we may derive the variance of the Poisson distribution.

Thus

$$\text{Var}(x) = \sum_{x=0}^{\infty} x^2 \frac{m^x}{x!} e^{-m} - m^2$$

$$= me^{-m} \sum_{x=1}^{\infty} x \frac{m^{x-1}}{(x-1)!} - m^2$$

$$= me^{-m} \left\{ \frac{1}{0!} + \frac{2m}{1!} + \frac{3m^2}{2!} + \frac{4m^3}{3!} \cdots \right\} - m^2. \qquad (1)$$

To sum the series in braces, notice that it can be rewritten as

$$\left\{ 1 + \left[\frac{m}{1!} + \frac{m}{1!} \right] + \left[\frac{m^2}{2!} + \frac{2m^2}{2!} \right] + \left[\frac{m^3}{3!} + \frac{3m^3}{3!} \right] + \cdots \right\}.$$

Rearranging terms this becomes

$$\left\{ \left[1 + \frac{m}{1!} + \frac{m^2}{2!} + \frac{m^3}{3!} + \cdots \right] + \left[\frac{m}{1!} + \frac{2m^2}{2!} + \frac{3m^3}{3!} + \cdots \right] \right\}$$

$$= \left\{ e^m + m \left[1 + \frac{m}{1!} + \frac{m^2}{2!} + \frac{m^3}{3!} + \cdots \right] \right\}$$

$$= \{ e^m + me^m \}.$$

Thus, from (1),

$$\text{Var}(x) = me^{-m} \{ e^m + me^m \} - m^2$$

$$= m(1 + m) - m^2$$

$$= m \qquad \qquad \text{q.e.d.}$$

The χ^2 Test for Goodness of Fit

SUPPOSE WE WISH to judge how well a theoretical frequency distribution fits an observed frequency distribution. (Note: we test whether a theory fits the facts: *not* whether the facts fit a theory). The best test is the χ^2-test for goodness of fit. This appendix gives instructions, without explanation, for carrying out the best. The underlying theory is beyond the level of this book; a full acount will be found in, for example, Kendall and Stuart (1967). The data of Table 7.1, reproduced here in Table 7.3, will be used as an example.

TABLE 7.3   To illustrate the χ^2 test for goodness of fit (Data as in Table 7.1, page 138)

| Variate value | Frequency | | Deviations | $\Sigma \dfrac{(f_x - np_x)^2}{np_x}$ |
| | Observed | Expected | | |
x	f_x	np_x	$(f_x - np_x)$	
0	7	5.7	+1.3	0.296
1	16	16.4	−0.4	0.010
2	20	23.4	−3.4	0.494
3	24	22.3	+1.7	0.130
4	17	16.0	+1.0	0.063
5	9	9.1	−0.1	0.001
6 and over	7	7.1	−0.1	0.001
	100	100.0	0.0	0.995

$$X^2 = 0.995$$

The fourth column of the table, headed $(f_x - np_x)$, lists the differences between the observed frequencies, f_x, and their expected values, np_x.

Since $\Sigma f_x = \Sigma np_x = n$ the differences sum to zero, but it is intuitively clear that the greater their magnitudes (disregarding their signs), the less good the fit. An overall measure of the difference between the observed and expected distributions is given by the test statistic X^2, defined as

$$X^2 = \sum \frac{(f_x - np_x)^2}{np_x}.$$

If the two distributions differ only because of chance sampling variation, then it can be proved that the size of X^2 depends only on its number of degrees of freedom, v. We determine v as follows: if c is the number of terms summed to give X^2 (i.e. the number of observed frequencies compared with expectation), then v is obtained by subtracting from $(c-1)$ the number of constants (parameters) derived from the observations and then used to calculate the expected frequencies. In the present case only one such parameter was derived, namely the mean $\bar{x} = 2.86$. This was the only observational result needed to enable the np_x to be calculated.

Thus in the example, $c = 7$ and $v = 5$.

Now clearly, for given v, the larger the value of X^2 the smaller the probability that the discrepancies between observation and expectation are due solely to chance. If we choose to reject the hypothesis under test as untenable when this probability is less than 5%, say, then we need to know the value of X^2 such that its chance of being equalled or exceeded (if the hypothesis is correct) is 0.05. The values of X^2 required are known as the 95% points of the χ^2 distribution; (95%, since this is the probability that, given the hypothesis, the calculated X^2 will be less than the value shown in standard statistical tables). Table 7.4 gives values of the 95% points of the χ^2 distribution for $v = 1, 2, \ldots, 30$. Thus, when a goodness of fit test is done and we wish to know whether an observed distribution differs significantly from a hypothetical distribution at the 5% level of significance, the values tabulated here

TABLE 7.4 95% points of the χ^2 distribution (v is the number of degrees of freedom)

v	Percentage point	v	Percentage point
1	3.84	16	26.3
2	5.99	17	27.6
3	7.81	18	28.9
4	9.49	19	30.1
5	11.1	20	31.4
6	12.6	21	32.7
7	14.1	22	33.9
8	15.5	23	35.2
9	16.9	24	36.4
10	18.3	25	37.7
11	19.7	26	38.9
12	21.0	27	40.1
13	22.4	28	41.3
14	23.7	29	42.6
15	25.0	30	43.8

give the required critical values (as they are called) of X^2. If the calculated X^2 is less than the tabulated value, the hypothesis may be accepted; otherwise it should be rejected.

Alternatively we may obtain a closer estimate of the probability, $P(\chi^2)$, that an observed X^2 will be equalled or exceeded, by consulting more detailed statistical tables (e.g. Table R in Rohlf and Sokal, 1969). Thus in Table 7.3 (and also in Table 7.1), $X^2 = 0.995$ and $\nu = 5$ so that $P(\chi^2) > 0.9$. The hypothesis fits the observations very well indeed.

A final point to note is that when a χ^2 goodness of fit test is done, small expected frequencies should be pooled so that none is less than 5. This is the reason for pooling the expected frequencies of obtaining $x = 6$, $x = 7$, $x = 8$, ... in Table 7.3. The observed frequencies are pooled correspondingly.

The Expected Value of ω in a Random Pattern

CONSIDER A POPULATION in which the density of the individuals is λ per circle of unit radius. Suppose a point is located at random in the area and let the distance from the point to the individual nearest it be measured. It was proved on page 157 that the probability that the square of this distance is less than or equal to some preassigned value, say ω, is given by

$$F(\omega) = 1 - e^{-\lambda\omega} \quad \text{with} \quad 0 \leq \omega < \infty.$$

Here $F(\omega)$ is the *cumulative distribution function* of ω. Differentiating $F(\omega)$ with respect to ω gives $f(\omega)$, the *probability density function*.

Thus $f(\omega) = \dfrac{dF(\omega)}{d\omega} = \lambda e^{-\lambda\omega}$ with $0 \leq \omega < \infty$. The mean of the distribution, that is the expected value of ω, is

$$\mathscr{E}(\omega) = \int\limits_0^\infty \omega f(\omega)\, d\omega = \int\limits_0^\infty \omega \lambda e^{-\lambda\omega}\, d\omega.$$

Now integrate by parts using the formula

$$\int x\, dy = xy - \int y\, dx.$$

In the present case put

$$x = \omega \quad \text{and} \quad dy = \lambda e^{-\lambda\omega}\, d\omega$$

so that

$$dx = d\omega \quad \text{and} \quad y = -e^{-\lambda\omega}.$$

It is now seen that

$$\mathscr{E}(\omega) = \int\limits_0^\infty \omega \cdot \lambda e^{-\lambda\omega} \cdot d\omega = \left[-\omega e^{-\lambda\omega}\right]_0^\infty - \left(-\int\limits_0^\infty e^{-\lambda\omega}\, d\omega\right)$$

$$= 0 + \left[\frac{-e^{-\lambda\omega}}{\lambda}\right]_0^\infty$$

$$= \frac{1}{\lambda}.$$

Table of Values of $\dfrac{1}{k} - \dfrac{e^{-k}}{1 - e^{-k}}$

k	$\dfrac{1}{k} - \dfrac{e^{-k}}{1 - e^{-k}}$	k	$\dfrac{1}{k} - \dfrac{e^{-k}}{1 - e^{-k}}$	k	$\dfrac{1}{k} - \dfrac{e^{-k}}{1 - e^{-k}}$
0.1	0.4917	1.8	0.3575	3.5	0.2546
0.2	0.4833	1.9	0.3504	3.6	0.2497
0.3	0.4750	2.0	0.3435	3.7	0.2449
0.4	0.4668	2.1	0.3366	3.8	0.2403
0.5	0.4585	2.2	0.3299	3.9	0.2358
0.6	0.4503	2.3	0.3234	4.0	0.2313
0.7	0.4421	2.4	0.3169	4.1	0.2271
0.8	0.4340	2.5	0.3106	4.2	0.2229
0.9	0.4260	2.6	0.3044	4.3	0.2188
1.0	0.4180	2.7	0.2983	4.4	0.2148
1.1	0.4101	2.8	0.2924	4.5	0.2110
1.2	0.4023	2.9	0.2866	4.6	0.2072
1.3	0.3946	3.0	0.2809	4.7	0.2036
1.4	0.3870	3.1	0.2754	4.8	0.2000
1.5	0.3795	3.2	0.2700	4.9	0.1966
1.6	0.3720	3.3	0.2647	5.0	0.1932
1.7	0.3647	3.4	0.2596		

L'Hospital's Rule

Suppose it is desired to evaluate the ratio $g(x)/h(x)$ when $x = a$, but that $g(a) = 0$ and $h(a) = 0$ so that $g(a)/h(a)$ is the indeterminate fraction 0/0. We can use the fact (proved in most textbooks of advanced calculus) that

$$\lim_{x \to a} \frac{g(x)}{h(x)} = \lim_{x \to a} \frac{\dfrac{d}{dx}[g(x)]}{\dfrac{d}{dx}[h(x)]}$$

and, so long as the expression on the right is not itself indeterminate, it will yield the desired limit. This is l'Hospital's rule.

As an example, we evaluate the right side of equation (10.3) (page 248) when $q = 0$.

The derivative of the numerator as $q \to 0$ is

$$\lim_{q \to 0} \frac{d}{dq}\left[Q \ln\left(1 - \frac{q}{1 - Q}\right)\right] = \lim_{q \to 0} Q\left(\frac{1 - Q}{1 - Q - q}\right)\left(\frac{-1}{1 - Q}\right) = \frac{-Q}{1 - Q}.$$

Similarly, the derivative of the denominator as $q \to 0$ is

$$\lim_{q \to 0} \frac{d}{dq}[q \ln(1 - Q)] = \lim_{q \to 0}[\ln(1 - Q)] = \ln(1 - Q).$$

Therefore, by l'Hospital's rule,

$$\lim_{q \to 0} \frac{Q \ln\left(1 - \dfrac{q}{1 - Q}\right)}{q \ln(1 - Q)} = \frac{-Q}{(1 - Q)\ln(1 - Q)}.$$

The rule can also be used to evaluate indeterminate fractions of the form ∞/∞.

Comparing Two 2×2 Tables

CONSIDER THE TWO 2×2 tables shown in perspective, one above the other, on the left in Figure 11.6. For concreteness, suppose the upper layer is a segregation table (see page 270) for species J and K in Area I and the lower layer is a segregation table for the same two species in Area II. Assume that $ad > bc$, implying a certain degree of segregation between the species in area I; and likewise $\alpha\delta > \beta\gamma$ implying segregation of the species in area II. We are not here concerned with separate tests for the significance of the segregations in the two areas; it is assumed that these have been done. We now wish to judge whether the degree of segregation is the same in the two areas.

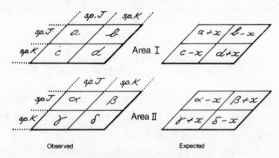

Observed   Expected

FIGURE 11.6   See text

Let the degree of segregation by measured by ad/bc in area I and by $\alpha\delta/\beta\gamma$ in area II. These are known as the cross-product ratios of the tables. The null hypothesis to be tested is that $ad/bc = \alpha\delta/\beta\gamma$, or, equivalently that

$$ad\beta\gamma = \alpha\delta bc. \tag{1}$$

Now imagine the two 2×2 tables combined to form a single, cubical $2 \times 2 \times 2$ table. Such a table has 12 marginal totals corresponding to the 12 edges of the cube. These totals are to remain unchanged when expected frequencies, under the null hypothesis, are substituted for observed frequencies in the eight cells of the table. Next suppose the expected frequency in, say, the upper left rear cell (in which the observed frequency is a) is set equal to $(a + x)$; then it follows that the expected frequencies in the other seven cells must have the values shown in the "expected" $2 \times 2 \times 2$ table on the right in the figure. No other values are possible if the 12 marginal totals are to keep their original values; we therefore see that the absolute magnitude of the deviation of the observed from the expected frequency is the same for all cells; only the signs alternate. And since, when one frequency is given all the others are automatically fixed, the table has one degree of freedom.

From equation (1) it is seen that, given the null hypothesis,

$$(a + x)(d + x)(\beta + x)(\gamma + x) = (\alpha - x)(\delta - x)(b - x)(c - x). \quad (2)$$

Solving this equation yields x, the deviation of the observed from the expected frequency in each cell of the table. Knowing x we can find the expected frequencies and can then judge their goodness of fit to the observed frequencies by a χ^2 test with one degree of freedom. If no allowance were made for the "continuity correction" (see page 191) the criterion for the goodness of fit test would be (see Appendix 7.2, page 389)

$$X^2 = \sum \frac{(\text{Observed frequency} - \text{Expected frequency})^2}{\text{Expected frequency}}$$

$$= x^2 \left\{ \frac{1}{a + x} + \frac{1}{d + x} + \frac{1}{\beta + x} + \cdots + \frac{1}{c - x} \right\}.$$

However, the fact that the table has only one degree of freedom makes a continuity correction possible. The correction is to ensure that the area under a smooth curve (the theoretical curve of the χ^2 variable) approximates as closely as possible the sum of discrete probabilities actually required. The closest approximation to the probability sought is given by calculating the adjusted criterion

$$X_c^2 = (|x| - 0.5)^2 \left\{ \frac{1}{a + x} + \frac{1}{d + x} + \frac{1}{\beta + x} + \cdots + \frac{1}{c - x} \right\} \quad (3)$$

and then, from a table of percentage points of the χ^2 distribution with one degree of freedom, finding the probability that the calculated X_c^2 would be exceeded if the null hypothesis were true.

The procedure can be demonstrated with the tables on page 270, for which equation (2) is

$$(26 + x)(149 + x)(11 + x)(9 + x) = (16 - x)(73 - x)(39 - x)(30 - x).$$

This simplifies to the cubic equation

$$353x^3 - 1006x^2 + 279{,}527x - 983{,}034 = 0$$

which must be solved by successive approximation (trial and error). It is found that, to one decimal place, $x = 3.5$. Therefore, the expected frequencies are as shown on the right in Figure 11.7 in which the observed $2 \times 2 \times 2$ table appears on the left. The test criterion for the goodness of fit test is

$$X_c^2 = (3.5 - 0.5)^2 \left\{ \frac{1}{29.5} + \frac{1}{152.5} + \cdots + \frac{1}{26.5} \right\}$$
$$= 3.147.$$

The probability of obtaining such a result, given the null hypothesis, is greater than 0.05. We therefore conclude that the data for the two segregation tables could have been drawn from a single population.

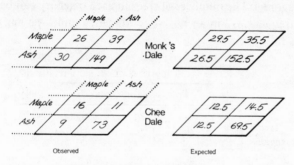

FIGURE 11.7   See text

An extension of the method, for use when more than two 2×2 tables are to be intercompared, has been described by Norton (1945), and also by Kastenbaum and Lamphiear (1959).

Contingency Table Tests and Homogeneity Tests

IMAGINE A SET of N things, or events, that can be classified in two distinct ways.

For example, the N things might be the trees in a forest and each could be classified first according to its species, and then according to its age class. Thus if the forest contains r species of trees, and the trees (regardless of species) are divided into c different age classes, each tree can be assigned to one of rc categories. The numbers of trees in each category can be tabulated as the "cell frequencies" in an $r \times c$ "contingency table" having r rows and c columns as follows:

		Age class			
		1	2	$\cdots$ c	Totals
	1	a_{11}	a_{12}	$\cdots$ a_{1c}	$a_1.$
Tree	2	a_{21}	a_{22}	$\cdots$ a_{2c}	$a_2.$
species	$\vdots$	$\vdots$	$\vdots$	$\vdots$ $\vdots$	$\vdots$
	r	a_{r1}	a_{r2}	$\cdots$ a_{rc}	$a_r.$
Totals		$a_{.1}$	$a_{.2}$	$a_{.c}$	N

Thus a_{ij} trees belonged to the ith species and the jth age class.

We now wish to test whether the two classifications are independent. The null hypothesis is that for any tree in the forest its age is independent of its species and vice versa. The hypothesis therefore states that the r separate frequency distributions represented by the rows of the table are homogeneous; each row represents the distribution of age classes in one species and it is hypotheisized that the proportions of trees in the c age classes are the same for all species. Equivalently, the hypothesis states that the c separate frequency distributions represented by the columns are homogeneous; that is that the proportions of trees in the r species are the same for all c age classes.

Then the expected frequency in the (i, j)th cell of the table is $(a_i.a_{.j})/N$. The goodness of fit of the expected to the observed frequencies is judged in the usual way by a χ^2 test. The test criterion is

$$X^2 = \sum \frac{(\text{Observed frequency} - \text{Expected frequency})^2}{\text{Expected frequency}}.$$

(Unless $r = c = 2$, a continuity correction cannot be made.) Under the null hypothesis, X^2 has the χ^2 distribution with $(r - 1)(c - 1)$ degrees of freedom and one may look up the probability that the calculated X^2 would be equalled or exceeded, if the null hypothesis were true, in a table of percentage points of the χ^2 distribution with the appropriate number of degrees of freedom.

The following convenient formula for calculating X^2 has been given by Maxwell (1961).

$$X^2 = N\left\{ \frac{1}{a_{1.}} \sum_{k=1}^{c} \frac{a_{1k}^2}{a_{.k}} + \frac{1}{a_{2.}} \sum_{k=1}^{c} \frac{a_{2k}^2}{a_{.k}} + \cdots + \frac{1}{a_{r.}} \sum_{k=1}^{c} \frac{a_{rk}^2}{a_{.k}} \right\} - N$$

$$= N \sum_{j=1}^{r} \left\{ \frac{1}{a_{j.}} \sum_{k=1}^{c} \frac{a_{jk}^2}{a_{.k}} \right\} - N.$$

A test of the homogeneity of two observed distributions (for example, those in Table 11.2 on page 274) is merely a special case of an $r \times c$ table test, with $r = 2$ if the distributions are written one above the other as horizontal rows, or $c = 2$ if they are written side by side as vertical columns. Thus Table 11.2 is a 3×2 table and has $(3 - 1)(2 - 1) = 2$ degrees of freedom.

The Distribution of s, the Number of Species per Sampling unit, when the Species are Independent

SUPPOSE WE WISH to test whether the occurrences of k species in N sampling units are independent. Conceptually it is possible to tabulate the observations to be tested in a 2^k table (the k-dimensional analog of a 2×2 table) and compare the observed and expected cell frequencies. This is rarely feasible, however. When k is at all large the table's cells are so numerous that nearly all of them have observed frequencies of zero and expected frequencies that are very small indeed. A practicable alternative is to compare the observed and expected distribution of s, the number of species per sampling unit. The procedure will be demonstrated using the data in Table 11.3 (page 281) as an example.

If the raw data are set out in the manner shown in Table 11.3, which is a typical example of a "data matrix", the observed values of s are given by the row totals of the matrix. The observed frequency distribution of the values of s for the cave animals is shown in Table 11.4 below.

The expected distribution of s, on the null hypothesis of independence of the k species, was derived by Barton and David (1959). Here we use their results without giving details of their derivation.

Denote by n_j the number of occurrences of the jth species for $j = 1, 2, ..., k$. These are the column totals of the data matrix.

Write $p_j = n_j/N$ for the proportion of the total number of observed occurrences that were of the jth species. Calculate the mean and variance of the k values of p_j.

The mean is

$$\bar{p} = \frac{1}{k}\Sigma p_j = \frac{1}{kN}\Sigma n_j.$$

$$= \frac{63}{7 \times 28} = 0.3214 \text{ in the example.}$$

The variance is

$$\text{var}(p) = \frac{1}{k}\Sigma(p_j - \bar{p})^2$$

$$= \frac{1}{kN^2}\left\{\Sigma n_j^2 - \frac{(\Sigma n_j)^2}{k}\right\}$$

$$= \frac{1}{7 \times 28^2}(651 - 63^2/7) = 0.01531$$

in the example.

Barton and David showed that if the species were mutually independent the number of different species per sampling unit, s, would have (approximately) a binomial distribution whose parameters, P and K, are given by

$$P = \bar{p} + \text{var}(p)/\bar{p} \quad \text{and} \quad K = \frac{k}{1 + \text{var}(p)/\bar{p}^2}.$$

In the present example, therefore

$$P = 0.37 \quad \text{and} \quad K = 6.29 \doteqdot 6.$$

Hence, writing f_r for the expected number of sampling units containing r different species, we have

$$f_r = \frac{NK!}{r!\,(K-r)!}P^r(1-P)^{K-r} \quad \text{and} \quad \sum_r f_r = N.$$

TABLE 11.4   The observed and expected frequencies of the number of animal species in 28 caves

Number of species	Frequencies		$\dfrac{(f_{obs} - f_{exp})^2}{f_{exp}}$
	Observed f_{obs}	Expected f_{exp}	
0	3⎤	1.8⎤	0.125
1	6⎦	6.2⎦	
2	8	9.1	0.133
3	5	7.1	0.621
4	4⎤	3.1⎤	
5	2⎬	0.7⎬	1.274
6 and over	0⎦	0.0⎦	
	28	28.0	2.153

$X^2 = 2.153$; with 3 degrees of freedom, $P(\chi^2) > 0.50$

Thus, in the example, direct calculation yields the expected number of caves with, for instance, two species as

$$f_2 = 28 \frac{6!}{2! \, 4!} (0.37)^2 (1 - 0.37)^4 = 9.06.$$

Alternatively, the required binomial probabilities (i.e., the values of f_r/N) may be looked up in statistical tables.

Table 11.4 shows the expected frequency distribution of s for the cave animals. The goodness of fit of the expected to the observed frequencies have been judged by a χ^2 test.

To do this the frequencies were first pooled as shown by the braces so that not more than one of the expected frequencies should be less than five.

The test statistic is $X^2 = 2.153$.

Since, after pooling, four pairs of observed and expected frequencies were available for comparison, there are three degrees of freedom. (None need be subtracted to allow for calculated parameters since in this case the parameters P and K were not derived from the observed distribution of s.) It follows that, given the null hypothesis, X^2 has the χ^2 distribution with 3 degrees of freedom, and tables show that the probability of obtaining a value of 2.153 or more is over 0.50. There is thus no reason to reject the hypothesis that the species of cave animals occurred independently of one another.

Bibliography

(The numbers in italics on the right refer to the pages on which each publication is mentioned)

Amadon, D. (1947). "Ecology and the evolution of some Hawaiian birds". *Evolution* **1**, 63–68. *332*

Anderson, D.J. (1967). "Studies on structure in plant communities. IV. Cyclical succession in *Dryas* communities from north-west Iceland". *J. Ecol.* **55**, 629 to 635. *170*

Andrewartha, H.G. (1961). *Introduction to the Study of Animal Populations.* University of Chicago Press. *63, 68, 69*

Andrewartha, H.G., and L.C.Birch (1954). *The Distribution and Abundance of Animals.* University of Chicago Press. *68*

Bainbridge, R. (1957). "The size, shape and density of marine phytoplankton concentrations". *Biol. Rev.* **32**, 91–115. *171*

Baker, F.S. (1950). *Principles of Silviculture.* McGraw Hill, New York. *62, 251*

Baker, H.G. (1970). "Differentiation of populations". *Science* **167**, 1637. *333*

Barclay-Estrup, P. (1971). "The description and interpretation of cyclical processes in a heath community. III. Microclimate in relation to the *Calluna* cycle". *J. Ecol.* **59**, 143–166. *170*

Barnes, H., and S.M.Marshall (1951). "On the variability of replicate plankton samples and some applications of 'contagious' series to the statistical distribution of catches over restricted periods". *J. Mar. Biol. Assoc. U. K.* **30**, 233–263. *145*

Barnett, V.D. (1962). "The Monte Carlo solution of a competing species problem". *Biometrics* **18**, 76–103. *225*

Bartholomew, B. (1970). "Bare zone between California shrub and grassland communities: the role of animals". *Science* **170**, 1210–1212. *169*

Bartlett, M.S. (1960). *Stochastic Population Models in Ecology and Epidemiology.* Wiley, New York. *226*

Barton, D.E., and F.N.David (1959). "The dispersion of a number of species". *J. R. Statist. Soc. B.* **21**, 190–194. *400*

Birch, L.C. (1948). "The intrinsic rate of natural increase of an insect population". *J. Anim. Ecol.* **17**, 15–26. *26*

Briggs, D., and S.M.Walters (1969). *Plant Variation and Evolution.* McGraw Hill, New York. *326*

Brillouin, L. (1962). *Science and Information Theory.* 2nd. ed. Academic Press, New York. *304*

Brown, J. H. (1971). "Mechanisms of competitive exclusion between two species of chipmunks." *Ecology* **52**, 305–311. *236*

Brown, W. L. Jr., and E. O. Wilson (1956). "Character displacement." *Syst. Zool.* **5**, 49–64. *332*

Buck, D. H., and C. F. Thoits III (1965). "An evaluation of Petersen estimation procedures employing seines in one-acre ponds." *J. Wildl. Mgmt.* **29**, 598 to 621. *124*

Calhoun, J. B. (1962). "Population density and social pathology." *Sci. American* **206**, 139–150. *62*

Cassie, R. M. (1962). "Frequency distribution models in the ecology of plankton and other organisms." *J. Anim. Ecol.* **31**, 65–92. *146*

Caughley, G. (1966). "Mortality patterns in mammals." *Ecology* **47**, 906–917. *27, 35*

Caughley, G. (1967). "Parameters for seasonally breeding populations." *Ecology* **48**, 834–838. *14, 15, 18*

Chitty, D. (1960). "Population processes in the vole and their relevance to general theory." *Can. J. Zool.* **38**, 99–113. *66, 148*

Chitty, D., and E. Phipps (1966). "Seasonal changes in survival in mixed populations of two species of vole." *J. Anim. Ecol.* **35**, 313–332. *95*

Chorley, R. J., and P. Haggett (1965). "Trend-surface mapping in geographical research." In *Spatial Analysis* (B. J. L. Berry and D. F. Marble, Eds.). Prentice-Hall, N. J. 1968. *197*

Clark, J. P. (1971). "The second derivative and population modeling." *Ecology* **52**, 606–613. *91*

Clark, P. J., and F. C. Evans (1955). "On some aspects of spatial pattern in biological populations." *Science* **121**, 397–398. *272*

Clarke, B. (1969). "The evidence for apostatic selection." *Heredity* **24**, 347–352. *330, 348*

Clausen, C. P. (1958). "Biological control of insect pests." *Ann. Rev. Entomol.* **3**, 291–310. *61*

Cochran, W. G. (1963). *Sampling Techniques*. 2nd. ed. Wiley, New York. *104, 107*

Cole, L. C. (1954). "The population consequences of life history phenomena." *Quart. Rev. Biol.* **29**, 103–137. *45, 340, 341*

Cole, L. C. (1960). "Competitive exclusion." *Science* **132**, 348–349. *234*

Coleman, M. J., and H. B. N. Hynes (1970). "The life histories of some Plecoptera and Ephemeroptera in a southern Ontario stream." *Can. J. Zool.* **48**, 1333 to 1339. *321*

Colwell, R. N. (1951). "The use of radioactive isotopes in determining spore distribution patterns." *Amer. J. Botany* **38**, 511–523. *333*

Connell, J. H. (1961a). "The influence of interspecific competition and other factors on the distribution of the barnacle *Chthamalus stellatus*." *Ecology* **42**, 710–723. *235*

Connell, J. H. (1961b). "Effects of competition, predation by *Thais lapillus* and other factors on natural populations of the barnacle *Balanus balanoides*." *Ecol. Monogr.* **31**, 61–104. *235, 240*

Connell, J. H., and E. Orias (1964). "The ecological regulation of species diversity." *Amer. Natur.* **98**, 399–414. *342*

Cooper, C. F. (1961). "Pattern in ponderosa pine forests." *Ecology* **42**, 493–499. *167*

Cooper, D.W. (1968). "The significance level in multiple tests made simultaneously." *Heredity* **23**, 614–617. *262*

Cormack, R.M. (1968). "The statistics of capture-recapture methods." *Oceanogr. Mar. Biol. Ann. Rev.* **6**, 455–506. *123, 124*

Crowell, K.L. (1968). "Rates of competitive exclusion by the Argentine ant in Bermuda." *Ecology* **49**, 551–555. *237*

Culver, D.C. (1970). "Analysis of simple cave communities. I. Caves as islands." *Evolution* **24**, 463–474. *281, 282*

Dasmann, R.F. (1964). *Wildlife Biology.* Wiley, New York. *67*

Davis, P.H., and V.H.Heywood (1963). *Principles of Angiosperm Taxonomy.* Van Nostrand, New York. *326*

DeBach, P. (1966). "The competitive displacement and coexistence principles." *Ann. Rev. Entomol.* **11**, 183–212. *234*

Deevey, E.S. (1947). "Life tables for natural populations of animals." *Quart. Rev. Biol.* **22**, 283–314. *28, 32, 35*

Doyle, R.W. (1972). "Genetic variability in *Ophiomusium lymani* (Echinodermata) populations in the deep sea." *Deep Sea Res.* 661–664 *331*

Ehrlich, P.R. (1965). "The population structure of the butterfly *Euphydryas editha.* II. The structure of the Jasper Ridge colony." *Evolution* **19**, 327–336. *333*

Ehrlich, P.R., and P.H.Raven (1969). "Differentiation of populations." *Science* **165**, 1228–1232. *333*

Eisenberg, R.M. (1966). "The regulation of density in a natural population of the pond snail *Lymnaea elodes.*" *Ecology* **47**, 889–906. *70*

Elton, C. (1927). *Animal Ecology.* Tenth Impression, 1966, October House Inc. New York. *316*

Elton, C. (1942). *Voles, Mice and Lemmings: Problems in Population Dynamics.* Oxford University Press. *95*

Elton, C., and M.Nicholson (1942). "The ten year cycle in numbers of the lynx in Canada." *J. Anim. Ecol.* **11**, 215–244. *88*

Erickson, J.A. *et al.* (1970). "Estimating ages of mule deer: an evaluation of technique accuracy." *J. Wildl. Mgmt.* **34**, 523–531. *31*

Erskine, A.J. (1972). "Buffleheads." *Canadian Wildlife Service, Monograph series No. 4*, Ottawa. *63*

Feller, W. (1943). "On a general class of contagious distributions." *Ann. Math. Statist.* **14**, 389–400. *146*

Fischer, A.G. (1960). "Latitudinal variation in organic diversity." *Evolution* **14**, 64–81. *336*

Flyger, V.F. (1959). "A comparison of methods for estimating squirrel populations." *J. Wildl. Mgmt.* **23**, 220–223. *122*

Garb, S. (1961). "Differential growth-inhibitors produced by plants." *Bot. Rev.* **27**, 422–443. *164*

Garfinkel, D., and R.Sack (1964). "Digital computer simulation of an ecological system, based on a modified mass action law." *Ecology* **45**, 502–507. *231*

Gause, G.F. (1934). *The Struggle for Existence.* Hafner, New York. *217, 237*

Giesel, J.T. (1970). "On the maintenance of shell pattern and behavior polymorphism in *Acmaea digitalis*, a limpet." *Evolution* **24**, 98–119. *330*

Gilmour, J.S.L., and J.W.Gregor (1939). "Demes: a suggested new terminology." *Nature, Lond.* **144**, 333–334. *326, 334*

Godfrey, W.E. (1966). *The Birds of Canada.* Information Canada, Ottawa. *320*

Goodall, D.W. (1963). "Pattern analysis and minimal area: some further comments." *J. Ecol.* **51**, 705–710. *178*

Goodall, D.W. (1967). "Computer simulation of changes in vegetation subject to grazing." *Jour. Indian Bot. Soc.* **46**, 356–362. *231*

Greenbank, D.O. (1957). "The role of climate and dispersal in the initiation of outbreaks of the spruce budworm in New Brunswick. I. The role of dispersal." *Can. J. Zool.* **35**, 385–403. *95*

Greenbank, D.O. (1963). "Climate and the spruce budworm." In "The Dynamics of Epidemic Spruce Budworm Populations" R.F.Morris, Ed. *Mem. Entomol. Soc. Canada* No. **31**, 174–180. *65*

Greig-Smith, P. (1952). "The use of random and contiguous quadrats in the study of the structure of plant communities." *Ann. Bot. Lond. N. S.* **16**, 293–316. *174, 178*

Greig-Smith, P. (1964). *Quantitative Plant Ecology.* 2nd. ed. Butterworths, London. *168*

Grinnell, J. (1904). "The origin and distribution of the chestnutbacked chickadee." *Auk* **21**, 364–382. *233*

Grinnell, J. (1917). "The niche-relationships of the California thrasher." *Auk* **34**, 427–433. *316*

Hadley, G. (1970). "Non-specificity of symbiotic infection in orchid mycorrhiza." *New Phytol.* **69**, 1015–1023. *64*

Hairston, N.G., F.E.Smith, and L.B.Slobodkin (1960). "Community structure, population control and competition." *Amer. Natur.* **94**, 421–425. *71*

Hannan, E.J. (1960). *Time Series Analysis.* Wiley, New York. *91*

Hansen, M.H., W.N.Hurwitz, and G.W.Madow (1953). *Sample Survey Methods and Theory. Volumes I and II.* Wiley, New York. *104*

Harcourt, D.G. (1969). "The development and use of life tables in the study of natural insect populations." *Ann. Rev. Ent.* **14**, 175–196. *39, 113*

Harper, J.L. (1964). "Establishment, aggression, and cohabitation in weedy species." In *The Genetics of Colonizing Species*, H.G.Baker and G.L.Stebbins, Eds. Academic Press. *241*

Harper, J.L. (1967). "A Darwinian approach to plant ecology." *J. Ecol.* **55**, 247–270. *63, 71*

Harper, J.L. (1968). "The regulation of numbers and mass in plant populations." In *Population Biology and Evolution*, R.C.Lewontin, Ed., Syracuse University Press. *286*

Harper, J.L. (1969). The role of predation in vegetational diversity. In "Diversity and Stability in Ecological Systems", *Proc. Brookhaven Symposium in Biology*, No. 22. *61*

Harper, J.L., J.N.Clatworthy, I.H.McNaughton, and G.R.Sagar (1961). "The evolution and ecology of closely related species living in the same area." *Evolution* **15**, 209–227. *286*

Harper, J.L., and I.H.McNaughton (1962). "The comparative biology of closely related species living in the same area. VII. Interference between individuals in pure and mixed populations of *Papaver* species." *New Phytol.* **61**, 175–188.
233, 241

Heatwole, H., and D.M.Davis (1965). "Ecology of three sympatric species of parasitic insects of the genus *Megarhyssa* (Hymenoptera: Ichneumonidae)." *Ecology* **46**, 140–150.   *319*

Hessler, R.R., and H.L.Sanders (1967). "Faunal diversity in the deep-sea." *Deep-Sea Research* **14**, 65–78.   *343, 344*

Hett, J.M., and O.L.Loucks (1968). "Application of life table analyses to tree seedlings in Quetico Provincial Park, Ontario." *Forestry Chron.* **44**, 29–32. *31, 32*

Horn, E.G. (1971). "Food competition among the cellular slime molds (Acrasieae)." *Ecology* **52**, 475–484.   *237*

Hurd, L.E., M.V.Mellinger, L.L.Wolf, and S.J.McNaughton (1971). "Stability and diversity at three trophic levels in terrestrial successional ecosystems." *Science* **173**, 1134–1136.   *348*

Hutchinson, G.E. (1957). Concluding remarks. *Cold Spring Harbor Symp. Quant. Biol.* **22**, 415–427.   *317*

Hutchinson, G.E. (1959). "Homage to Santa Rosalia, or why are there so many kinds of animals." *Amer. Natur.* **93**, 145–159.   *332*

Hutchinson, G.E. (1961). "The paradox of the plankton." *Amer. Natur.* **95**, 137 to 145.   *239*

Hutchinson, G.E. (1967). "*A Treatise on Limnology. Volume II.*" Wiley, New York.   *289*

Inger, R.F., and B.Greenberg (1966). "Annual reproductive pattern of lizards from a Bornean rain forest." *Ecology* **47**, 1007–1020.   *16*

International Commission for the Northwest Atlantic Fisheries (1963). Northwest Atlantic Fish Marking Symposium. *Special Publication No. 4.*   *123*

Jaeger, R.G. (1970). "Potential extinction through competition between two species of terrestrial salamanders." *Evolution* **24**, 632–642.   *236*

Janzen, D.H. (1967). "Why mountain passes are higher in the tropics." *Amer. Natur.* **101**, 233–249.   *341*

Janzen, D.H. (1970). "Herbivores and the number of tree species in tropical forests." *Amer. Natur.* **104**, 501–528.   *284*

Jensen, A.L. (1971). "The effect of increased mortality on the young in a population of brook trout, a theoretical analysis." *Trans. Amer. Fish Soc. 1971 No. 3*, 456–459.   *232*

Jones, D.A. (1966). "On the polymorphism of cyanogenesis in *Lotus corniculatus*. I. Selection by animals." *Can J. Genetics and Cytology* **8**, 556–567.   *327*

Kastenbaum, M.A., and D.E.Lamphiear (1959). "Calculation of chi-square to test the no three-factor interaction hypothesis." *Biometrics* **15**, 107–115.   *397*

Keast, A. (1966). "Trophic interrelationships in the fish fauna of a small stream." *Univ. Mich. Great Lakes Res. Div. Publ.* **15**, 51–79.   *321*

Keith, L.B. (1963). *Wildlife's Ten-Year Cycle.* University of Wisconsin Press.
92, 94, 95

Kendall, M.G., and A.Stuart (1966). *The Advanced Theory of Statistics, Volume 3*
Griffin, London.   *180*

Kendall, M.G., and A.Stuart (1967). *The Advanced Theory of Statistics, Volume 2.*
2nd. ed. Griffin, London.   *389*

Kershaw, K.A. (1957). "The use of cover and frequency in the detection of pattern
in plant communities." *Ecology* 38, 291–299.   *174, 178*

Kershaw, K.A. (1958). "An investigation of the structure of a grassland community. I. The pattern of *Agrostis tenuis*." *J. Ecol.* 46, 571–592.   *168*

Kershaw, K.A. (1964). *Quantitative and Dynamic Ecology*. Arnold, London.   *168*

Keyfitz, N., and E.M.Murphy (1967). "Matrix and multiple decrement in population analysis." *Biometrics* 23, 485–503.   *10, 24*

Khinchin, A.I. (1957). *Mathematical Foundations of Information Theory*. Dover,
New York.   *291*

Klopfer, P.H., and R.H.MacArthur (1961). "On the causes of tropical species
diversity: niche overlap." *Amer. Natur.* 95, 223–226.   *338, 339*

Knight Jones, E.W., and J.Moyse (1961). "Intraspecific competition in sedentary
marine animals. In "Mechanisms in Biological Competition", F.L.Milthorpe,
Ed. *Symposium of the Society of Experimental Biologists, No. 15.* Academic
Press.   *252*

Koehn, R.K. (1969). "Esterase heterogeneity: dynamics of a polymorphism."
Science 163, 943–944.   *331*

Kohn, A.J. (1968). "Microhabitats, abundance and food of *Conus* on atoll reefs in
the Maldive and Chagos Islands." *Ecology* 49, 1046–1062.   *336*

Krebs, C.J. (1970). "*Microtus* population biology: behavioral changes associated
with the population cycle in *M. ochrogaster* and *M. pennsylvanicus*." *Ecology* 51,
34–52.   *95*

Lack, D. (1954). "Cyclic mortality." *J. Wildl. Mgmt.* 18, 25–37.   *92*

Lack, D. (1966). *Population Studies of Birds*. Oxford University Press.   *68, 69*

Larkin, P.A., and A.S.Hourston (1964). "A model for simulation of the biology of
Pacific salmon." *J. Fish. Res. Bd. Canada* 25, 1245–1265.   *232*

Laughlin, R. (1965). "Capacity for increase: a useful population statistic." *J.
Anim. Ecol.* 34, 77–91.   *14, 18*

Lefkovitch, L.P. (1965). "The study of population growth in organisms grouped by
stages." *Biometrics* 21, 1–18.   *24*

Lerner, I.M., and E.R.Dempster (1962). "Indeterminism in interspecific competition." *Proc. Nat. Acad. Sci. Wash.* 48, 821–826.   *227, 229*

Le Roux, E.J. *et al.* (1963). Major mortality factors in the population dynamics of
the eye-spotted bud moth, the pistol casebearer, the fruit-tree leafroller and the
European cornborner. In "Population Dynamics of Agricultural and Forest
Insect Pests." E.J.Le Roux, Ed. *Mem. Entomol. Soc. Canada No. 32.*   *39*

Leslie, P.H. (1945). "The use of matrices in certain populationma thematics."
Biometrika 33, 183–212.   *24, 78*

Leslie, P.H. (1948). "Some further notes on the use of matrices in population
mathematics." *Biometrika* 35, 213–245.   *24, 78*

Leslie, P.H. (1958). "A stochastic model for studying the properties of certain
biological systems by numerical methods." *Biometrika* 45, 16–31.   *83*

Leslie, P.H. (1959). "The properties of a certain lag type of population growth and the influence of an external random factor on a number of such populations." *Physiol. Zool.* **32**, 151–159. *78*

Leslie, P.H. (1966). "The intrinsic rate of increase and the overlap of successive generations in a population of guillemots (*Uria aalge* Pont.)." *J. Anim. Ecol.* **35**, 291–301. *18, 24, 28, 35*

Leslie, P.H., and D.H.S.Davis (1939). "An attempt to determine the absolute number of rats on a given area." *J. Anim. Ecol.* **8**, 94–113. *129*

Leslie, P.H., and J.C.Gower (1960). "The properties of a stochastic model for the predator-prey type of interaction between two species." *Biometrika* **47**, 219 to 234. *83*

Leslie, P.H., T.Park, and D.B.Mertz (1968). "The effect of varying initial numbers on the outcome of competition between two *Tribolium* species." *J. Anim. Ecol.* **37**, 9–23. *226*

Levins, R. (1968). *Evolution in Changing Environments.* Princeton University Press. *352*

Lewis, E.G. (1942). "On the generation and growth of a population." *Sankhya* **6**, 93–96. *24*

Lloyd, M. (1967). "Mean crowding". *J. Anim. Ecol.* **36**, 1–30. *150, 151, 266*

Lloyd, M., R.F.Inger, and F.W.King (1968). "On the diversity of reptile and amphibian species in a Bornean rain forest." *Amer. Nat.* **102**, 497–515. *305, 309, 343*

MacArthur, R.H. (1964). "Environmental factors affecting bird species diversity." *Amer. Natur.* **98**, 387–397. *335*

MacArthur, R.H. (1965). "Patterns of species diversity." *Biol. Rev.* **40**, 510–533. *342*

MacArthur, R.H. (1968). "The theory of the niche." In *Population Biology and Evolution*, R.C.Lewontin, Ed. Syracuse University Press. *335*

MacArthur, R.H. (1969). "Species packing and what interspecies competition minimizes." *Proc. Nat. Acad. Sci. Wash.* **64**, 1369–1371. *354*

MacArthur, R.H., and J.H.Connell (1966). *The Biology of Populations.* Wiley, New York. *71*

MacArthur, R.H., and R.Levins (1967). "The limiting similarity, convergence and divergence of coexisting species." *Amer. Natur.* **101**, 377–385. *352*

MacArthur, R.H., and E.O.Wilson (1963). "An equilibrium theory of insular zoogeography." *Evolution* **17**, 373–387. *354, 356*

MacArthur, R.H., and E.O.Wilson (1967). *The Theory of Island Biography.* Princeton University Press. *340, 354*

McIntosh, R.P. (1967). "An index of diversity and the relation of certain concepts to diversity." *Ecology* **48**, 392–404. *290*

McIntosh, R.P. (1970). "Community, competition and adaptation." *Quart. Rev. Biol.* **45**, 259–280. *203, 287, 351*

McMillan, C. (1960). "Ecotypes and community function." *Amer. Natur.* **94**, 245–255. *328*

McNaughton, S.J. (1968). "Autotoxic feedback in the regulation of *Typha* populations." *Ecology* **49**, 367–369. *168*

McNeilly, T. (1968). "Evolution in closely adjacent plant populations. III. *Agrostis tenuis* on a small copper mine." *Heredity* **23**, 99–108. *327*

MacPherson, A.H. (1969). "The dynamics of the Canadian arctic fox population."
    Can. Wildl. Serv. Report Series No. 8.   78
McPherson, J.K., and C.H.Muller (1969). "Allelopathic effects of *Adenostoma
    fasciculatum*, "chamise", in the California chaparral." *Ecol Monogr.* **39**, 177
    to 198.   169
Maelzer, D.A. (1970). "The regression of log N_{n+1} on log N_n as a test of density
    dependence: an exercise with computer-constructed density-independent popu-
    lations." *Ecology* **51**, 810–822.   56
Malthus, T.R. (1798). *An Essay on the Principle of Population.* Reprinted by Mac-
    Millan & Co., 1894.   9, 46
Margalef, D.R. (1958). "Information theory in ecology." *General Systems* **3**,
    36–71.   290
Margalef, R. (1969). Diversity and stability: a practical proposal and a model of
    interdependence. In "Diversity and Stability in Ecological Systems" (G.M.Wood-
    well and H.H.Smith, Eds.). *Proc. Brookhaven Symposium in Biology No. 22.*
                                                                                     346
Matern, B. (1960). "Spatial variation. Stochastic models and their applications to
    some problems in forest surveys and other sampling investigations." *Medd. fran
    Statens Skogsforskningsinstitut.* **49**, 1–144.   187, 188
Matthewman, W.G., and D.P.Pielou (1971). "Arthropods inhabiting the sporo-
    phores of *Fomes fomentarius* (Polyporaceae) in Gatineau Park, Quebec." *Can.
    Entomol.* **103**, 775–847.   336
Maxwell, A.E. (1961). *Analysing Qualitative Data.* Methuen, London.   398
Maynard Smith, J. (1970). "Genetic polymorphism in a varied environment."
    Amer. Natur, **104**, 487–490.   324
Mayr, E. (1963). *Animal Species and Evolution.* Belknap Press, Harvard University.
                                                                                    333
Mayr, E. (1966). Review of G.E.Hutchinson's "The Ecological Theater and the
    Evolutionary Play" in *Bioscience* **16**, 555–556.   350
Mead, R. (1971). "A note on the use and misuse of regression models in ecology."
    J. Ecol. **59**, 215–219.   351
Mergen, F. (1959). "A toxic principle in the leaves of *Ailanthus*." *Bot. Gaz.* **121**,
    32–36.   169
Mergen, F. (1963). "Ecotypic variation in *Pinus strobus*." *Ecology* **44**, 716–727.
                                                                                    328
Miller, C.A. (1963). The spruce budworm. In "The Dynamics of Epidemic Spruce
    Budworm Populations" (R.F.Morris, Ed.) *Mem. Entomol. Soc. Can. No. 31.*
                                                                                   37, 38
Miller, R.B. (1958). "The role of competition in the mortality of hatchery trout."
    J. Fish. Res. Bd. Canada **15**, 27–45.   317
Miller, R.S. (1967). "Pattern and process in competition." *Advances in Ecol.
    Research* **4**, 1–74.   234, 235, 252, 318
Miller, R.S. (1968). "Conditions of competition between redwinged and yellow-
    headed blackbirds." *J. Anim. Ecol.* **37**, 43–62.   235
Milne, A. (1957). "The natural control of insect populations." *Can. Entomol.* **89**,
    193–213.   68
Milne, A. (1961). Definition of competition among animals. In "Mechanisms of
    Biological Competition" (F.Milthorp, Ed.) *Symposium of the Soc. of Exptal
    Biol. No. 25.* Academic Press.   202

Mook, L.J. (1963). Birds and the spruce budworm. In "The Dynamics of Epidemic Spruce Budworm Populations" (R.F.Morris, Ed.) *Mem. Entomol. Soc. Canada No. 31,* 268–271.   64

Moran, P.A.P. (1949). "The statistical analysis of the sunspot and lynx cycles." *J. Anim. Ecol.* **18,** 115–116.   87, 88

Moran, P.A.P. (1952). "The statistical analysis of game bird records." *J. Anim. Ecol.* **21,** 154   87, 89

Moran, P.A.P. (1953a). "The statistical analysis of the Canadian lynx cycle. I. Structure and prediction." *Austr. J. Zool.* **1,** 163–173.   87, 89, 90

Moran, P.A.P. (1953b). "The statistical analysis of the Canadian lynx cycle. II. Synchronization and meteorology." *Austr. J. Zool.* **1,** 291–298.   87, 91

Morris, R.F., (1963). "Predictive population equations based on key factors" *Mem. Entomol. Soc. Canada,* **32.**   55

Morrison, R.G. and G.A.Yarranton (1970). "An instrument for rapid and precise point sampling of vegetation." *Can. J. Botany* **48,** 293–297.   174

Mott, D.G. (1967). The analysis of determination in population systems. In *Systems Analysis in Ecology,* (K.E.F.Watt, Ed.) Academic Press.   40, 44

Mountford, M.D. (1961). "On E.C.Pielou's index of non-randomness." *J. Ecol.* **49,** 271–275.   160

Muller, C.H. (1966). "The role of chemical inhibition (allelopathy) in vegetational composition." *Bull. Torrey Bot. Club* **93,** 332–351.   165, 168

Muller, C.H. (1970). The role of allelopathy in the evolution of vegetation. In "Biochemical Evolution" (K.L.Chambers, Ed.). *Proc. 29th Annual Biology Colloquium.* Oregon State University Press.   168

Murdoch, W.W. (1969). "Switching in general predators: experiments on predator specificity and stability of prey populations." *Ecol. Monogr.* **39,** 335–354.   348

Murdoch, W.W. (1970). "Population regulation and population inertia." *Ecology* **51,** 497–502.   70, 346

Neave, F. (1953). "Principles affecting the size of pink and chum salmon populations in British Columbia." *J. Fish. Res. Bd. Canada* **9,** 450–491.   66

Nero, R.W. (1964). "Comparative behavior of the yellowheaded blackbird, redwinged blackbird and other icterids." *Wilson Bull.* **75,** 376–413.   235

Nicholson, A.J. (1954). "An outline of the dynamics of animal populations." *Austr. J. Zool.* **2,** 9–65.   59, 68, 81, 251, 252

Nicholson, A.J. (1957). "Comments on paper of T.B.Reynoldson." *Cold Spring Harbor Symp. on Quant. Biol.* **22,** 326.   68, 70

Norton, H.W. (1945). "Calculation of chi-square for complex contingency tables." *Jour. Amer. Statist. Ass.* **40,** 251–258.   397

Ogawa, R.E. (1969). "Lake Ontario phytoplankton, September 1964." *Great Lakes Fish Comm. Tech. Rept. No.* **14,** 27–38.   108

Okali, D.U.U. (1966). "A comparative study of the ecologically related tree species *Acer pseudoplatanus* and *Fraxinus excelsior.* II. The analysis of adult tree distribution." *J. Ecol.* **54,** 419–425.   270, 273

O'Neill, R.V. (1967). "Niche segregation in seven species of diplopods." *Ecology* **48,** 983.   320

Orians, G.H. (1961). "The ecology of blackbird (*Agelaius*) social systems." *Ecol. Monogr.* **31,** 285–312.   235

Paine, R.T. (1966). "Food web complexity and species diversity." *Amer. Natur.* **100**, 65–75.   *60, 240*

Park, T., P.H.Leslie, and D.B.Mertz (1964). "Genetic strains and competition in populations of *Tribolium*." *Physiol. Zool.* **37**, 97–162.   *226*

Paulik, G.J. (1967). Digital simulation of natural animal communities. In *Pollution and Marine Ecology*, T.A.Olson and F.J.Burgess, Eds., Wiley-Interscience. *232*

Pianka, E.R. (1966). "Latitudinal gradients in species diversity: a review of concepts." *Amer. Natur.* **100**, 33–46.   *337*

Pianka, E.R. (1970). "On *r* and *K* selection." *Amer. Natur.* **104**, 592–597.   *340*

Pielou, D.P., and A.N.Verma (1968). "The arthropod fauna associated with the birch bracket fungus, *Polyporus betulinus*, in eastern Canada." *Can. Entomol.* **100**, 1179–1199.   *343, 344*

Pielou, E.C. (1961). "Segregation and symmetry in two-species populations as studied by nearest neighbour relations." *J. Ecol.* **49**, 255–269.   *252, 269, 272*

Pielou, E.C. (1962*a*). "The use of plant-to-neighbour distances for the detection of competition." *J. Ecol.* **50**, 357–367.   *162, 163*

Pielou, E.C. (1962*b*). "Runs of one species with respect to another in transects through plant populations." *Biometrics* **18**, 579–593.   *274, 276*

Pielou, E.C. (1965). The concept of segregation pattern in ecology: some discrete distributions applicable to the run lengths of plants in narrow transects. In *Classical and Contagious Discrete Distributions* (G.P.Patil, Ed.) Statistical Publishing Society, Calcutta.   *269*

Pielou, E.C. (1966*a*). "Shannon's formula as a measure of specific diversity: its use and misuse. *Amer. Natur.* **100**, 463–465.   *305*

Pielou, E.C. (1966*b*). "Species-diversity and pattern-diversity in the study of ecological succession." *J. theor. Biol.* **10**, 370–383.   *303*

Pielou, E.C. (1966*c*). "The measurement of diversity in different types of biological collections." *J. theor. Biol.* **13**, 131–144.   *307*

Pielou, E.C. (1967*a*). "A test for random mingling of the phases of a mosaic." *Biometrics* **23**, 657–670.   *286*

Pielou, E.C. (1967*b*). "The use of information theory in the study of the diversity of biological populations." *Proc. 5th. Berkeley Symposium on Math Stat. and Prob.* **4**, 163–177.   *294*

Pielou, E.C. (1969). *An Introduction to Mathematical Ecology*. Wiley-Interscience, New York.   *79, 152, 290, 294, 301, 362*

Pielou, E.C. (1972*a*). Measurement of structure in animal communities. In "Ecosystem Structure and Function" (J.A.Wiens, Ed.). *Proc. 31st. Annual Biology Colloquium*, Oregon State University Press.   *278*

Pielou, E.C. (1972*b*). "2^k contingency tables in ecology." (With an appendix by D.S.Robson). *J. theor. Biol.* **34**, 337–352.   *278, 280*

Pielou, E.C. (1972*c*). "Niche width and niche overlap: a method for measuring them." *Ecology.* **53**, 687–692.   *239*

Pimentel, D., E.H.Feinberg, P.W.Wood, and J.T.Hayes (1965). "Selection, spatial distribution, and the coexistence of competing fly species." *Amer. Natur.* **99**, 97–109.   *243*

Pitelka, F.A. (1967). Some characteristics of microtine cycles in the Arctic. In "Arctic Biology" (H.P.Hansen, Ed.), Oregon State University Press.   *93*

Platt, T., L.M.Dickie, and R.W.Trites (1970). "Spatial heterogeneity of phytoplankton in a near-shore environment." *J. Fish. Res. Bd. Canada* **27**, 1453 to 1473. *187*

Pollard, J.H. (1966). "On the use of the direct matrix product in analysing certain stochastic population models." *Biometrika* **53**, 397–415. *25*

Poulson, T.L., and D.C.Culver (1969). "Diversity in terrestrial cave communities." *Ecology* **50**, 153–158. *343, 344*

Powell, J.R. (1971). "Genetic polymorphism in varied environments." *Science* **174**, 1035–1036. *331*

Preston, F. (1968). "On modeling islands." (Review of "The Theory of Island Biogeography" by R.H.MacArthur and E.O.Wilson). *Ecology* **49**, 592–594. *356*

Raj, D. (1968). *Sampling Theory*. McGraw Hill. *104*

Renyi, A. (1960). "On measures of entropy and information." *Proc. 4th Berkeley Symp. on Math. Stat. and Prob.* **1**, 547–561. *298*

Reynoldson, T.B. (1964). Evidence for intraspecific competition in field populations of triclads. In "Brit. Ecol. Soc. Jubilee Symposium" (A.Macfadyen and P.J.Newbould, Eds.). Blackwell Sci. Publications, Oxford. *252*

Roberts, M.R., and H.Lewis (1955). Subspeciation in *Clarkia biloba*. *Evolution* **9**, 445–454. *333*

Robertson, R.J. (1972). "Optimal niche space of the redwinged blackbird *(Agelaius phoeniceus)*. I. Nesting success in marsh and upland habitat." *Can. J. Zool.* **50**, 247–263. *235*

Rohlf, F.J., and R.R.Sokal (1969). *Statistical Tables*. W.H.Freeman and Co. San Francisco. *103, 138*

Rowan, W. (1954). "Reflections on the biology of animal cycles." *J. Wildl. Mgmt.* **18**, 52–60. *93*

Sadleir, R.M.F.S. (1969). *The Ecology of Reproduction in Wild and Domestic Animals*. Methuen, London. *16, 62*

Samarasinghe, S. (1965). "The biology and dynamics of the oystershell scale *Lepidosaphes ulmi* (L.) (Homoptera: Coccidae) on apple in Quebec." Ph.D.Thesis. McGill University. *42*

Samarasinghe, S., and E.J.LeRoux (1966). "The biology and dynamics of the oystershell scale on apple in Quebec." *Annals Ent. Soc. Quebec* **11**, 206–292. *42*

Sammeta, K.P.V., and R.Levins (1970). "Genetics and ecology." *Ann. Rev. Genetics* **4**, 469–488. *351*

Sampford, M.R. (1962). *An Introduction to Sampling Theory*. Oliver and Boyd, Edinburgh. *104, 107*

Samuel, E. (1969). "Comparison of sequential rules for estimation of the size of a population." *Biometrics* **25**, 517–527. *125*

Sanders, H.L. (1968). "Marine benthic diversity: a comparative study." *Amer. Natur.* **102**, 243–282. *341*

Savile, D.B.O. (1960). "Limitations of the competitive exclusion principle." *Science* **132**, 1761. *338, 339*

Schoener, T.W. (1968). "The *Anolis* lizards of Bimini: resource partitioning in a complex fauna." *Ecology* **49**, 704–726. *320*

Seber, G.A.F. (1970). "The effects of trap response on tag recapture estimates."
  Biometrics **26**, 13–22.   *125*
Seber, G.A.F., and E.D.Le Cren (1967). "Estimating population parameters from
  catches large relative to the population." *J. Anim. Ecol.* **36**, 631–643.   *127*
Selander, R.K. (1966). "Sexual dimorphism and differential niche utilization in
  birds." *Condor* **68**, 113–151.   *330*
Shapiro, A.M. (1970). "Differentiation of populations." *Science* **167**, 1636–1637.
                                                                                                  333
Sheppard, D.H. (1971). "Competition between two chipmunk species (*Eutamias*)."
  Ecology **52**, 320–329.   *236*
Simberloff, D.S. (1969). "Experimental zoogeography of islands: a model for insu-
  lar colonization." *Ecology* **50**, 296–314.   *357*
Simberloff, D.S., and E.O.Wilson (1969). "Experimental zoogeography of is-
  lands: the colonization of empty islands." *Ecology* **50**, 278–296.   *357*
Simpson, E.H. (1949). "Measurement of diversity." *Nature* **163**, 688.   *297, 298*
Simpson, G.G. (1964). "Species density of North American Recent mammals."
  Syst. Zool. **13**, 57–73.   *337*
Skellam, J.G. (1951). "Random dispersal in theoretical populations." *Biometrika*
  38, 196–218.   *244*
Slobodkin, L.B. (1962). *Growth and Regulation of Animal Populations*. Holt, Rine-
  hart and Winston.   *18*
Slobodkin, L.B. (1964). Experimental populations of Hydra. In "British Ecol. Soc.
  Jubilee Symposium" (A.Macfadyen and P.J.Newbould, Eds.).   *237, 240*
Smith, F.E. (1963). "Population dynamics in *Daphnia magna*." *Ecology* **44**, 651
  to 663.   *52*
Solomon, M.E. (1957). "Dynamics of insect populations." *Ann. Rev. Entomol.* **2**,
  121–142.   *62, 68, 69*
Southwood, T.R.E. (1966). *Ecological Methods*. Methuen, London.   *123*
St Amant, J.L.S. (1970). "The detection of regulation in animal populations."
  Ecology **51**, 823–828.   *56*
Stehli, F.G., R.G.Douglas, and N.D.Newell. (1969). "Generation and mainten-
  ance of gradients in taxonomic diversity." *Science* **164**, 947–950.   *337*
Stephen, W.J.D. (1967). "Bionomics of the sandhill crane." *Can. Wildl. Serv.*
  Rept. Series No. 2.   *99*
Swynnerton, C.F.M. (1936). "The tsetse flies of East Africa." *Trans. Roy. Ent.*
  Soc. Lond. **84**, 1–579.   *123*

Taber, R.D. (1971). "Criteria of sex and age." In *Wildlife Management Techniques*
  (R.G.Hiles, Jr. Ed.) 3rd. ed. Wildlife Soc. Ann Arbor.   *30*
Thomas, D.C. (1969). "Population estimates of barren ground caribou; March to
  May, 1967." *Can. Wildl. Serv. Rept. Ser. No. 9.*   *100*
Thompson, W.R. (1956). "The fundamental theory of natural and biological con-
  trol." *Ann. Rev. Entomol.* **1**, 379–402.   *68*
Tuck, L.M. (1960). "The Murres". *Can. Wildl. Serv. Rept. No.* 1.   *28, 320, 321*

Usher, M.B. (1966). "A matrix approach to the management of renewable resources, with special reference to selection forests." *J. Appl. Ecol.* **3**, 355–367.   *232*

Usher, M.B. (1969). "A matrix model for forest management." *Biometrics* **25**, 309–315. *232*

Utida, S. (1957). "Cyclic fluctuations of population density intrinsic to the host-parasite system." *Ecology* **38**, 442–449. *85*

Vandermeer, J.H. (1969). "The competitive structure of communities: an experimental approach with protozoa." *Ecology* **50**, 362–371. *217*

Vesey-Fitzgerald, D.F. (1960). "Grazing succession among East African game animals." *J. Mammalogy* **41**, 161–172. *324*

Wangersky, P.J., and W.J.Cunningham (1956). "On time lags in equations of growth." *Proc. Nat. Acad. Sci. Wash.* **42**, 699–702. *221*

Wangersky, P.J. and W.J. Cunningham (1957). "Time lag in population models." *Cold Spring Habor Symp. Quant. Biol.* **22**, 329–338. *220*

Watt, A.S. (1947). "Pattern and process in the plant community." *J. Ecol.* **35**, 1–22. *169*

Watt, K.E.F. (1961). "Mathematical models for use in insect pest control." *Can. Entomol.* **93**, Suppl. 19. *40*

Watt, K.E.F. (1964). "Comments on fluctuations of animal populations and measures of community stability." *Can. Entomol.* **96**, 1434–1442. *347*

Watt, K.E.F. (1968). *Ecology and Resource Management.* McGraw Hill, New York. *67, 95*

Weaver, W. (1948). "Probability, rarity, interest, and surprise." *Sci. Monthly* **67**, 390–392. *297*

Wellington, W.G. (1957). "Individual differences as a factor in population dynamics: the development of a problem." *Can. J. Zool.* **35**, 293–323. *147*

Wellington, W.G. (1960). "Qualitative changes in natural populations during changes in abundance." *Can. J. Zool.* **38**, 289–314. *147*

Wellington, W.G. (1965). "Some maternal influences on progeny quality in the western tent caterpillar *Malacosoma pluviale* (Dyar)." *Can. Entomol.* **97**, 1–14. *148*

Whittaker, R.H. (1960). "Vegetation of the Siskiyou Mountains, Oregon and California." *Ecol. Monogr.* **30**, 279–338. *310, 314, 342, 343*

Whittaker, R.H. (1972). Evolution and measurement of species diversity. In "Origin and Measurement of Diversity", Summer Institute in Systematics V, Smithsonian Institution, Wash. *310, 311, 313*

Williams, L.G. (1964). "Possible relationships between plankton-diatom species numbers and water quality estimates." *Ecology* **45**, 809–823. *339*

Wilson, E.O. (1955). "A monographic revision of the ant genus *Lasius.*" *Bull. Mus. Comp. Zool. Harvard Coll.* **113**, 1–199. *332*

Wilson, E.O. (1969). The species equilibrium. In "Diversity and Stability in Ecological Systems" (G.M.Woodwell and H.H.Smith, Eds.) *Proc. Brookhaven Symposium in Biology No. 22.* *357, 358*

Wilson, E.O., and D.S.Simberloff (1969). "Experimental zoogeography of islands: defaunation and monitoring techniques." *Ecology* **50**, 267–278. *357*

Woodwell, G.M. (1962). "Effects of ionizing radiations on terrestrial ecosystems." *Science* **138**, 572–577. *65*

Woodwell, G.M. (1963). "The ecological effects of radiation." *Sci. American* **208**, 40–49. *65*

Yates, F. (1960). *Sampling Methods for Censuses and Surveys.* 3rd. ed. Griffin, London.   *104*

Zippin, C. (1956). "An evaluation of the removal method of estimating animal populations". *Biometrics* **12**, 163–189.   *128*

Zippin, C. (1958). "The removal method of population estimation." *J. Wildl. Mgmt.* **22**, 82–90.   *128, 129*

Subject Index

417